Ex Libris
Robert Wojtysiak

ELEMENTARY THEORY OF

STRUCTURES

YUAN-YU HSIEH

Professor of Civil Engineering
National Taiwan University

PRENTICE-HALL, INC.
Englewood Cliffs, New Jersey

PRENTICE-HALL INTERNATIONAL, INC., *London*
PRENTICE-HALL OF AUSTRALIA, PTY. LTD., *Sydney*
PRENTICE-HALL OF CANADA, LTD., *Toronto*
PRENTICE-HALL OF INDIA PRIVATE LTD., *New Delhi*
PRENTICE-HALL OF JAPAN, INC., *Tokyo*

13-261552-5

Current printing (last digit):
10 9 8 7 6 5 4 3 2 1

Printed in the United States of America

To
My Parents

PREFACE

This book is intended for elementary courses in the structural theory of civil engineering. In preparing material for it, the author has assumed that the reader is not familiar with the subject. The first seven chapters contain the basic concepts of structure and an analysis of statically determinate structures. Chapter 8 deals with elastic deformations. Chapters 9 through 15 are concerned with statically indeterminate structures, including the method of consistent deformations, least work, slope deflection, and moment distribution. Chapter 16 is devoted to an introductory discussion of matrix algebra. The scope is wide enough to provide adequate background for reading the remainder of the book. Finally, the last two chapters present a unified treatment of structures by matrix methods based on the finite-element approach. Since this is an elementary treatment, emphasis is on the development of general theory in terms of matrix operations rather than on the particular details of computer programs.

The elementary theory of structures is not difficult, involving as it does only a limited amount of higher mathematics. For most of the book, the only preparation needed is a knowledge of arithmetic and high school algebra. We may also say, however, that this theory is difficult, in that it generally requires careful study in order to achieve a thorough understanding of its basic philosophy. The author has endeavored to present clearly and lucidly the fundamentals of structural theory. These fundamentals are arranged in a systematic order and are supplemented by examples illustrating their application in some of the common structures—namely, beams, trusses, and rigid frames.

While written primarily for use as a textbook in the classroom, this book can also be of help to structural and architectural engineers in their independent study.

The author wishes to express his appreciation to those who have assisted in the preparation of this book. He is especially grateful to Dr. Y. C. Fung, professor at the University of California, San Diego, for his enthusiastic encouragement throughout the writing of this volume and for his constructive criticism of the manuscript. Particular acknowledgment is also due Dr. Z. A. Lu of the University of California, Berkeley, who made many valuable suggestions for improving the contents, and Dr. John Yao, who read part of the manuscript. Thanks also are due Pamela Fischer and Cordelia Thomas of Prentice-Hall, who offered their competent knowledge in producing and editing this book. Finally, the author is indebted to his wife, Nelly, who typed all of the manuscript.

Y. Y. H.

CONTENTS

ix

15 ANALYSIS OF STATICALLY INDETERMINATE BEAMS

AND RIGID FRAMES COMPOSED OF NONPRISMATIC

MEMBERS

16 MATRIX ALGEBRA FOR STRUCTURAL ENGINEERS

17 MATRIX ANALYSIS OF STRUCTURES BY THE FINITE

ELEMENT METHOD   PART I: THE FORCE METHOD

ELEMENTARY THEORY OF
STRUCTURES

1

INTRODUCTION

1-1. ENGINEERING STRUCTURES

The word *structure* has various meanings. By an *engineering structure* we mean roughly something constructed or built. The principal structures of concern to civil engineers are bridges, buildings, walls, dams, towers, and shell structures. Structures as such are composed of one or more solid elements so arranged that the whole structures as well as their components are capable of holding themselves without appreciable geometric change during loading and unloading.

To design a structure involves many considerations among which are two major objectives that must be satisfied:

1. The structure must meet the performance requirement.
2. The structure must carry loads safely.

Consider, for example, the roof truss resting on columns shown in Fig. 1-1. The purposes of the roof truss and of the columns are, on the one hand, to hold in equilibrium their own weights, the load of roof covering, and the wind and snow (if any); and, on the other hand, to provide rooms for housing a family, for a manufacturing plant, or for other uses. In this book, we are concerned only with the load-carrying function of structures.

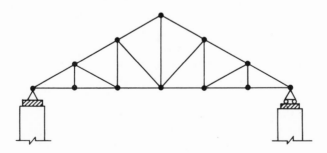

Fig. 1-1

1

1-2. THEORY OF STRUCTURES DEFINED

The complete design of a structure is contained in the following stages:

1. Establishing the general layout. The general layout for a structure is selected from many possible alternatives. The primary concern is the performance requirement of the structure whether it is to house, to convey, or to support in space. Many secondary considerations should also be examined including the economic, aesthetic, legal, and financial aspects.

2. Investigating the loadings. General information about the loadings imposed on a structure is usually given in the specifications and codes. Basically it is part of the designer's responsibility to specify the loading conditions and to take care of exceptional cases. Loadings based on *static* consideration may be classified as:

a. Dead load. Dead load is the weight of the structure itself and is regarded as fixed in magnitude and location. Since the dead load must be assumed before the structure is designed, the original data are only tentative. Revision must be made if the estimate is not satisfactory.

b. Live load. Live load may be classified into movable load and moving load. *Movable loads* are loads that may be transported from one location to another on a structure, for example, people and furniture on a building floor. *Moving loads* are loads that move continuously over the structure, such as railway trains or trucks on a bridge.

c. Impact load. The effects of impact are usually associated with moving live loads. In structural design the impact load is treated as an increase in the live load if the live load is taken as a gradually applied static load.

3. Stress analysis. Once the external loads are defined, a *stress analysis* must be made to determine the internal forces, sometimes referred to as *stresses*, that will be produced in the various members. When live loads are involved, emphasis should be put on the maximum possible stresses in each member under consideration. To obtain this we must know not only what loading is imposed but also where it is placed.

4. Selection of members. The choice of materials and proportioning of the sizes for structure members are based on the results of Stage 3 together with the design provisions given by the specifications or codes.

5. Drawing and detailing. After the make-up of each part of the structure is determined, the final stage, drawing and detailing, provides the necessary information for construction.

The subject matter of the theory of structures is stress analysis with occasional reference to loadings. The emphasis of structural theory is usually on the fundamentals rather than on the details of design.

1-3. THEORIES OF STRUCTURES CLASSIFIED

Structural theories may be classified from various points of view. For convenience, we shall characterize them by the following aspects:

1. Statics versus dynamics. Ordinary structures are usually designed under static loads. Dead load and snow load are static loads that cause no dynamic effect on structures. Some live loads, such as trucks and locomotives moving on bridges, are also assumed as concentrated static load systems. They do cause impact on structures; however, the dynamic effects are treated as a fraction of the moving loads to simplify the design.

The particular, specialized branch that deals with the dynamic effects on structures of accelerated moving loads, earthquake loads, wind gusts, or bomb blasts is *structural dynamics*.

2. Plane versus space. No structure is really planar. However, structural analyses for beams, trussed bridges, or rigid frame buildings are usually treated as plane problems, although they are never two-dimensional in themselves. On the other hand, in some structures, such as towers and framing for domes, the stresses are interrelated between members not lying in a plane in such a way that the analysis cannot be simplified on the basis of component planar structures. Such structures must be considered as space frameworks under non-co-planar force system.

3. Linear versus nonlinear structures. Linear structure means that a linear relationship is assumed to exist between the applied loads and the resulting displacements in a structure. This assumption is based on the following conditions:

a. The material of the structure is elastic and obeys Hooke's law at all points and throughout the range of loading considered.

b. The changes in the geometry of the structure are so small that they can be neglected when the stresses are calculated.

Note that if the principle of superposition is to apply, a linear relationship must exist, or be assumed to exist, between loads and displacements.

A nonlinear relationship between the applied actions and the resulting displacements exists under either of two conditions:

a. The material of the structure is inelastic.

b. The material is within the elastic range, but the geometry of the structure changes significantly during the application of loads.

The study of nonlinear behavior of structures includes *plastic analysis of structures* and *buckling of structures*.

4. Statically determinate versus statically indeterminate structures. The term *statically determinate structure* means that structural analysis can be

carried out by statics alone. If this is not so, the structure is statically indeterminate.

A statically indeterminate structure is solved by the equations of statics together with the equations furnished by the geometry of the elastic curve of the structure in linear analysis. We note that the elastic deformations of the structure are not only associated with the applied loads on the structure but are also affected by the material properties (e.g., the modulus of elasticity E) and by the geometric properties of the member section (e.g., the cross-sectional area A or the moment of inertia I). Thus, loads, material properties, and geometric properties are all involved in solving a statically indeterminate structure, while load factor alone dominates in a statically determinate problem.

1-4. ACTUAL AND IDEAL STRUCTURES

All analyses are based on some assumptions not quite in accordance with the facts. It is impossible for an actual structure to correspond fully to the idealized structure on which the analysis is based. The materials of which the structure is built do not follow exactly the assumed properties, and the dimensions of the actual structure do not coincide with their theoretical values.

To illustrate, let us take a simple example. In designing a reinforced concrete beam of rectangular section, the values of E and I are usually assumed to be constant. However, the amount of reinforcing steel placed in the beam varies with the stresses; therefore the values of E and I are not constant throughout the span. Besides, there is great uncertainty involved even in choosing a constant E or I. Even without considering other factors, such as the supports, the connections, and the working dimensions of the structure, we find that the behavior of an actual structure often deviates from that of an idealized structure by a considerable amount. However, it does not follow from this that the results of analysis are not useful for practical purposes. We must idealize a structure in order to carry out practical analysis, and from practical analyses we make the idealization more and more consistent with actuality.

1-5. SCOPE OF THIS BOOK

Three major types of basic structures are thoroughly discussed in this book:

1. *Beam*
2. *Truss*
3. *Rigid frame*

A *beam*, in its narrow sense, is a straight member subjected only to transverse loads. A beam is completely analyzed when the values of bending moment and shear are determined.

A *truss* is composed of members connected by frictionless hinges or pins. The loads on a truss are assumed to be concentrated at the joints. Each member of a truss is considered as a two-force member subjected to axial forces only.

A *rigid frame* is built of members connected by rigid joints capable of resisting moment. Members of a rigid frame, in general, are subjected to bending moment, shear, and axial forces.

The present book is confined exclusively to the static, planar, and linear aspects of structure. The first seven chapters deal with the basic concepts of structure and an analysis of statically determinate structures. The remaining chapters deal primarily with an analysis of statically indeterminate structures.

2

STABILITY AND DETERMINACY OF

STRUCTURES

2-1. EQUATIONS OF EQUILIBRIUM FOR A COPLANAR FORCE SYSTEM

The first and major function of a structure is to carry loads. Beams, trusses, and rigid frames all have this element in common: Each sustains the burden of certain loads without showing appreciable distortions. In structural statics all force systems are assumed to act on rigid bodies. Actually there are always some small deformations that may cause some small change of dimension in structure and a shifting of the action lines of the forces. However, such deviations are neglected in stress analysis.

A structure is said to be *in equilibrium* if, under the action of external forces, it remains at rest relative to the earth. Also, each part of the structure, if taken as a free body isolated from the whole, must be at rest relative to the earth under the action of the internal forces at the cut sections and of the external forces thereabout. If such is the case, the force system is balanced, or in equilibrium, which implies that the resultant of the force system (either a resultant force or a resultant couple) imposed on the structure, or segment thereof, must be zero.

Since this book is confined to planar structures, all the force systems are coplanar. The generally balanced coplanar force system must then satisfy the following three simultaneous equations:

$$\sum F_x = 0 \qquad \sum F_y = 0 \qquad \sum M_a = 0 \qquad (2\text{-}1)$$

where $\sum F_x$ = summation of x component of each force in the system

$\sum F_y$ = summation of y component of each force in the system

The subscripts x and y indicate two mutually perpendicular directions in the Cartesian coordinate system;

$\sum M_a$ = summation of moment about any point a in the plane due to each force in the system

Note that $\sum F_x$ also represents the x component of the resultant of the

6

force system, $\sum F_y$ the y component of the resultant of the force system, and $\sum M_a$ the moment about a of the resultant of the force system.

The alternative to Eq. 2-1 may be given by

$$\sum F_y = 0 \qquad \sum M_a = 0 \qquad \sum M_b = 0 \qquad (2\text{-}2)$$

provided that the line through points a and b is not perpendicular to the y axis, a and b being two arbitrarily chosen points and the y axis being an arbitrarily chosen axis in the plane. Or

$$\sum M_a = 0 \qquad \sum M_b = 0 \qquad \sum M_c = 0 \qquad (2\text{-}3)$$

provided that points a, b, and c are not collinear, a, b, and c being three arbitrarily chosen points in the plane.

Explanation of Eq. 2-2 is as follows:

1. Let R denote the resultant of the force system. Assume that $R \neq 0$. Since $\sum M_a = 0$ and $\sum M_b = 0$, the resultant R cannot be a couple. It must be a force through a and b and by assumption is not perpendicular to the y axis.

2. By $\sum F_y = 0$, we mean that the resultant has no y-axis component and must therefore be perpendicular to the y axis.

The above contradictory statements lead to the conclusion that the force is also zero. Therefore Eq. 2-2 is the condition for $R = 0$.

Similar explanation is given here for Eq. 2-3:

1. Assume that $R \neq 0$. Since $\sum M_a = 0$, $\sum M_b = 0$, and $\sum M_c = 0$, the resultant R cannot be a couple. It must be a force through a, b, and c.

2. But by assumption a, b, and c are not collinear.

The above lead us to conclude that the force is also zero. Therefore Eq. 2-3 is the condition for $R = 0$.

Two special cases of the coplanar force system in equilibrium are worth noting:

1. Concurrent forces. If a system of coplanar, concurrent forces is in equilibrium, then the forces of the system must satisfy the following equations:

$$\sum F_x = 0 \qquad \sum F_y = 0 \qquad (2\text{-}4)$$

Another set of independent equations necessary and sufficient for the equilibrium of the forces of a coplanar, concurrent force system is

$$\sum F_y = 0 \qquad \sum M_a = 0 \qquad (2\text{-}5)$$

provided that point a is not on the line through the concurrent point of forces and perpendicular to y axis.

A third set of equations of equilibrium for a coplanar, concurrent force system is

$$\sum M_a = 0 \qquad \sum M_b = 0 \tag{2-6}$$

where a and b are any two points in the plane of the forces, provided that the line through a and b does not pass through the concurrent point of forces.

2. Parallel forces. If a coplanar, parallel force system is in equilibrium, the forces of the system must satisfy the equations

$$\sum F_y = 0 \qquad \sum M_a = 0 \tag{2-7}$$

where the y axis is in the direction of the force system and a is any point in the plane.

Another set of independent equations of equilibrium for a system of coplanar, parallel forces may be given as

$$\sum M_a = 0 \qquad \sum M_b = 0 \tag{2-8}$$

where a and b are any two points in the plane, provided that the line through a and b is not parallel to the forces of system.

There are two simple, special cases of equilibrium that deserve explicit mention:

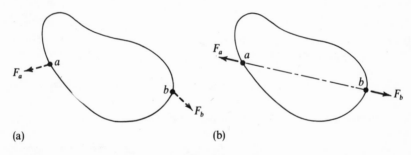

(a)   (b)

Fig. 2-1

1. Two-force member. In Fig. 2-1 is shown a body subjected to two external forces applied at a and b. If the body is in equilibrium, then the two forces cannot be in random orientation, as shown in Fig. 2-1(a), but must be directed along ab, as shown in Fig. 2-1(b). Furthermore, they must be equal in magnitude and opposite in sense. This can be proved by using first the equations $\sum M_a = 0$ and $\sum M_b = 0$. In order for the moment a to vanish, the force F_b must pass through a. Similarly, the force F_a must pass through b. Next, since $\sum F = 0$, it is readily seen that $F_a = -F_b$.

2. Three-force member. Figure 2-2 shows a body subjected to the action of three external forces applied at a, b, and c. If the body is in equilibrium, then the three cannot be in random orientation, as shown in Fig. 2-2(a). They must be concurrent at a common point O, as shown in Fig. 2-2(b); otherwise the total moment about the intersection of any two forces could not vanish. A

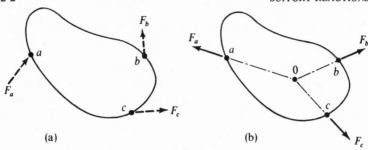

Fig. 2-2

limiting case occurs when point O moves off at infinite distance from a, b, and c, in which case the forces F_a, F_b, and F_c are parallel.

2-2. SUPPORT REACTIONS

Structures are either partially or completely restrained so that they cannot move freely in space. Such restraints are provided by supports that connect the structure to some stationary body, such as the ground or another structure. The first step in structural analysis is to take the structure without the supports and calculate the forces, known as *reactions*, exerted on the structure by the supports. The reactions are considered part of the external forces other than the loads on the structure and are to balance the other external loads in a state of equilibrium.

Certain symbols used to designate supports must first be described. There are generally three different types of support: the *hinge*, the *roller*, and the *fixed support*. The distribution of the reactive forces of a support may be very complicated, but in an idealized state the resultant of the forces may be represented by a single force completely specified by three elements—the *point of application*, the *direction*, and the *magnitude*. It may be noted that, in analysis, the direction simply means the slope of the action line, while the magnitude of force may be positive or negative, thus indicating not only its numerical size but also the sense of the action line.

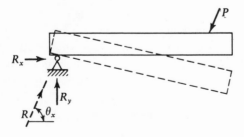

Fig. 2-3

Hinge Support. A hinge support is represented by the symbol ⚓ or

⬇ . It can resist a general force P in any direction but cannot resist the moment of the force about the connecting point, as illustrated in Fig. 2-3.

The reaction of a hinge support is assumed to be through the center of the connecting pin; its magnitude and slope of action line are yet to be determined. It is therefore a reaction with two unknown elements which could equivalently be represented by the unknown magnitudes of its horizontal and vertical components both acting through the center of the hinge pin. This is justified by the following equations from statics:

$$|R| = \sqrt{R_x^2 + R_y^2} \qquad \theta_x = \tan^{-1}\frac{R_y}{R_x} \qquad (2\text{-}9)$$

where $|R| =$ magnitude of the reaction R

$R_x = x$ component of R

$R_y = y$ component of R

$\theta_x =$ angle that R makes with x direction

The magnitude and direction of R can be determined if the unknown magnitudes of R_x and R_y are found.

From this, a hinge support can also be replaced by two links along the horizontal and vertical directions through the center of the connecting pin as shown in Fig. 2-4(a). Each link is a two-force member, the axial force of

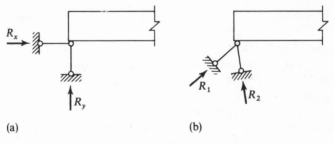

(a)   (b)

Fig. 2-4

which represents an element of reaction (R_x or R_y). In general, a hinge support is equivalent to two supporting links provided in any two different directions, not necessarily an orthogonal set, through the connecting point as shown in Fig. 2-4(b), where R_1 and R_2 indicate the axial forces in two links. The reaction R at the pin can always be determined if the magnitudes of R_1 and R_2 are obtained.

Roller Support. A roller support is represented by either the symbol

⚬ or ▭ The support mechanism used is such that the reaction acts normal to the supporting surface through the center of the connecting pin, as

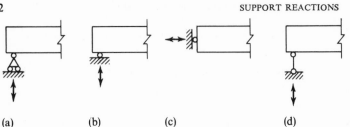

(a)   (b)   (c)   (d)

Fig. 2-5

shown in Figs. 2-5(a), (b), and (c). The reaction may be either away from or toward the supporting surface. As such, the roller support is incapable of resisting moment and lateral force along the surface of support.

A roller support supplies a reactive force, fixed at a known point and in a known direction, the magnitude of which is unknown. It is therefore a reaction with one unknown element.

A link support, shown in Fig. 2-5(d), is also of this type since the link is a two-force member and the reaction must be along the link.

Fixed Support. A fixed support is designated by the symbol ⫽⊢ . It is capable of resisting force in any direction and moment of force about the connecting end, thus preventing the end of the member from both translation and rotation. The reaction supplied by a fixed support may be represented by the unknown magnitudes of a moment called M_o, a horizontal force R_x, and a vertical force R_y acting through the centroid of the end cross section O, as

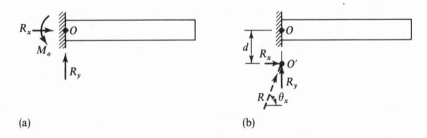

(a)   (b)

Fig. 2-6

shown in Fig. 2-6(a). These three unknown elements can be expressed as equal to a single force R with its three elements—the magnitude, direction, and point of application—yet to be determined, as shown in Fig. 2-6(b). Now the magnitude and direction of R can be related to its components R_x and R_y by Eq. 2-9

$$|R| = \sqrt{R_x^2 + R_y^2} \qquad \theta_x = \tan^{-1}\frac{R_y}{R_x}$$

and the point of application O' can be located by the distance d from O, which, in turn, is related to M_o by

$$d = \frac{M_o}{R_x}$$

Since fixed support provides moment resistance, it is one step beyond the hinge support in rigidity.

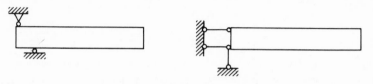

Fig. 2-7

Two devices equivalent to the fixed support are shown in Fig. 2-7. Each is composed of a hinge and a roller and represents three elements of reaction capable of resisting both force and moment.

2-3. INTERNAL FORCES AT A CUT SECTION OF A
STRUCTURE

A truss structure is composed of pin-connected members and is assumed to be pin loaded, as is shown in Fig. 2-8(a). Now if any one of the members is

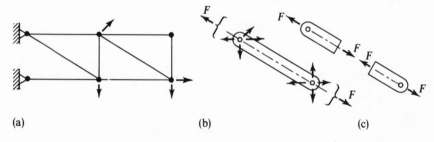

(a)   (b)   (c)

Fig. 2-8

taken from its connecting pins as a free body, the forces exerted on the member must be concentrated at the two ends of the member through the centers of pins. Furthermore, these two systems of concurrent forces can be combined into two resultant forces that must be equal, opposite, and collinear, as is indicated in Fig. 2-8(b). In other words, each member of a truss is a *two-force* member. Hence the internal forces existing in any cut section of a truss member (assumed straight and uniform) must be a pair of equal and opposite axial forces to balance the axial forces exerted on the ends, as is shown in Fig. 2-8(c).

The fact that each member of a truss represents an unknown element of internal force enables us to obtain the total number of unknown elements of internal force by counting the total number of members of which the truss is composed.

Members of structures, such as beam and rigid frame, are acted on by more than two forces. Let us investigate the elements of internal force in any cut section A-A of the beam in Fig. 2-9(a) or of the rigid frame in Fig. 2-9(b).

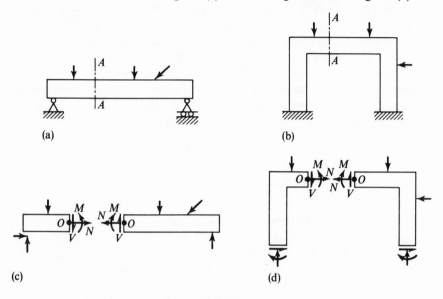

(a)   (b)

(c)   (d)

Fig. 2-9

We begin by taking the free bodies of the portions to the left and right of the section A-A, as shown in Figs. 2-9(c) and (d). It is obvious that forces of internal constraint must exist between these two portions in order to hold them together. Such internal forces, of course, always occur in pairs of equal and opposite forces. The actual distribution of these internal forces cannot be easily discovered. To maintain the equilibrium of the free body, however, the internal forces must be statically equal and opposite to the system of forces acting externally on the portion considered, and the internal forces can always be represented by a force applied at the centroid O of the cross section together with a couple of moment M. Furthermore, the force can, in turn, be resolved into a normal component N and a tangential component V. Thus in Figs. 2-9(c) and (d) we represent the stress resultant on any section A-A by the three unknown magnitudes of N, V, M, called, respectively, the *normal force*, the *shearing force*, and the *resisting moment* at that section.

From the foregoing discussion, we remember that to take a free body from a beam or rigid frame we must assume three unknown elements of internal force generally existing in the cut section.

2-4. EQUATIONS OF CONDITION OR CONSTRUCTION

Structures such as truss, beam, and rigid frame may sometimes be considered to be one rigid body sustained in space by a number of supports. Out of several such rigid bodies a compound form of structure may be built by means of connecting devices, such as hinges, links, or rollers, and mounted on a number of supports. In either the simple or the compound type of structure the external force system of the entire structure, consisting of the loads on the structure and the support reactions, must satisfy the equations of equilibrium if the structure is to remain at rest. However, in the compound type of structure the connecting devices enforce further restrictions on the force system acting on the structure, thus providing additional equations of statics to supplement the equations of equilibrium. Equations supplied by the method of special construction (other than external supports) are called *equations of condition or construction*. We shall discuss these further in Sec. 2-6.

2-5. STABILITY AND DETERMINACY OF A STRUCTURE
WITH RESPECT TO SUPPORTS WHEN THE STRUCTURE
IS CONSIDERED A MONOLITHIC RIGID BODY

When one considers the design of a structure, careful thought must be given to the number and arrangement of the supports directly related to the statical stability and determinacy of the structure. In the following discussions we shall treat the structure as one monolithic rigid body mounted on a number of supports. Thus there will be no internal condition involved, and the stability and determinacy of the structure will be judged solely by the stability and determinacy of supports.

1. Two elements of reaction supplied by supports, such as two forces each with a definite point of application and direction, are not sufficient to ensure the stability of a rigid body, because the two are either collinear, parallel, or concurrent. In each of these cases, the condition of equilibrium is violated, not because of the lack of strength of supports, but because of the insufficient number of support elements. This is referred to as *statical instability*.

If two reactive forces are collinear (see Fig. 2-10(a)), they cannot resist an external load that has a component normal to the line of reactions. If they are parallel (see Fig. 2-10(b)), they cannot prevent the body from lateral sliding. If they are concurrent (see Fig. 2-10(c) or (d)), they cannot resist the moment about the concurrent point O due to any force not through O.

Algebraically, in each of the above cases, one equilibrium condition is not satisfied. For instance, in Fig. 2-10(a) or (b) the condition $\sum F_x = 0$ is violated (x indicates the direction normal to the line of reaction); whereas in Fig.

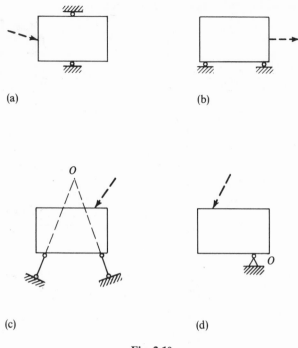

(a)   (b)

(c)   (d)

Fig. 2-10

2-10(c) or (d) the condition $\sum M_o = 0$ is not fulfilled. The body is, therefore, not in equilibrium. It is unstable.

Only under some very special conditions of loading can the body be stable, such as those shown in Fig. 2-11. In case (a) (Fig. 2-11) the applied loads acting on the body are themselves in equilibrium; therefore, no reaction is required. In case (b) the applied load is in the same direction as the reactions so that equilibrium can be maintained for the parallel force system; and in case (c) or (d) the applied load is through the concurrent point O; therefore, equilibrium can also be established.

Structures stable under special conditions of loading but unstable under general conditions of loading are said to be in a state of *unstable equilibrium* and are classified as unstable structures.

2. At least three elements of reaction are necessary to restrain a body in stable equilibrium. Consider each of the cases shown in Fig. 2-12. The rigid body is subjected to restraints by three elements of reaction, and the restraints can be solved by the three available equilibrium equations. The satisfaction of all the three equilibrium equations, $\sum F_x = 0$, $\sum F_y = 0$, and $\sum M = 0$, for loads and reactions acting on the body guarantees, respectively, that the body will neither move horizontally or vertically nor rotate. The system is said to be *statically stable and determinate*.

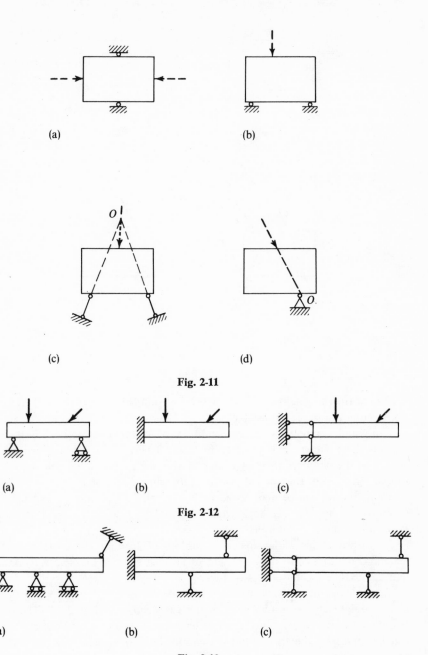

(a)

(b)

(c)

(d)

Fig. 2-11

(a)

(b)

(c)

Fig. 2-12

(a)

(b)

(c)

Fig. 2-13

3. If there are more than three elements of reaction, as in each of the cases shown in Fig. 2-13, the body is necessarily more stable, because of the additional restraints. Since the number of unknown elements of reaction is more than the number of equations for static equilibrium, the system is said to be *statically indeterminate* with regard to the reactions of support.

4. That the number of elements of reaction should be at least three is a necessary but not a sufficient condition for an externally stable structure. There are many cases that are obviously not stable with respect to the support system even though three or more than three elements of reaction are supplied. When, for example, the lines of reaction are all parallel, as in Fig. 2-14(a), the body

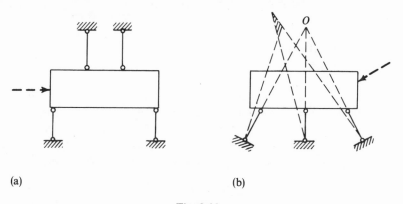

(a)   (b)

Fig. 2-14

is unstable, because it is vulnerable to lateral sliding. Another case is shown in Fig. 2-14(b), where lines of the three reaction elements are originally con-current at point O. The system is also unstable because, even though complete collapse probably will not result, a small initial rotation about O because of the moment caused by any force not through O will certainly occur until the three reaction lines form the triangle indicated by the cross-hatched lines.

The above-mentioned instability, which results from the inadequacy of arrangement of supports, is referred to as *external geometric instability*.

5. A monolithic rigid body is rigid by definition; hence it will have no problem of internal instability. Furthermore, at any cut section of a monolithic rigid body, the elements of internal force, which are no more than three in number, can always be determined by the equations of equilibrium, once the reactions are completely defined. Therefore the stability and determinacy of the entire system, mentioned in this section, are solely determined by the stability and determinacy of supports and reactions.

Let us sum up the main points of foregoing discussions as follows:

1. If the number of unknown elements of reaction is fewer than three, the

equations of equilibrium are generally not satisfied, and the system is said to be unstable.

2. If the number of unknown elements of reaction is equal to three and if no external geometric instability is involved, then the system is statically stable and determinate.

3. If the number of unknown elements of reaction is more than three, then the system is statically indeterminate; it is stable provided that no external geometric instability is involved. The excess number n of unknown elements designates the nth degree of indeterminacy. For example, in each case of Fig. 2-13 there are five unknown elements of reaction. Thus, $5 - 3 = 2$, which indicates an indeterminacy of second degree.

2-6. GENERAL STABILITY AND DETERMINACY OF STRUCTURES

Structural stability and determinacy must be judged by the number and arrangement of the supports as well as the number and arrangement of the members and the connections of the structure. They are determined by inspection or by formula. For convenience, we shall deal with the general stability and determinacy of beams, trusses, and rigid frames in separate sections.

2-6a. General Stability and Determinacy of Beams. If a beam is built up without any internal connections (internal hinge, roller, or link), the entire beam may be considered as one piece of monolithic rigid body placed on a number of supports, and the question of the stability and of the determinacy of the beam is settled solely by the number and arrangement of supports, as discussed in Sec. 2-5.

Now let us proceed to investigate what will happen if a certain connecting device is inserted in a beam. Let us suppose that a hinge is introduced into the statically stable and determinate beam of Fig. 2-15(a) or (b). The beam in each case will obviously become unstable under general loading as the result of a relative rotation between the left and the right portions of the beam at the internal hinge, as indicated in Fig. 2-15(c) or (d). That the hinge has no capacity to resist moment constitutes a restriction on the external forces acting on the structure; i.e.,

$$M = 0$$

about the hinge. In other words, the moment about the hinge calculated from the external forces on either side of the hinge must be zero in order to guarantee that these portions will not rotate about the hinge.

Referring to Fig. 2-15(c) or (d), we see that in each case there are three elements of reaction supplied by supports, whereas there are four conditions of statics to restrict the external forces—three from equilibrium plus one from

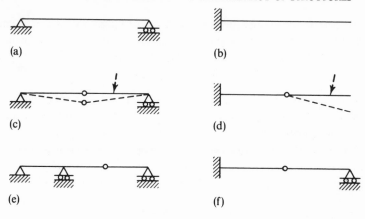

Fig. 2-15

construction. This means that the number of unknown elements of reaction is one fewer than the independent equations of statics available for their solution. Therefore, the equations of statics for the force system are generally not satisfied. The beam is unstable, unless we provide at least one additional element of reaction, such as the additional roller support shown in Fig. 2-15(e) or (f), which makes the total number of unknown elements of reaction equal to the total number of independent equations of statics needed to determine the elements. If this is done, the beam will be restored to a statically stable and determinate state.

Next, let us suppose that a link (or a roller) is introduced into a section of the statically stable and determinate beam of Fig. 2-15(a) or (b). We expect that this beam will be less stable than that of a hinged connection because the link (roller) cannot resist both moments about the link pin and forces normal to the link. The beam will collapse under general types of loading as a result of the relative rotation and the lateral translation of the left and right portions of the beam at the link, as indicated in Fig. 2-16(a) or (b).

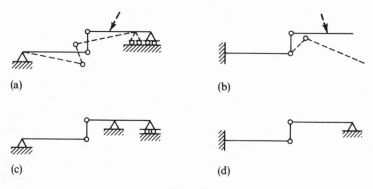

Fig. 2-16

That a link (or roller) has no capacity to resist lateral forces and moments constitutes two restrictions to the external forces acting on the structure, namely,

$$H = 0 \quad \text{and} \quad M = 0$$

about the link (either of two pins). H is the sum of forces on either side of the link in the direction normal to the link. The satisfaction of condition $H = 0$ for the portion of the structure on either side of the link prevents movement in the direction normal to the link of one portion of the structure relative to the other. Satisfying condition $M = 0$ for the portion of the structure on either side of the link ensures that these portions will not rotate about the pins of the link.

Referring to Fig. 2-16(a) or (b), we find that in each case there are three elements of reaction supplied by the support system, while there are five conditions of statics to restrict them—three from equilibrium and two from construction. Since the number of elements in reaction is two fewer than the number of statical equations to determine them, the beam is, therefore, quite unstable unless we supply at least two more elements of reaction, such as the hinged support shown in Fig. 2-16(c) or (d), to balance the situation. This done, the beam will be restored to its statically stable and determinate state.

There are beams for which the number of reaction elements is greater than the total number of independent equations of statics available. The beams are then classified as *statically indeterminate*; and the excess number of unknown elements indicates the degree of indeterminacy.

Geometric instability is most likely to occur whenever internal connections are introduced into an originally stable structure. Consider, for example, Fig. 2-17(a). The beam is statically indeterminate to the first degree. Now if a

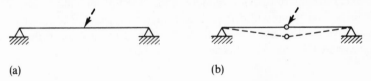

(a)   (b)

Fig. 2-17

hinge is inserted into the beam, as shown in Fig. 2-17(b), it seems to be statically determinate. However, when a load is applied, a small initial displacement will result and will not be resisted elastically by the structure. In such a case, the beam is unstable not because of the inadequacy of the supports but because of the inadequacy of the arrangement of members. This is referred to as *internal geometric instability*. Very often when this occurs, the structure will collapse. In the present case collapse will not occur; the beam will come to rest in a position such as that marked by the dashed lines shown in Fig. 2-17(b).

From the foregoing discussions, a criterion may be established for the stability and determinacy of beams. Let r denote the number of reaction elements and c the number of equations of condition ($c = 1$ for a hinge; $c = 2$ for a roller; $c = 0$ for a beam without internal connection).

1. If $r < c + 3$, the beam is unstable.
2. If $r = c + 3$, the beam is statically determinate provided that no geometric instability (internal and external) is involved.
3. If $r > c + 3$, the beam is statically indeterminate.

Further illustrations are given in Table 2-1.

TABLE 2-1

Beam	r	c	$r \gtreqless c + 3$	Classification
	5	2	$5 = 5$	Stable and determinate
	6	2	$6 > 5$	Stable and indeterminate to the first degree
	5	2	$5 = 5$	Unstable*
	4	3	$4 < 6$	Unstable
	6	3	$6 = 6$	Stable and determinate
	7	2	$7 > 5$	Unstable*

*Internal geometric instability; a possible form of displacement is indicated by the dotted lines.

2-6b. General Stability and Determinacy of Trusses. A truss is composed of a number of bars connected at their ends by a number of pinned joints so as to form a network, usually a series of triangles, and mounted on a number of supports, such as the one shown in Fig. 2-18(a). Each bar of a truss is a two-force member; hence, each represents one unknown element of internal force (see Sec. 2-3). The total number of unknown elements for the entire

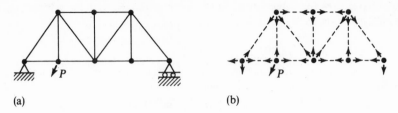

Fig. 2-18

system is counted by the number of bars (internal) plus the number of independent reaction elements (external). Thus if we let b denote the number of bars and r the number of reaction components, the total number of unknown elements of the entire system is $b + r$. Now if the truss is in equilibrium, every isolated portion must likewise be in equilibrium. For a truss having j joints, the entire system may be separated into j free bodies, as illustrated in Fig. 2-18(b), in which each joint yields two equilibrium equations, $\sum F_x = 0$ and $\sum F_y = 0$, for the concurrent force system acting on it. From this a total of $2j$ independent equations, involving $(b + r)$ unknowns, is obtained. We may thus establish a criterion for the stability and the determinacy of truss by counting the total unknowns and the total equations.

1. If $b + r < 2j$, the system is unstable.
2. If $b + r = 2j$, the system is statically determinate provided that it is also stable.
3. If $b + r > 2j$, the system is statically indeterminate.

The satisfaction of condition $b + r \gtrless 2j$ does not ensure a stable truss. For the truss to be stable requires fulfillment of further conditions. First, the value of r must be equal to or greater than the three required for statical stability of supports. Next, there must be no inadequacy in the arrangement of supports and bars so as to avoid both external and internal geometric instability.

Basically, a stable truss can usually be obtained by starting with three bars pinned together at their ends in the form of a triangle and then by extending from it by adding two new bars for each new joint, as shown in Fig. 2-18(a). Since this truss satisfies $b + r = 2j$ ($b = 13$, $r = 3$, $j = 8$), it is therefore statically determinate.

Suppose that this truss form is changed, as shown in Fig. 2-19. The number of bars and joints remains the same; the criterion equation is still satisfied. But it is geometrically unstable, since there is no bar to carry

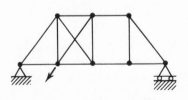

Fig. 2-19

the vertical force (shear) in the panel where the diagonal is omitted. Other examples are given in Table 2-2.

TABLE 2-2

Truss	b	r	j	$b + r \lessgtr 2j$	Classification
	7	3	5	$10 = 10$	Stable and determinate
	7	3	5	$10 = 10$	Unstable*
	7	3	5	$10 = 10$	Unstable**
	6	3	5	$9 < 10$	Unstable
	6	4	5	$10 = 10$	Stable and determinate
	8	4	5	$12 > 10$	Stable and indeterminate to the second degree
	6	4	5	$10 = 10$	Unstable***

*Internal geometric instability due to three pins a, b, c on a line; possible displacement as indicated by dotted lines.

**External geometric instability due to parallel lines of reaction.

***Internal geometric instability due to lack of lateral resistance in panel $abcd$.

Figure 2-20 shows a long-span trussed bridge, which we may consider to be composed of three rigid trusses connected by a hinge A and a link BC and

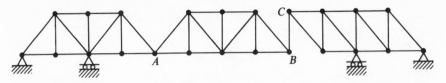

Fig. 2-20

mounted on a number of supports. These connections are not completely rigid, so that certain equations of condition are introduced to restrict the external forces acting on the structure. In this case the hinge at A provides one condition equation $M_A = 0$, which means that the moment about A of the forces on either side of A must be zero. The hanger BC provides two condition equations, $M_B = 0$ (or $M_C = 0$) and $H = 0$, which means that the moment about B (or C) of the forces on either side of B (or C) must be zero and also that the sum of the horizontal forces on either side of the hanger must be zero since the vertical hanger is incapable of resisting horizontal forces.

The stability and determinacy of the truss may be investigated by first counting the number of bars, joints, and reaction elements. It is found that the equation $b + r = 2j$ is satisfied by the truss since $b = 40$, $r = 6$, and $j = 23$. Thus the necessary condition for the system to be statically determinate is fulfilled. Next, there is no obvious instability either in the formation of the truss or in the supports. Both the portions to the left of A and to the right of BC are rigidly formed and adequately supported. The portion in the center span is also rigidly formed. Its connection to the side portions by a hinge and a hanger constitutes three elements of support. As regards reactions, there are a total of six elements which can just be determined by six statical equations, three from equilibrium and three from construction. Thus the entire system is stable and statically determinate; furthermore, it is stable and statically determinate as regards support reactions.

There are certain cases in which the stability or instability of a truss is not obvious. One way of determining stability is to attempt a stress analysis and to discover whether the results are consistent or not. An inconsistent result indicates that the answer is not unique but infinite and indeterminate. If such is the case, then the truss is said to be *unstable*. We shall discuss this further in Sec. 4-4.

2-6c. General Stability and Determinacy of Rigid Frames. A rigid frame is built of beams and columns connected rigidly, such as the one shown in Fig. 2-21(a). The stability and determinacy of a rigid frame may also be investigated by comparing the number of unknowns (internal unknowns and reaction unknowns) with the number of equations of statics available for their solution. Like a truss, a rigid frame may be separated into a number of free bodies of joints, as shown in Fig. 2-21(b), which requires that every member of the frame be taken apart. As discussed in Sec. 2-3, there are usually three unknown magnitudes (N, V, M) existing in a cut section of a member. However, if these quantities are known at one section of a member, similar quantities for any other section of the same member can be determined. Hence there are only three independent, internal, unknown elements for each member in a frame. If we let b denote the total number of members and r the reaction elements, then the total number of independent unknowns in a rigid frame is $(3b + r)$.

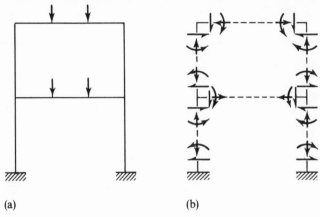

(a)   (b)

Fig. 2-21

Next, a rigid joint isolated as a free body will generally be acted upon by a system of forces and couples, as indicated in Fig. 2-21(b), since a rigid joint is capable of resisting moments. For equilibrium of such a joint, this system, therefore, must satisfy three equilibrium equations, $\sum F_x = 0$, $\sum F_y = 0$, and $\sum M = 0$. Thus if the total number of rigid joints is j, then $3j$ independent equilibrium equations may be written for the entire system.

It may happen that hinges or other devices of construction are introduced into the structure so as to provide additional equations of statics, say a total of c. Then the total number of equations of statics available for the solution of the $(3b + r)$ unknowns is $(3j + c)$. The criteria for the stability and the determinacy of the rigid frame are thus established by comparing the number of unknowns $(3b + r)$ with the number of independent equations $(3j + c)$:

1. If $3b + r < 3j + c$, the frame is unstable.
2. If $3b + r = 3j + c$, the frame is statically determinate provided that it is also stable.
3. If $3b + r > 3j + c$, the frame is statically indeterminate.

It should be recalled from the similar discussion dealing with the criterion for trusses that satisfaction of the condition $3b + r \gtreqless 3j + c$ does not warrant a stable frame unless $r \gtreqless 3$ and, also, that no geometric instability is involved in the system.

TABLE 2-3

Frame	b	r	j	c	$3b + r \lessgtr 3j + c$	Classification
	10	9	9	0	$39 > 27$	Indeterminate to the 12th degree
	10	9	9	4	$39 > 31$	Indeterminate to the eighth degree
	10	9	9	1	$39 > 28$	Indeterminate to the 11th degree
	10	9	9	3*	$39 > 30$	Indeterminate to the ninth degree
	10**	6	9	0	$36 > 27$	Indeterminate to the ninth degree

*If a pin is inserted in a rigid frame, generally, c = the number of members meeting at the pin minus one. In this case $c = 4 - 1 = 3$.

**The overhanging portions, such as ab and cd on the right side of the frame, should not be counted in the number of members.

Consider the frame in Fig. 2-21(a). There are six joints (including those at supports), six members, and six reaction elements, but no condition of construction. Thus $3b + r = 18 + 6 > 3j + c = 18 + 0$. The excess number six in unknowns indicates that the frame is statically indeterminate to the sixth degree. Further examples for classifying frame stability and determinacy are given in Table 2-3.

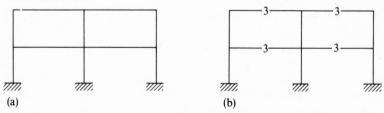

(a)   (b)

Fig. 2-22

Criteria, such as the above, are general and useful; but many problems, which may be investigated by a formula, can readily be settled by inspection through cutting frame members and reducing the structure to several simple parts.

Suppose that we wish to analyze the degree of indeterminacy of the frame shown in Fig. 2-22(a). The best approach is to cut members as indicated in Fig. 2-22(b) so that the structure is separated into three statically determinate and stable parts. The number of restraints removed to accomplish this result gives the degree of indeterminacy of the frame. Since each cut involves three internal unknown elements, the total number of restraints removed by four cuts is $(4)(3) = 12$; the frame is statically indeterminate to the 12th degree.

The advantage of this approach over counting the number of bars and joints and reaction elements will easily be seen when we come to determine the degree of indeterminacy of the frame of a tall building, such as the one shown in Fig. 2-23. Since the building can be separated into 12 stable and deter-

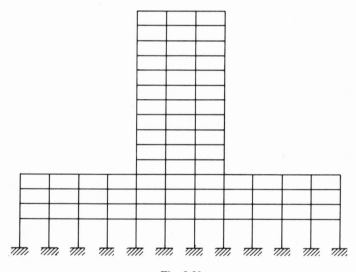

Fig. 2-23

minate parts by 77 cuts in the beams, it is statically indeterminate to the 231st degree.

PROBLEMS

2-1. Discuss the stability and determinacy of the beams shown in Fig. 2-24.

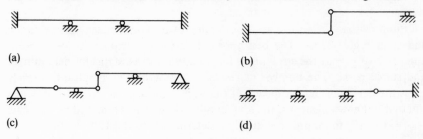

(a)

(b)

(c)

(d)

Fig. 2-24

2-2. Discuss the stability and determinacy of the trusses shown in Fig. 2-25.

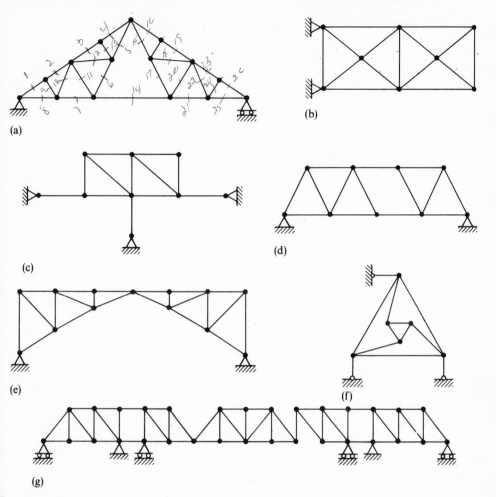

Fig. 2-25

2-3. Discuss the stability and determinacy of the rigid frames shown in Fig. 2-26.

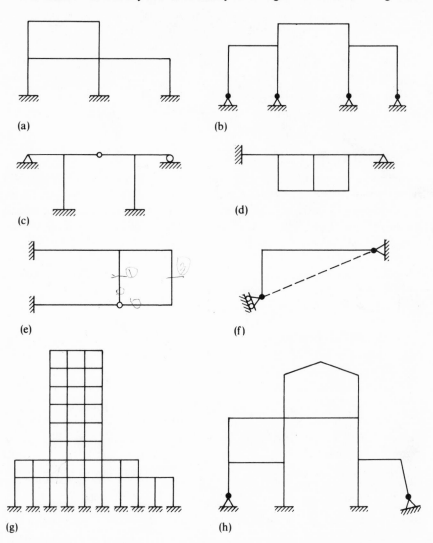

Fig. 2-26

3

STATICALLY DETERMINATE BEAMS

3-1. GENERAL

A *beam* is defined as a structural member predominantly subjected to bending moment. In this chapter and later chapters we shall limit our discussions to beams of symmetrical section, the centroidal axis of such beams being a straight line. Furthermore, we assume that the beam is acted on by only transverse loading and moment loading and that all the loads and reactions lie in the plane of symmetry. From these, it follows that such a beam will be subjected to bending and shear in the plane of loading without axial stretching and twisting.

Statically determinate beams may be classified as:

1. Simple beam. A beam which is supported at its two ends with a hinge and a roller (see Fig. 3-1(a)), or their equivalents (see Fig. 3-1(b)), is termed a *simply supported beam* or *simple beam*.

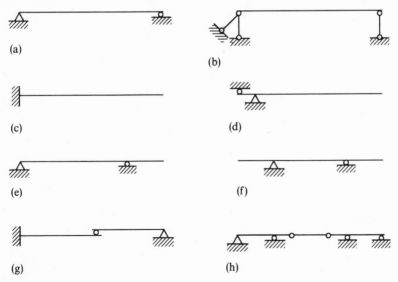

Fig. 3-1

31

2. Cantilever beam. In Fig. 3-1(c) shows a *cantilever beam*, which is fixed or built-in at one end and free at the other end. Another form of cantilever is shown in Fig. 3-1(d) in which the three elements of reaction of fixed support are provided by a hinge and a roller closely placed at one end segment.

3. Simple beam with overhang. A beam may be simply supported at two points and have its end portion (or portions) extend beyond the support, as in Fig. 3-1(e) or (f); It is then called a *simple beam with overhang*.

4. Compound beam. The beams indicated above may be connected by internal hinges or rollers to form a *compound beam*, such as those shown in Figs. 3-1(g) and (h). Care must be used in providing the connections so that instability is not produced.

3-2. ANALYSIS OF STATICALLY DETERMINATE BEAMS

To illustrate the general procedure in analyzing a beam, let us consider the loaded beam in Fig. 3-2(a).

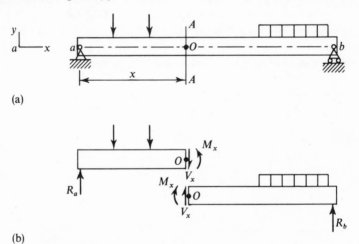

(a)

(b)

Fig. 3-2

The first step in the analysis is to find the reactions at ends a and b denoted by R_a and R_b respectively. This can readily be accomplished by applying the equilibrium equations:

$$\sum M_a = 0 \qquad \sum M_b = 0$$

or $\qquad\qquad \sum M_b = 0 \qquad \sum F_y = 0$

Next we investigate the shear force and bending moment at each transverse cross section of the beam. The *shear force* at any transverse cross section of the beam, say section $A\text{-}A$, at a distance x from the left end (see Fig. 3-2(a)), is the algebraic sum of the external forces (including those of reaction) applied to the portion of the beam on either side of *A-A*. The *bending moment* at section *A-A* of the beam is the algebraic sum of the moments taken about an axis through *O* (the centroid of section *A-A*) and normal to the plane of loading of all the external forces applied to the portion of the beam on either side of *A-A*. By considering either the left or the right portion as the free body, as shown in Fig. 3-2(b), we readily see that the *shear resisting force* at the section V_x is equal and opposite to the shear force for that section just defined; and the *resisting moment* at the section M_x is equal and opposite to the bending moment for the section just defined. The values of V_x and M_x can be found from the two equations of equilibrium

$$\sum F_y = 0 \quad \text{and} \quad \sum M_o = 0$$

for the portion considered.

Since the shear and bending moment in a transversely loaded beam will, in general, vary with the distance *x* defining the location of the cross section on which they occur, both are therefore the functions of *x*. It is advisable to plot curves or diagrams from which the value of functions (V_x and M_x) at any cross section may readily be obtained. To do this, we let one axis, the *x* axis, coincide with the centroidal axis of beam, indicating the position of beam section, and the other axis, the *y* axis, indicate the value of function V_x or M_x. The graphic representation is called *shear* or *moment curve*.

Our sign conventions for beam shear and moment are as follows:

1. Shear is considered positive at a section when it tends to rotate the portion of the beam in the clockwise direction about an axis through a point inside the free body and normal to the plane of loading; otherwise it is negative (see Fig. 3-3(a)).

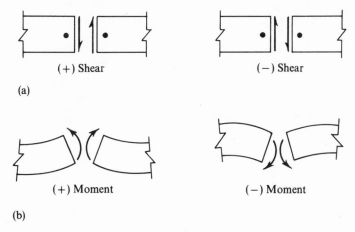

(a)   (+) Shear   (−) Shear

(b)   (+) Moment   (−) Moment

Fig. 3-3

2. Bending moment is considered positive at a section when it tends to bend the member concave upward; otherwise it is negative (see Fig. 3-3(b)).

Such sign conventions, although arbitrary, must be carefully observed to avoid confusion.

The analysis of statically determinate beams is illustrated in the following examples.

Example 3-1. In Fig. 3-4(a) is shown a simple beam under a concentrated load P acting at C.

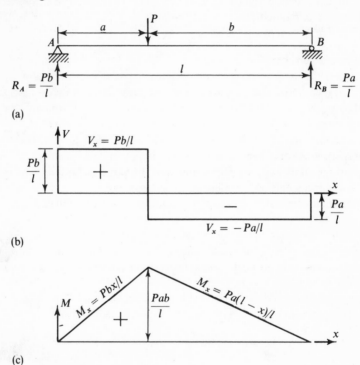

Fig. 3-4

The reactions R_A and R_B are readily found to be

$$R_A = \frac{Pb}{l} \qquad R_B = \frac{Pa}{l}$$

from $\sum M_B = 0$ and $\sum M_A = 0$.

The shear at any section to the left of P is equal to R_A; i.e.,

$$V_x = \frac{Pb}{l} \qquad (0 < x < a)$$

The shear at any section to the right of P is found to be equal to R_B but with negative sign, according to our sign convention. Thus

$$V_x = -\frac{Pa}{l} \qquad (a < x < l)$$

There is a change process of shear occuring at section C, the total change being $-P$ (from Pb/l to $-Pa/l$). In this connection, we note that at a concentrated load (including reaction) there is, in general, an abrupt change in the shear equal to the load. Consider the shear in the immediate vicinity of each support. The shear on a section an infinitesimal distance to the right of point A is Pb/l; therefore the shear curve rises abruptly from zero to Pb/l at A. Similarly, the shear goes to zero from the value $-Pa/l$ at B. In general, *the shear curve always starts at zero and ends at zero* (see Fig. 3-4(b) for the shear curve).

The bending moment at any section distance x from A is given by

$$M_x = \frac{Pb}{l}x \qquad (0 \le x \le a)$$

$$M_x = \frac{Pa}{l}(l - x) \qquad (a \le x \le l)$$

Both are of linear variation and are plotted in Fig. 3-4(c).

If there are several concentrated loads on the beam, we need as many linear equations to represent the shear or moment as the number of segments involved. The shear or moment diagram is then composed of a series of line segments.

It is customary to drop the notion of coordinate axes in the diagram unless the origin of coordinate system is otherwise specified.

Example 3-2. Fig. 3-5(a) shows a simple beam subjected to a uniform load of intensity w.

Because of symmetry, the reactions are each equal to $wl/2$ as shown. Then at any section distance x from the left end A, we have

$$V_x = \frac{wl}{2} - wx$$

$$M_x = \frac{wl}{2}x - \frac{wx^2}{2}$$

These are shown in Fig. 3-5(b) and (c) respectively.

Example 3-3. In Fig. 3-6(a) is shown a simple beam subjected to an external couple of M applied at C.

The reactions R_A and R_B must be such as to form a couple to balance M. They must be equal to M/l and opposite in sense, as indicated in Fig. 3-6(a).

In this case the shear is of constant value equal to $-M/l$ in the range $0 < x < l$, as shown in Fig. 3-6(b).

The moments vary linearly from A to C and from C to B and are given by

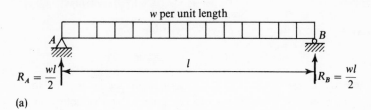

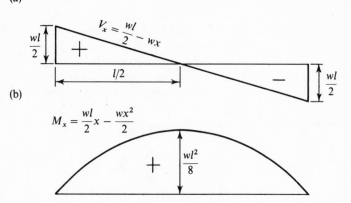

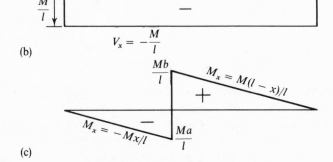

Fig. 3-5

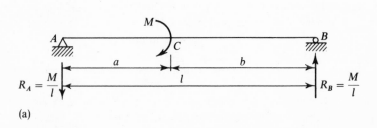

Fig. 3-6

$$M_x = -\frac{M}{l}x \qquad (0 \leq x < a)$$

$$M_x = \frac{M}{l}(l - x) \qquad (a < x \leq l)$$

The process of moment changes at $C(x = a)$, the total change being M (from $-Ma/l$ to Mb/l). The moment diagram is shown in Fig. 3-6(c), and the point of inflection (zero moment) is at C where the moment curve passes the x-axis.

If the beam is subjected to several loads (or several groups of loads), the

TABLE 3-1

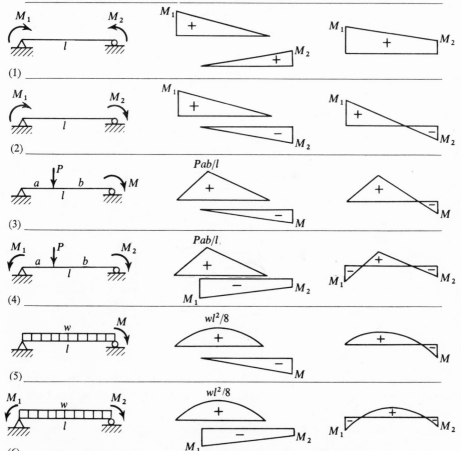

Case	Separate Moment Diagram	Combined Moment Diagram*

*With large end moment (or moments) the combined moment diagrams for Cases (3), (4), (5), or (6) could be all negative without point of inflection.

shear or the moment diagram may be plotted separately for each load (or each group of loads) and then combined into one diagram by the principle of super-position.

Table 3-1 shows the application of this principle in drawing qualitative moment diagram for a simple beam (or a general member) restrained by end moments.

From Table 3-1, we see that if the beam carries no load but end moments, the moment curve is a straight line with one point of inflection or no point of inflection. If the beam carries a concentrated load or a uniform load together with the end moment on one end or both ends, the moment curve may pass zero at one, or two, or no point on the beam axis.

It may be worth mentioning here that if the beam is made of elastic ma-terial, the beam will be deformed under load; the elastic deformations of beams are primarily caused by bending. With reference to the moment diagram and points of zero moment, we can easily sketch the deflected elastic curve.

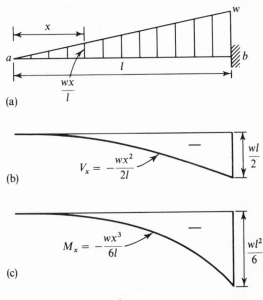

Fig. 3-7

Example 3-4. In Fig. 3-7(a) is shown a cantilever carrying a distributed load the intensity of which varies linearly from w per unit length at the fixed end to zero at free end.

At any section distance x from the free end a,

$$V_x = -\left(\frac{wx}{l}\right)\left(\frac{x}{2}\right) = -\frac{wx^2}{2l}$$

$$M_x = -\left(\frac{wx^2}{2l}\right)\left(\frac{x}{3}\right) = -\frac{wx^3}{6l}$$

These are plotted in Figs. 3-7(b) and (c) respectively.

Example 3-5. In Fig. 3-8(a) is shown a simple beam with an overhang. We wish to plot the shear and moment diagrams.

It may be convenient to analyze the beam by separating it into two parts, a simple beam and the overhanging portion and then by superposing the results. To do this, we cut the beam into two free bodies by passing a section at the immediate right side of support B, as shown in Fig. 3-8(b). The analysis of the overhanging portion, similar to that of a cantilever, illustrated in Example

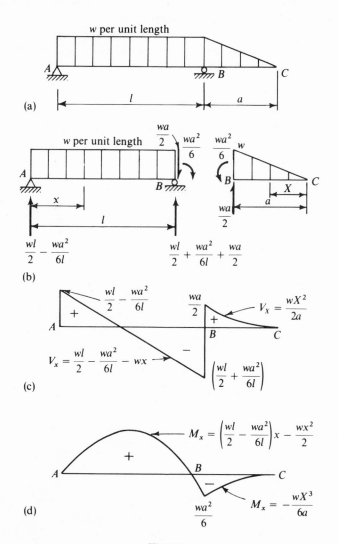

Fig. 3-8

3-4, is first laid out. At any section distance X from end C (origin is taken at C), we have

$$V_X = \frac{wX^2}{2a} \qquad M_X = -\frac{wX^3}{6a}$$

The shear and moment at the cut section just to the right of support B are thus found to be:

$$V_B = \frac{wa}{2} \quad \text{(up)} \qquad M_B = \frac{wa^2}{6} \quad \text{(counter-clockwise)}$$

Next, consider the portion AB, which is subjected to a uniform load over the entire span together with an end shear of $wa/2$ (down) and an end moment of $wa^2/6$ (clockwise) at B exerted on by the portion BC. The value of $wa/2$ will go to support B and cause no bending and shear on AB. At any section distance x from A (origin is taken at A),

$$V_x = \left(\frac{wl}{2} - \frac{wa^2}{6l}\right) - wx$$

which is of linear variation; and

$$M_x = \left(\frac{wl}{2} - \frac{wa^2}{6l}\right)x - \frac{wx^2}{2}$$

which represents a parabola. Note that the construction of the moment curve for the portion of the simple beam may be done by superposing the positive moment and the negative moment, as suggested in Table 3-1.

The shear and moment diagrams for the entire beam are plotted in Figs. 3-8(c) and (d).

Example 3-6. Analyze the loaded compound beam of Fig. 3-9(a), which is composed of a simple beam and a cantilever connected by an internal hinge.

We note that the number of unknown reaction elements, including the horizontal reaction of support e, is four. These are to be solved by four independent equations of statics, i.e., three equilibrium equations and one condition equation:

$$\sum F_x = H_e = 0$$
$$\sum F_y = V_a + V_e - 2P = 0$$
$$\sum M_e = (V_a)(4p) - P(3p + p) + M_e = 0$$
$$M_c = (V_a)(2p) - Pp = 0$$

From the above, we obtain $V_a = P/2$, $V_e = 3P/2$, $H_e = 0$, and $M_e = 2Pp$.

After the reactions are determined, it is not difficult to construct the shear and moment diagrams, shown in Figs. 3-9(b) and (c) respectively. Note that the moment curve must pass zero at c as required.

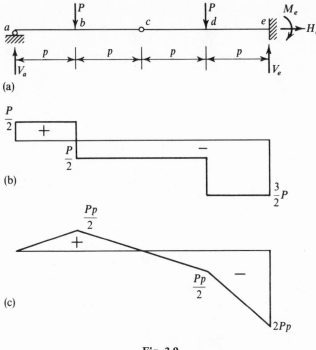

Fig. 3-9

3-3. PRINCIPLE OF VIRTUAL WORK AND BEAM
ANALYSIS

The principle of virtual work as applied to statically determinate plane structures may be stated as follows:

If a small virtual displacement consistent with the constraints is given to an ideal system in equilibrium under the action of a balanced force system, the total work done by all active forces is equal to zero.

By an *ideal system* we mean a system of rigid bodies connected and supported smoothly and rigidly so that no energy is stored or dissipated during a small displacement.

Note that if such a system is in equilibrium under the action of a balanced force system, no displacement can actually occur, and no work can actually be done by the forces. However, an arbitrary, fictitious displacement, called *virtual displacement*, for the system may be assumed to occur. This has nothing to do with the real displacement of the body but only serves as a technique in solving equilibrium problems. Virtual displacement has no finite value, otherwise it might cause some shifting of the lines of action of the forces so that the system would be no longer in equilibrium. For this reason, the virtual displacement is generally assumed to be vanishingly small. It is customary to use δs to denote the linear virtual displacement and $\delta\theta$ the angular virtual displacement, as distinguished from real infinitesimal displacements ds and $d\theta$. The corresponding work done by the force system is called *virtual work*, and is usually denoted by δW.

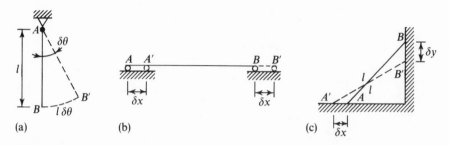

Fig. 3-10

Also note that virtual displacements for a system should be so chosen that they are compatible with the constraints on the system. Thus for the bar in Fig. 3-10(a) possible rotation about the supporting hinge at A may occur so that bar AB will turn through an angle $\delta\theta$ at A as indicated. For the beam in Fig. 3-10(b) horizontal displacement may take place as a result of the lack of lateral resistance of the roller supports so that end A will move to A' and B to B'; and $AA' = BB' = \delta x$. Likewise, for the beam shown in Fig. 3-10(c) end A may move to the left from A to A', $AA' = \delta x$, and end B to B' downward, $BB' = \delta y$. In all cases, the structure must be in an unstable form, otherwise the displacement is impossible.

Active force is the opposite of *workless force*. The work done by a constantly applied force P during a displacement s is defined as the product of the magnitude of the force, the magnitude of the displacement, and the cosine of the angle θ between the force and the displacement, i.e.,

$$W = Ps \cos \theta \qquad (3\text{-}1)$$

And the work done by a couple of constant moment M is defined as the product of the magnitude of the moment and of the magnitude of the angular displacement ϕ of the couple; i.e.,

$$W = M\phi \qquad (3\text{-}2)$$

A force produces positive work when the displacement and the projection of the force along the displacement are in the same direction; a couple does positive work when the angular displacement and the couple have the same sence. From this we see that such forces as the following are termed *workless*.

1. the force which is normal to the displacement

2. the force which does not move or the couple which does not turn

3. the internal forces or moments which exist in pairs with equal magnitude but opposite direction so that they do an equal amount of positive work and negative work for any displacement assigned to the system

Note that the internal forces do produce work in an elastic body known as *strain energy*, as will be discussed in Sec. 8-3.

To prove the principle of virtual work, we begin with the case of single particle A under a balanced coplanar concurrent force system $P_1, P_2, \cdots, P_n$ as shown in Fig. 3-11. Imagine that A is given a small arbitrary displacement δs as shown. Then the total virtual work done by this force system is

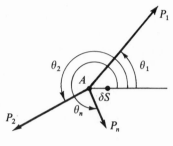

Fig. 3-11

$$\delta W = (P_1 \cos \theta_1 + P_2 \cos \theta_2 + \cdots + P_n \cos \theta_n)\delta s$$

The expression in parenthesis is the projection of resultant force in the direction of the displacement, and it must be zero since the particle is in equilibrium. Therefore,

$$\delta W = 0 \tag{3-3}$$

meaning that the virtual work done on a particle during any virtual displacement is zero.

Next, let us consider a rigid body in equilibrium. A rigid body may be regarded as a collection of particles, say n particles, and the system is in equilibrium if every one of its n constituent particles is in equilibrium. Therefore, for an isolated particle i, the total work done by the forces acting on it due to any virtual displacement given to the body is zero according to Eq. 3-3; i.e.,

$$\delta W_i = 0$$

For the system considered as the summation of the total particles we may write the expression of virtual work as the sum of the amount of virtual work done on all the particles of the body; i.e.,

$$\delta W = \delta W_1 + \delta W_2 + \cdots + \delta W_n = \sum_{i=1}^{n} \delta W_i = 0 \tag{3-4}$$

Now the forces involved in the above expression are of two kinds, external and internal. The internal forces occur in self-canceling pairs; only the active part of external forces contributes to the virtual work. Thus we narrow down to the statement that the total virtual work done by all active forces acting on a rigid body during any virtual displacement is zero.

Finally, let us extend the principle to a balanced ideal system which consists of a number of rigid bodies, say m bodies, joined together and supported smoothly and rigidly. For any isolated rigid body j in the system, the total work done by the forces acting on it because of any small compatible virtual displacement assigned to the system is zero according to Eq. 3-4

$$\delta W_j = 0$$

Since there is no energy loss or gain in each ideal connection or means of support, we can write an expression of virtual work for the system based on the summation of virtual works done on all of the individual rigid bodies; i.e.,

$$\delta W = \delta W_1 + \delta W_2 + \cdots + \delta W_m = \sum_{j=1}^{m} \delta W_j = 0 \qquad (3\text{-}5)$$

However, of the forces involved in the above expressions, some are internal forces exerted by connections which produce positive virtual work on one body and the same amount of negative virtual work on the other of each connected pair; others are workless support reactions, and only the active part of the external forces on the system contributes to the virtual work. This completes our proof for the statement.

In solving certain problems of statics, we find that methods of solution based on the principle of virtual work have a decided advantage over the equilibrium equations based on the procedure of taking a free body. The principle is of general use. However, here we shall limit its application to the analysis of statically determinate beams, as will be illustrated in the following examples.

Example 3-7. Figure 3-12(a) shows a simple beam AB subjected to a single concentrated load P acting at point C. We wish to find the reaction at end A by the method of virtual work.

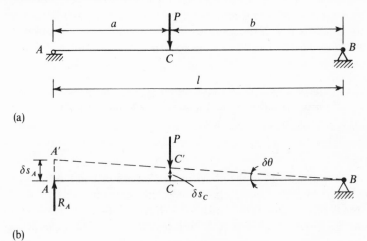

(a)

(b)

Fig. 3-12

To do this, we start by removing the constraint at A and substituting for it a vertical force R_A, as shown in Fig. 3-12(b). Note that in applying the principle of virtual work, we must remove certain constraints in a stable structure since otherwise no displacement is possible.

Now if an infinitesimal virtual angular displacement $\delta\theta$ of rigid beam AB about hinge at B is assumed to take place, the beam will deflect as shown by the dotted line $A'C'B$ in Fig. 3-12(b). The corresponding virtual displacements at A and C, denoted by δs_A and δs_C respectively, are found to be

$$\delta s_A = l\,\delta\theta$$

$$\delta s_C = b\,\delta\theta$$

Both δs_A and δs_C can be considered to be verticle, since the arcs can be taken as coincident with their own verticle tangents for a small rotation. Thus the equation of virtual work can be written as

$$(R_A)(l\,\delta\theta) - (P)(b\,\delta\theta) = 0$$

from which

$$R_A = \frac{Pb}{l}$$

Example 3-8. In Fig. 3-13(a) is shown a simple beam subjected to a concentrated load P applied at midspan. Let us find the shear and moment at section C by the method of virtual work.

Before starting, we should note that the internal forces of shear and moment at any section exist in pairs equal in magnitude but opposite in direction. They are workless when both acquire the same displacement; they produce work only when the section is cut and a relative displacement occurs between the two sides of the cut section.

To analyze the internal forces at section C of Fig. 3-13(a), we first let the beam be cut at section C and replaced by shearing forces V_C and resisting moments M_C in pairs, as shown in Fig. 3-13(b), thus placing the beam in a state of unstable equilibrium. The virtual displacement is so chosen that each time we introduce only one active unknown element into the system, to avoid solving simultaneous equations. Thus to find V_C we let portion AC turn about A through a virtual angular displacement $\delta\theta$ and the portion BC rotate about B with same virtual angular displacement $\delta\theta$. This leaves the internal moment forces workless, since their contributions ($M_C\,\delta\theta - M_C\,\delta\theta$) cancel each other. In other words, we allow a small, relative transverse sliding between the two portions AC and BC at C and at the same time prevent them from relative rotation. The displacement diagram for the beam consistent with the assumed angular displacements is shown by the dotted lines AC_1C_2B in Fig. 3-13(c). Based on the load diagram in Fig. 3-13(b) and the displacement diagram in Fig. 3-13(c), the equation of virtual work can be written

$$V_C(a\,\delta\theta + b\,\delta\theta) - P\left(\frac{l}{2}\,\delta\theta\right) = 0$$

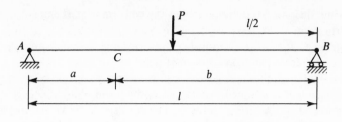

(a)

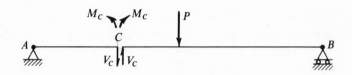

(b)

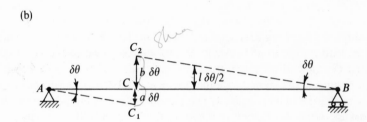

(c)

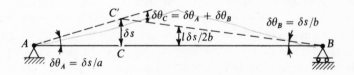

(d)

Fig. 3-13

Or

$$V_c l\,\delta\theta - P\frac{l}{2}\,\delta\theta = 0$$

from which

$$V_c = \frac{P}{2}$$

Similarly, to find M_c we let both portions AC and BC move transversely the same virtual displacement δs at C, thus leaveing the shearing forces inactive in work. In other words, we allow relative rotation between the two portions AC and BC at C but at the same time prevent them from relative transverse sliding. The displacement diagram for the beam, consistent with the assumed δs, is given in Fig. 3-13(d) by dotted line $AC'B$. Based on the load diagram in Fig. 3-13(b) and the displacement diagram in Fig. 3-13(d), we may write the equation of virtual work as

$$M_C \left(\frac{\delta s}{a} + \frac{\delta s}{b} \right) - P \left(\frac{l\,\delta s}{2b} \right) = 0$$

Or
$$M_C \left(\frac{l\,\delta s}{ab} \right) - P \left(\frac{l\,\delta s}{2b} \right) = 0$$

from which
$$M_C = \frac{Pa}{2}$$

Example 3-9. The technique demonstrated in the previous example can be used to solve more complicated cases, like the compound beam loaded as shown in Fig. 3-14(a). Let us determine the reactions at supports A and E.

To find the reaction at A, called R_A, we remove the support and introduce R_A at A. Furthermore, we let end A move vertically a small virtual displacement δs_A. (Note that, as explained before, this is equivalent to rotating the portion AC through a small angle about the hinge at C.) The whole beam undergoes deflections as shown in Fig. 3-14(b) by $A'B'CDE$. Applying the principle of virtual work, we have

$$(R_A)(\delta s_A) + (Pa) \left(\frac{\delta s_A}{2a} \right) = 0$$

from which
$$R_A = -\frac{P}{2}$$

Next, to determine the reaction at E, we remove the constraint at E and introduce in its place the reaction components V_E and M_E. Solving V_E, we let point E move a small vertical distance δs_E without rotating the beam axis at E; the beam will deflect as shown by the dotted line $AB'C'D'E'$ in Fig. 3-14(c). Applying the virtual work equation gives

$$(V_E)(\delta s_E) - (P)(\delta s_E) - (Pa) \left(\frac{\delta s_E}{2a} \right) = 0$$

from which
$$V_E = \frac{3P}{2}$$

Note that in this case M_E is inactive in work.

Solving M_E, we let CE rotate about the point E through a small angular displacement $\delta\theta_E$ without producing linear movement at E. The deflected beam

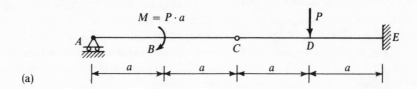

(a)

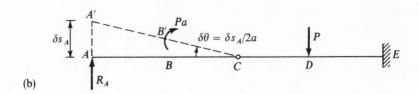

(b)

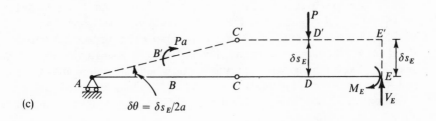

(c)

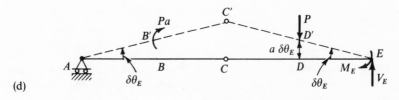

(d)

Fig. 3-14

consistent with this is shown by the dotted line $AB'C'D'E$ in Fig. 3-14(d). Thus

$$(M_E)(\delta\theta_E) - (P)(a\,\delta\theta_E) - (Pa)(\delta\theta_E) = 0$$

Or

$$M_E = 2Pa$$

Note that in this case V_E is inactive in work.

From the above illustration, we see that the value of the principle of virtual work in solving problems of statics is that no free bodies need be taken and that solving simultaneous equations can be avoided.

3-4. RELATIONSHIPS BETWEEN LOAD, SHEAR, AND
BENDING MOMENT

There exist at any cross section of a loaded beam certain relationships between load, shear, and bending moment that are tremendously helpful in constructing the shear and bending moment curves.

Consider a portion of a beam of any type subjected to transverse loading and moment loading, such as the one shown in Fig. 3-15(a). To investigate the relationships between load, shear, and bending moment in a beam, we may classify the beam segments in the following way (as partly illustrated in Fig. 3-15(a)):

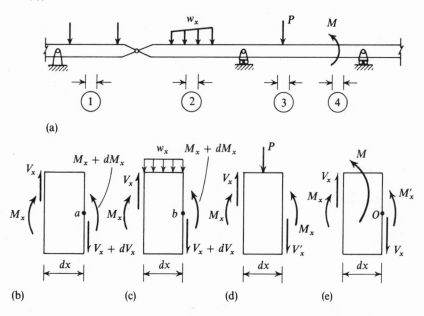

Fig. 3-15

1. Segment under no load
2. Segment under distributed load
3. Segment under concentrated load
4. Segment under moment load

We shall deal with each of the above four cases as follows:

1. Segment under no load. As indicated in Fig. 3-15(a), a segment between two concentrated loads is an example of a segment under no load. Let us take an element cut out by two adjacent cross sections distance dx apart, as shown in Fig. 3-15(b). On the left-hand face of this element, we represent the shear

force and bending moment by V_x and M_x and on the right-hand face of the element by $V_x + dV_x$ and $M_x + dM_x$ in which dV_x and dM_x are changes of shear and moment in dx. We assume that x increases from left to right. Since the element is in equilibrium, we have from $\sum F_y = 0$

$$V_x - (V_x + dV_x) = 0$$

that is,

$$dV_x = 0$$

or

$$V_x = \text{constant} \qquad (3\text{-}6)$$

Also from $\sum M_a = 0$,

$$M_x + V_x\,dx - (M_x + dM_x) = 0$$

Reducing and using Eq. 3-6, we arrive at

$$\frac{dM_x}{dx} = \text{constant} \qquad (3\text{-}7)$$

Equation 3-6 states that no change of shear takes place, and Eq. 3-7 states that the rate of change of bending moment at any point with respect to x is constant.

2. Segment under distributed load. Let us take an element subjected to a distributed load cut out by two adjacent cross sections distance dx apart, as shown in Fig. 3-15(c). Assume a downward distributed load in a positive direction. From $\sum F_y = 0$,

$$V_x - (V_x + dV_x) - w_x\,dx = 0$$

$$dV_x = -w_x\,dx$$

or

$$\frac{dV_x}{dx} = -w_x \qquad (3\text{-}8)$$

From $\sum M_b = 0$, $\qquad M_x + V_x\,dx - w_x\,dx\,\dfrac{dx}{2} - (M_x + dM_x) = 0$

Neglecting the small term $w_x(dx)^2/2$ and reducing, we find

$$\frac{dM_x}{dx} = V_x \qquad (3\text{-}9)$$

Equation 3-8 states that the rate of change of shear with respect to x at any point is equal to the intensity of the load at that point but with the opposite sign. Equation 3-9 states that the rate of change of bending moment with respect to x at any point is equal to the shear force at that point.

3. Segment under concentrated load. Fig. 3-15(d) shows an element subjected to a concentrated load P. Now P is assumed to be acting at a point. As the distance between two adjacent sections becomes infinitesimal, there will be no moment difference between the sections to the immediate left of P and to

the immediate right of P. However, an abrupt change in the shear force equal
to P between the two sections takes place, since from $\sum F_y = 0$,

$$V_x - P - V'_x = 0$$

or

$$V'_x = V_x - P \qquad (3\text{-}10)$$

as indicated in Fig. 3-15(d). Accordingly, there will be an abrupt change in
the derivative dM_x/dx at the point of application of concentrated force.

4. Segment under moment load. In Fig. 3-15(e) is shown an element subjected to a couple of M. Now M is assumed to be acting at a point. As the
distance between the two adjacent sections becomes infinitesimal, there will be
no shear difference between the sections to the immediate left of M and to the
immediate right of M. However, there will be an abrupt change of moment
equal to M between the two sections, for from $\sum M_o = 0$ we have

$$M_x - M - M'_x = 0$$

or

$$M'_x = M_x - M \qquad (3\text{-}11)$$

as indicated in Fig. 3-15(e).

Construction of the shear and moment diagrams is facilitated by the relationships previously stated. For instance, the equation

$$\frac{dV_x}{dx} = -w_x$$

implies that the slope of the shear curve at any point is equal to the negative
value of the ordinate of load diagram applied to the beam at that point. There
are cases worth noting:

1. For a segment under no load, the slope of the shear curve is zero, i.e.,
parallel to the beam axis. The shear curve is therefore a line parallel to the
beam axis.

2. For a segment under a uniform load of intensity w, the slope of the
shear curve is constant. The shear curve is therefore a sloping line.

3. At a point of concentrated load, the intensity of the load is infinite, and
the slope of the shear curve will thus be infinite, i.e., vertical to the beam axis.
There will be a discontinuity in the shear curve, and a change process of shear
equal to the applied force occurs between the two sides of the loaded point.

4. Under distributed load the change in shear between two cross sections
a differential distance dx apart is

$$dV_x = -w_x \, dx$$

Thus the difference in the ordinates of the shear curve between any two points
a and b is given by

$$V_b - V_a = -\int_{x_a}^{x_b} w_x \, dx$$

$$= -(\text{area of load diagram between } a \text{ and } b) \qquad (3\text{-}12)$$

Suppose that there are additional concentrated forces $\sum P$ acting between a and b. The result of the shear difference between the two points must include the effect due to $\sum P$; i.e.,

$$V_b - V_a = -\int_{x_a}^{x_b} w_x \, dx - \sum P$$

$$= -(\text{area of load diagram between } a \text{ and } b + \sum P) \qquad (3\text{-}13)$$

in which $\sum P$ has been assumed to act downward.

Similarly, from the equation

$$\frac{dM_x}{dx} = V_x$$

the slope of the bending moment curve at any point equals the ordinate of the shear curve at that point. We note the following:

1. If the shear is constant in a portion of the beam, the bending moment curve will be a straight line in that portion.

2. If the shear varies in any manner in a portion of the beam, the bending moment curve will be a curved line.

3. At a point where a concentrated force acts there will be an abrupt change in the ordinate of shear curve and, therefore, an abrupt change in the slope of the bending moment curve at the point. In fact, the moment curve will have two different slopes at that point.

4. Maximum and minimum bending moments occur at the points where a shear curve goes through the x axis—the maximum where shear changes from positive (at the left) to negative (at the right); the minimum in the reverse manner.

5. For a concentrated force system the maximum bending moment must occur under a certain concentrated force, since change of shear from positive to negative must occur at a certain point where a concentrated force is applied.

6. Referring to the equation $dM_x/dx = V_x$, we find that under transverse loading the change in bending moment between two cross sections a differential distance dx apart is given by

$$dM_x = V_x \, dx$$

Therefore the difference in the ordinates of the bending moment curve between any two points a and b is given by

$$M_b - M_a = \int_{x_a}^{x_b} V_x \, dx$$

$$= \text{(area of shear diagram between } a \text{ and } b\text{)} \qquad (3\text{-}14)$$

If there are external moments $\sum M$ acting between a and b, then the result of the moment difference between the two points must include the effect due to these moments; i.e.,

$$M_b - M_a = \int_{x_a}^{x_b} V_x \, dx - \sum M$$

$$= \text{(area of shear diagram between } a \text{ and } b\text{)} - \sum M \qquad (3\text{-}15)$$

in which $\sum M$ has been assumed to act in a counter-clockwise direction.

3-5. NUMERICAL EXAMPLES

The following examples will serve to illustrate the construction of shear and bending moment diagrams of transversely loaded beams by using the principles stated in Sec. 3-4.

Example 3-10. Consider a simple beam with an overhang loaded as shown in Fig. 3-16(a).

From $\sum M_b = 0$ and $\sum M_d = 0$, the support reactions are found to be

$$R_b = R_d = 16 \text{ kips}$$

We may now regard the beam as being in equilibrium under the balanced system of applied loads and reactions and present the load diagram as shown in Fig. 3-16(b).

A freehand sketch of the shear diagram can then be drawn, as in Fig. 3-16(c). In connection with this diagram, we note the following facts:

1. The shear at a goes from 0 dropping to -2 kips; also the shear at e is 0. Recall that the shear curve always starts at zero and ends up at zero.

2. There will be constant shear in portion ab since it is not loaded. As a result, the shear curve in this portion is a horizontal line parallel to the beam axis.

3. There will be abrupt changes in shear at b, c, and d corresponding to the concentrated forces acting at these points. The total change of shear at each of these points is equal to the force at the point.

4. From b (right) to c (left), c (right) to d (left), and d (right) to e, the shear curves are sloping lines, the slope being given by

$$\frac{dV}{dx} = -w = -1$$

i.e., $1 : 1$ downward to the right as indicated in Fig. 3-16(c).

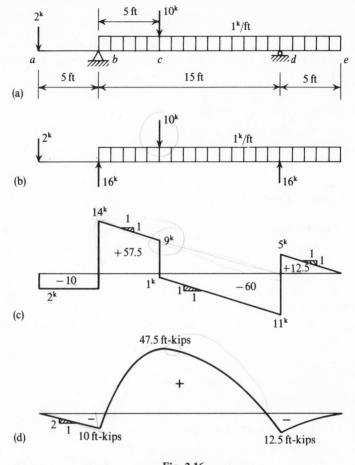

Fig. 3-16

A freehand sketch of a bending moment diagram can be drawn, as in Fig. 3-16(d). In connection with it, we note the following:

1. Moments at a and e are null. The moment curve from a to b is a sloping line with the slope given by

$$\frac{dM}{dx} = V = -2$$

i.e., $2 : 1$ downward to the right, as indicated in Fig. 3-16(d).

2. There are extreme values of moment at points b, c, and d where the shear curve goes through the x axis. Minimum bending moments occur at b and d since abrupt changes in the slope of moment curve from negative to positive

take place at these points corresponding to the abrupt changes in shear from negative to positive. Maximum bending moment occurs at c where an abrupt change in the slope of the moment curve from positive to negative takes place, corresponding to the shear change from positive to negative at c.

3. Since the shear curve between bc or cd or de decreases from left to the right, the slope of the moment curve in each portion also decreases from left to right. This means that the moment curve in each portion is concave downward.

4. One way to obtain the ordinates of the moment diagram at b, c, and d is to compute the areas of the shear diagram (see the values indicated in Fig. 3-16(c)) from which we may find the moment difference between any two points; i.e.,

$$M_b - M_a = -10 \text{ ft-kips} \qquad M_c - M_b = 57.5 \text{ ft-kips}$$
$$M_d - M_c = -60 \text{ ft-kips} \qquad M_e - M_d = 12.5 \text{ ft-kips}$$

From the above and using $M_a = M_e = 0$, we find

$$M_b = -10 \text{ ft-kips} \qquad M_c = 47.5 \text{ ft-kips} \qquad M_d = -12.5 \text{ ft-kips}$$

as indicated in Fig. 3-16(d).

It may be noted that the algebraic sum of the total area of the shear diagram for the beam is zero in this example, since the difference in moment between a and e is zero.

Example 3-11. Fig. 3-17(a) shows a simple beam carrying a distributed load that varies linearly from 2 kips per ft at the left end to 0 at the right end.

Applying $\sum M_b = 0$ and $\sum M_a = 0$, we obtain the vertical reaction at a equal to 16 kips and that at b equal to 8 kips, respectively. The load diagram for the beam is shown in Fig. 3-17(b).

In connection with the sketch of the shear diagram shown in Fig. 3-17(c), we note:

1. The shear ordinates at a and b are from zero to 16 kips and -8 kips, respectively.
2. The shear curve has a slope of $-(2:1)$ at a and zero at b; the slope of the shear curve increases continuously from left to right, which corresponds to a continuously decreasing load intensity from left to right. Hence the shear curve is concave upward.

In connection with the sketch of the moment diagram shown in Fig. 3-17(d), we note:

1. $M_a = M_b = 0$. Between a and b the moment curve is concave downward

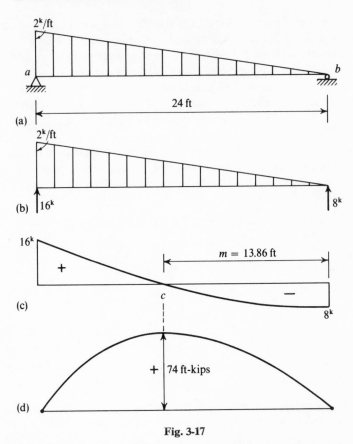

Fig. 3-17

corresponding to the continuously decreasing shear, i.e., the continuously decreasing slope of the moment curve from the left end to the right end.

2. There will be a maximum bending moment corresponding to the zero shear, the location of which may be determined by writing the expression for shear at point c in terms of its distance m from end b (see Fig. 3-17(c)), equating this shear to zero, and then solving for m. Thus,

$$V_c = \left(\frac{1}{2}\right)\left(\frac{2m}{24}\right)(m) - 8 = 0$$

$$m = 8\sqrt{3} = 13.86 \text{ ft}$$

In this particular case the maximum bending moment can be more conveniently obtained by conventional calculation than from the area of shear diagram. Using the portion to the right of c as free body, we obtain

$$M_{\max} = (8)(13.86) - \frac{(2)(13.86)^3}{(2)(3)(24)} = 74 \text{ ft-kips}$$

PROBLEMS

3-1. In each part of Fig. 3-18 qualitative loadings are shown. Draw the shear and moment diagrams consistent with these loadings; give the equation for each curve.

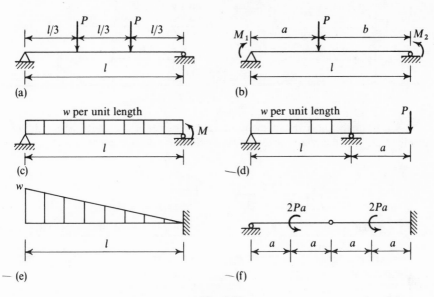

Fig. 3-18

3-2. Calculate the reactions at supports A, B, and D, the moment at B, and the shear at C of the compound beam shown in Fig. 3-19 by the method of virtual work.

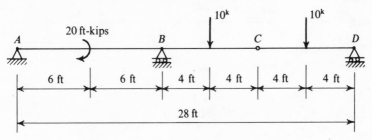

Fig. 3-19

3-3. Use the method of virtual work to find the reactions induced at support A and F and the shear and moment at B of the compound beam shown in Fig. 3-20.

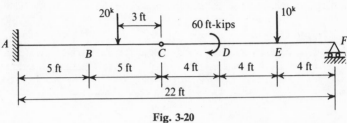

Fig. 3-20

3-4. Sketch the shear and moment diagrams for each of the loaded beams shown in Fig. 3-21. Use the relationships between load, shear, and moment.

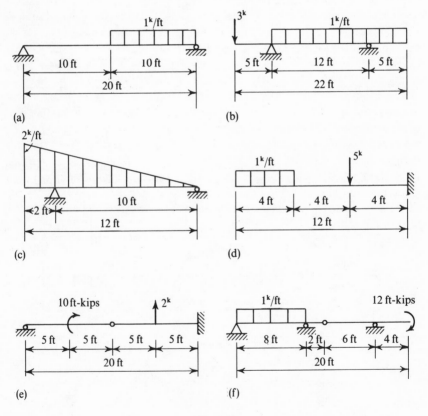

Fig. 3-21

4

STATICALLY DETERMINATE TRUSSES

4-1. GENERAL

A *truss*, such as the one shown in Fig. 4-1, may be defined as a plane structure composed of a number of members joined together at their ends by smooth pins so as to form a rigid framework the external forces and reactions of which are assumed to lie in the same plane and to act only at the pins. Furthermore, we assume that the centroidal axis of each member coincides with the line connecting the joint centers at the ends of member and that the weight of each member is negligible in comparison to the other external forces acting on the truss. From these conditions it follows that each member in a truss is a *two-force* member and is subjected only to direct axial forces (tension or compression). A truss is completely analyzed when the internal axial forces in all members have been determined. It is customary to use a plus sign to designate tension and a minus sign to designate compression. Thus $+5$ kips means a tension of 5 kips, and -5 kips a compression of 5 kips.

A modern truss made of riveted or welded joints is not really a truss by a strict interpretation of this definition. However, since a satisfactory stress analysis may usually be worked out by assuming that such a structure acts as if it were pin-connected, it may still be called a truss.

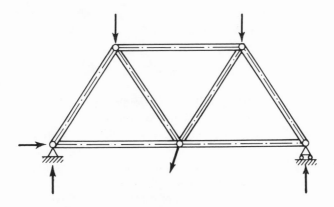

Fig. 4-1

Common trusses may be classified according to their formations as *simple*, *compound*, and *complex*.

1. Simple truss. A rigid plane truss can always be formed by beginning with three bars pinned together at their ends in the form of a triangle and then extending from this two new bars for each new joint, as explained in Sec. 2-6. Of course the new joint and the two joints to which it is connected should never lie along the same straight line in order to avoid geometric instability. Trusses the members of which have been so arranged are called *simple trusses*, for they are the simplest type of bar arrangement encountered in practice.

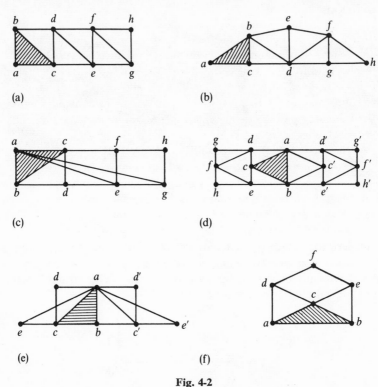

Fig. 4-2

The trusses shown in Fig. 4-2 are all simple trusses. The shaded triangle *abc* in each truss diagram is the base figure from which we extended the form by using two additional bars to connect each of the new joint in alphabetical order.

It can easily be shown that there exists a very definite relationship between the number of bars b and the number of joints j in a simple truss. Since the base triangle of a simple truss consists of three bars and three joints, the additional bars and joints required to complete the truss are $(b - 3)$ and $(j - 3)$, respectively. These two numbers should be in a 2 : 1 ratio. Thus

$$b - 3 = 2(j - 3)$$

or
$$b + 3 = 2j$$

Comparing the above equation with the necessary condition for a statically determinate truss (see Sec. 2-6) given by

$$b + r = 2j$$

we find that, if the supports of a simple truss are so arranged that they are composed of three elements of reaction neither parallel nor concurrent, then the structure is stable and statically determinate under general conditions of loading.

2. Compound truss. If two or more simple trusses are connected together to form one rigid framework, the composite truss is called a *compound truss*. One simple truss can be rigidly connected to another simple truss at certain joints by three links neither parallel nor concurrent or by the equivalent of this type of connection. Additional simple trusses can be joined in a similar manner

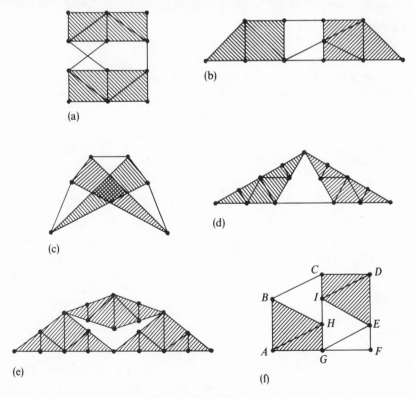

Fig. 4-3

to the framework already constructed to obtain a more elaborate compound truss.

Figure 4-3 shows several examples of compound trusses in which the simple trusses that have been connected together are crosshatched. In cases (a), (b), and (c) two simple trusses are connected by three links. In case (d) two simple trusses are connected by one hinge and one link, which constitute three elements of connection. In case (e) two simple trusses may be connected by a hinge and a link (bar) similar to that of case (d); however, the link is replaced by a third simple truss. Sometimes a compound truss is not apparent. The truss shown in Fig. 4-3(f) is such a case. This truss is composed of two simple trusses *ABHG* and *CDEI* connected by two bars *BC* and *HI* and a third truss *EFG*; therefore, it is a compound truss.

Note that the relationship between the number of bars and the number of joints for a simple truss,

$$b + 3 = 2j$$

still holds for a compound truss. Therefore, a compound truss may be supported in the same way as a simple truss to obtain a stable and statically determinate structure.

3. Complex truss. Trusses that cannot be classified as either simple or compound are called *complex trusses*.

In Fig. 4-4(a) is shown a simple truss. Rearranging the bars results in a compound truss such as the one shown in Fig. 4-4(b). However, the truss

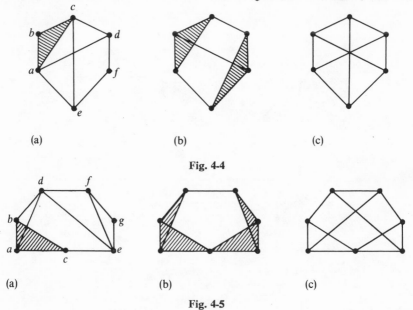

(a)   (b)   (c)

Fig. 4-4

(a)   (b)   (c)

Fig. 4-5

shown in Fig. 4-4(c), made of the same number of bars and joints, does not belong to either of the above categories and may be termed a complex truss. Similarly, we find that the truss shown in Fig. 4-5(a) is a simple truss, that in Fig. 4-5(b) a compound truss, and that in Fig. 4-5(c) a complex truss.

4-2. ANALYSIS OF SIMPLE TRUSSES: METHOD OF
JOINT AND METHOD OF SECTION

The method of joint and the method of section are the most fundamental tools in the stress analysis of trusses. These procedures may be explained by considering a specific example, such as the simple truss shown in Fig. 4-6.

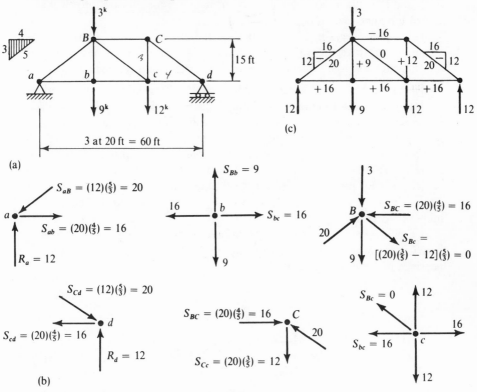

(c)

(a)

(b)

Fig. 4-6

Method of Joint. The reactions

$$R_a = R_d = 12 \text{ kips}$$

are first obtained by taking the whole truss as a free body.

The two equations of equilibrium $\sum F_x = 0$ and $\sum F_y = 0$ are then applied to the *free body of each joint* in such an order that not more than two unknowns are involved in each free body. This can always be done for a simple truss. In this example we start with joint a at the left end and proceed in succession to joints b and B; then we turn to joint d at the right end and proceed to joints C and c. We thus provide three checks for the analysis by obtaining the internal forces in members BC, Bc, and bc from two directions.

The stress analysis for each joint is given briefly in Fig. 4-6(b). Usually when the slopes of the members are in simple ratios, the solution for unknowns can readily be obtained by inspection rather than by using equations. The arrows in each free body of the joint indicate the directions of the member forces acting on the joint, not the actions of the joint on the member. Note that the internal force in the member is a tensile force if it acts outward such as S_{ab} and that the internal force in the member is a compressive force if it acts toward the joint such as S_{aB}.

The *answer diagram* (Fig. 4-6(c)) gives the results obtained from the preceding analysis together with the horizontal and vertical components. A plus sign indicates a tensile force, and a minus sign indicates a compressive force.

Method of Section. Sometimes when only the stresses in certain members are desired or when the method of joint is less convenient for solving stresses, it is expedient to use the method of section, which involves isolating a portion of the truss by cutting certain members and then solving the stresses on these members with the equilibrium equations. Consider the truss in Fig. 4-6(a). Let us determine the internal forces in the members BC, Bc, and bc.

We start by passing a section m-m through these members and treating either side of the truss as a free body (see Fig. 4-7). Note that the sense of the unknown force in each cut member is assumed to be tensile and if this is done, a plus sign in the answer indicates that the assumed sense is correct, and therefore tension; whereas a minus sign indicates that the assumed sense is incorrect, and therefore compression.

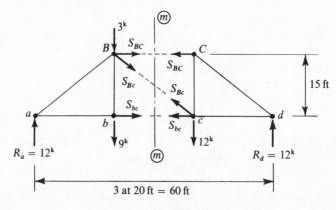

Fig. 4-7

Since in each free body only three unknowns are involved, the unknowns can be solved by three equilibrium equations. In this example, it is convenient to solve S_{BC} by $\sum M_c = 0$; S_{bc} by $\sum M_B = 0$; and S_{Bc} by $\sum F_y = 0$. Thus if we consider the left portion of Fig. 4-7 as a free body, we have

$$\sum M_c = 0 \qquad (12)(40) - (9 + 3)(20) + 15S_{BC} = 0$$

$$S_{BC} = -16 \text{ kips (compression)}$$

$$\sum M_B = 0 \qquad (12)(20) - 15S_{bc} = 0$$

$$S_{bc} = +16 \text{ kips (tension)}$$

$$\sum F_y = 0 \qquad 12 - (9 + 3) - V_{Bc} = 0$$

$$V_{Bc} = 0 \quad \text{or} \quad S_{Bc} = 0$$

in which V_{Bc} represents the vertical component of S_{Bc}. Since $S_{Bc} = (5/3)V_{Bc}$, the zero value of V_{Bc} evidently implies the nonexistence of S_{Bc}.

In connection with the method of joint and the method of section we may note the following:

1. For an unloaded joint, if two bars are collinear, then the force in the third member is zero.

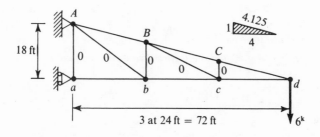

Fig. 4-8

Example 4-1. For the loaded truss shown in Fig. 4-8, we find that the internal forces in bars Cc, cB, Bb, and bA are zero successively from taking free bodies at joints C, c, B, and b and applying $\sum F_V = 0$. V indicates the direction perpendicular to the collinear bars. The remaining forces are

$$S_{ab} = S_{bc} = S_{cd} = 24 \text{ kips} \qquad S_{AB} = S_{BC} = S_{Cd} = 24.75 \text{ kips} \qquad S_{Aa} = 0$$

2. In applying the method of section, we note that by proper choice of moment centers we can often determine the stresses on certain members, such as the members BC and bc of Fig. 4-7, directly from the moment equations and avoid solving simultaneous equations. This technique is called the *method of moment* and can best be illustrated in the following example.

Example 4-2. In Fig. 4-9(a) is shown a simple nonparallel chord truss. Find the stresses in chord members cd and CD and in the diagonal Cd.

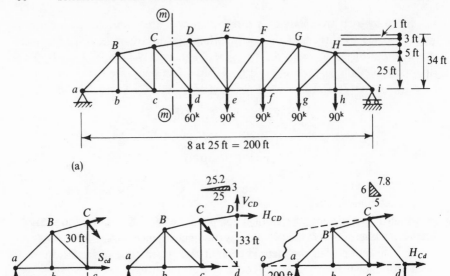

Fig. 4-9

First, from $\sum M_i = 0$ for the entire structure, the reaction at a is found to be

$$R_a = \frac{(5)(60) + (4 + 3 + 2 + 1)(90)}{8} = 150 \text{ kips}$$

Next, to find the internal force in member cd, we pass a section m-m through members CD, Cd, and cd, as indicated by the dotted line in Fig. 4-9(a), and take the left portion of the truss as a free body, as shown in Fig. 4-9(b). From $\sum M_c = 0$

$$S_{cd} = \frac{(150)(50)}{30} = 250 \text{ kips (tension)}$$

To find the internal force in member CD, we use the same free body and resolve S_{CD} into a vertical component V_{CD} and a horizontal component H_{CD} at D, as shown in Fig. 4-9(c). From $\sum M_d = 0$

$$H_{CD} = -\frac{(150)(75)}{33} = -341 \text{ kips} \quad \text{(compression)}$$

Thus, $\quad S_{CD} = (-341)\left(\frac{25.2}{25}\right) = -344 \text{ kips} \quad \text{(compression)}$

Similarly, to find the internal force in member Cd, we resolve S_{Cd} into a vertical component V_{Cd} and a horizontal component H_{Cd} at d, as shown in Fig. 4-9(d). Note that the moment center is chosen at o where the extending lines of members CD and cd intersect. The distance of oa is found to be 200 ft.

From $\sum M_o = 0$,

$$V_{Cd} = \frac{(150)(200)}{275} = 109 \text{ kips} \qquad \text{(tension)}$$

Thus, $\qquad S_{Cd} = (109)\left(\frac{7.8}{6}\right) = 142 \text{ kips} \qquad \text{(tension)}$

3. Note also that in a cut section if all the members except one are parallel, such as the diagonal Bc in the truss of Fig. 4-7, then this one carries the shear, as we say; i.e., the vertical component of the member force equals the shear of the cut section. This technique is called the *method of shear*.

4. It is sometimes possible to find some of unknown forces by the method of section even when there are *more than three bars cut through*, as illustrated in the following example.

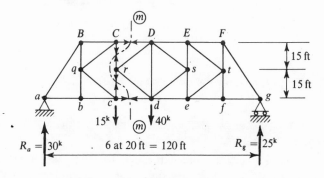

Fig. 4-10

Example 4-3. Consider the K truss loaded as shown in Fig. 4-10. The reactions are found to be

$$R_a = 30 \text{ kips} \quad \text{and} \quad R_g = 25 \text{ kips}$$

To obtain chord member stresses, such as the internal force in member cd, we pass a section m-m through four bars, as indicated by the dotted line, use either side of the truss as a free body, and apply $\sum M_c = 0$. Thus, if the left portion of the truss is taken as the free body,

$$S_{cd} = \frac{(30)(40)}{30} = 40 \text{ kips} \qquad \text{(tension)}$$

Similarly, we apply $\sum M_c = 0$ to obtain the internal force in chord member CD,

$$S_{CD} = -\frac{(30)(40)}{30} = -40 \text{ kips} \qquad \text{(compression)}$$

The above results can be checked by $\sum F_x = 0$; i.e., in this case the internal forces in the chord members cd and CD must be equal and opposite.

5. In general no truss is analyzed by one method alone. Instead, it is often analyzed by a *mixed method* based on knowledge from both the joint and section methods combined as illustrated in the following example.

Example 4-4. Suppose we wish to find S_{Dr} and S_{dr} in the truss of Fig. 4-10. We first fix our attention on the free body of joint r, as shown in Fig. 4-11(a). Assume that S_{Dr} is compression; then obviously S_{dr} is tension as indicated. Furthermore,

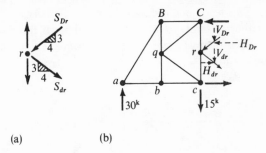

(a)   (b)

Fig. 4-11

$$S_{Dr} = S_{dr} \quad \text{or} \quad V_{Dr} = V_{dr}$$

since the slopes of members Dr and dr are equal and their horizontal components of bar forces must be balanced.

Next, by passing a vertical section through bars CD, Dr, dr, and cd, and by taking the left portion of truss as a free body, as shown in Fig. 4-11(b), we obtain from $\sum F_y = 0$,

$$30 - 15 - V_{Dr} - V_{dr} = 0$$

Using $V_{Dr} = V_{dr}$ and reducing, we arrive at

$$V_{Dr} = V_{dr} = +7.5 \text{ kips}$$

or $\qquad\qquad S_{Dr} = S_{dr} = (7.5)(\tfrac{5}{3}) = +12.5 \text{ kips}$

The positive sign indicates that S_{Dr} is compression and S_{dr} is tension as assumed.

4-3. ANALYSIS OF COMPOUND TRUSSES: MIXED
METHOD

A compound truss is formed by interconnecting two or more simple trusses usually by three elements of connection neither parallel nor concurrent as ex-

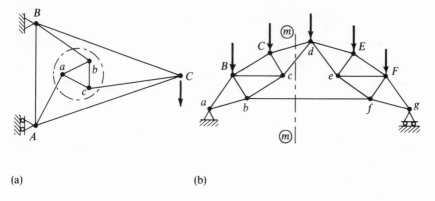

(a)   (b)

Fig. 4-12

plained in Sec. 4-1. Such a truss, if adequately sustained by three elements of support, will result in a stable and statically determinate structure. However, it frequently happens that a compound truss cannot be completely analyzed by the method of joint alone. For instance, Figure 4-12(a) shows a loaded compound truss that consists of two triangles connected by three bars. As a first step in the analysis of this truss, we apply the equilibrium equations to determine the reactions at A and B by taking the entire structure as a free body. After this, we find that the method of joint cannot be applied at the very beginning because there is no joint to which less than three bars are connected. In order to proceed we must resort to the method of section, and we cut through the three connecting members as indicated by the dotted line in Fig. 4-12(a), thus isolating triangle ABC. From the three equilibrium equations we obtain the internal forces for members Aa, Bb, and Cc. As soon as this is done, the rest of the analysis can be carried out by the method of joint.

As another illustration, let us consider the compound truss in Fig. 4-12(b).

After the reactions are found, we start the stress analysis from the left end joint a by the method of joint. This done, each of the joints next met has more than two unknowns which cannot be solved by the method of joint, and the difficulty is not overcome by going to the other side of truss. However, we can remove the trouble by taking section m-m, as shown, and by the method of moment we can easily obtain S_{bf} from $\sum M_d = 0$. After that, we go to joint b and proceed successively to joints $B, c, C, d. . .$ without difficulty.

A specific application of the foregoing procedure of the mixed method is illustrated in the following example.

Example 4-5. In Fig. 4-13(a) we have a compound truss consisting of two simple trusses connected by bar IJ and two triangle trusses BCD and FGH.

The first step in the analysis is to obtain the reactions at A and G by considering the entire truss as a free body. Thus

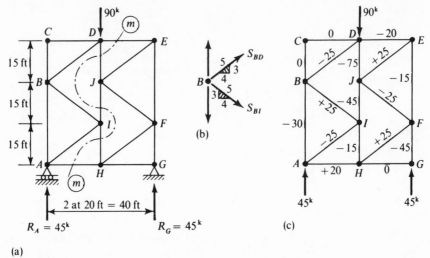

Fig. 4-13

$$R_A = R_G = 45 \text{ kips}$$

Next, from joint G, we find that

$$S_{FG} = -45 \text{ kips} \quad \text{and} \quad S_{GH} = 0$$

and from joint C, $\qquad\qquad S_{BC} = S_{CD} = 0$

After this the method of joint fails since each remaining joint involves more than two unknowns. It also appears at first glance that it is not possible to apply the method of section, since we cannot take any section that cuts only three bars that are not concurrent. However, if we pass section m-m through five bars, as indicated in Fig. 4-13(a), we can easily obtain from $\sum M_D = 0$ that

$$S_{AH} = \frac{(45)(20)}{45} = 20 \text{ kips}$$

by taking either portion of the truss as a free body and by assuming S_{AH} acts in positive direction. Similarly, from $\sum M_H = 0$ or $\sum F_x = 0$, we obtain that

$$S_{DE} = -20 \text{ kips}$$

Having done this, we can solve the stresses for the remaining bars by the method of joint without difficulty. The analysis may be facilitated by observing that in this particular case

$$S_{BI} = -S_{BD} = -S_{AI} = S_{EJ} = -S_{FJ} = S_{FH} = 25 \text{ kips}$$

To prove this, we first isolate joint B as in Fig. 4-13(b) where all bar forces are assumed to act in positive direction.

From $\sum F_x = 0$

$$\tfrac{4}{5}S_{BD} + \tfrac{4}{5}S_{BI} = 0 \quad \text{or} \quad S_{BI} = -S_{BD}$$

In a similar manner, we obtain that

$$S_{BI} = -S_{AI} \qquad \text{(joint } I\text{)}$$

$$S_{EJ} = -S_{FJ} \qquad \text{(joint } J\text{)}$$

$$S_{FH} = -S_{FJ} \qquad \text{(joint } F\text{)}$$

Next, from joint E we note that

$$S_{EJ} = 25 \text{ kips}$$

and from joint A $\qquad S_{AI} = -25 \text{ kips}$

This completes our proof. An answer diagram for the analysis is given in Fig. 4-13(c).

In some cases analysis may be facilitated by using the notion of *secondary truss* action. Let us consider the loaded truss in Fig. 4-14(a). We replace member *DE* with a secondary truss *DEFG*, as shown in Fig. 4-14(b). The function of this secondary truss, if it takes no external load itself as in the case

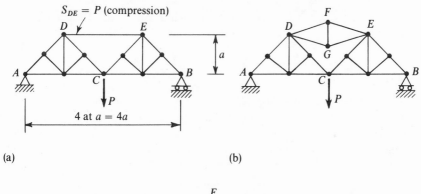

(a)   (b)

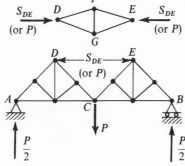

(c)

Fig. 4-14

shown, is nothing more than to act as a single member DE. To analyze the secondary truss, let us refer to Fig. 4-14(c) in which it is taken free from joints D and E. It is apparent that the truss shown below can be in equilibrium only if the forces exerted by the secondary truss on joints D and E equal S_{DE} of the truss of Fig. 4-14(a). Note that S_{DE} is a compressive force equal to P in this case. Therefore, the forces imposed on the secondary truss at D and E by the main truss will have the same value with signs as indicated.

Now if the secondary truss also carries a load, as shown in Fig. 4-15(a), then it not only performs the duty of DE but, at the same time, functions as a truss to transfer its own load to the joints at its ends as shown in Fig. 4-15(b).

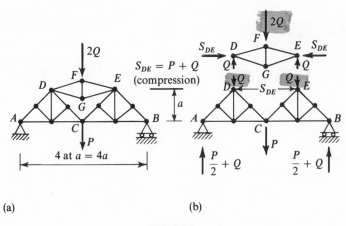

(a)   (b)

Fig. 4-15

Referring to Fig. 4-15(b), we obtain in this case

$$S_{DE} = \frac{(P/2 + Q)(2a) - Qa}{a} = P + Q \qquad \text{(compression)}$$

by the method of section ($\sum M_C = 0$). This done, the rest of the analysis can be carried out without difficulty if complete dimensions are given.

Sometimes panels of a simple truss may be subdivided by adding sub-members and by thus forming a compound truss, such as the one shown in Fig. 4-16(a) in which the dashed lines indicate the submembers. Each shaded portion acts as a secondary truss which performs dual functions: that of taking the duty of the main members of an unsubdivided truss under equivalent load-ing; and that of receiving load at the panel.

To explain this in an easily understood way, let us refer to Fig. 4-16 in which (a) is thought of being equivalent to the sum of the effects of (b) and of (c) which, in turn, are equivalent to the sum of the effects of (d) and of (e). Note that in (b) each submember carries no stress. This is readily seen from the method of joint. In (c) each member not belonging to the secondary trusses carries no stress. This seen from the method of section and the method of

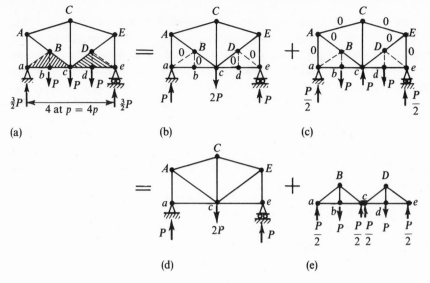

Fig. 4-16

joint. The truss in Fig. 4-16(a) may thus be considered as consisting of a main truss shown in Fig. 4-16(d), upon which two small secondary trusses, shown in Fig. 4-16(e), are superposed.

In the design of long-span trussed bridges, we often subdivide each panel with secondary bars for economical reasons. It is advantageous to carry through a complete analysis of these trusses by making use of this simple division into main and secondary elements.

4-4. ANALYSIS OF COMPLEX TRUSSES: METHOD OF
SUBSTITUTE MEMBER

In analyzing a complex truss, we frequently find that the method of joint and the method of section, described in previous sections, are not directly applicable. For example, let us consider the loaded complex truss shown in Fig. 4-17(a). After the reactions at A and E are found, we observe that no further progress can be made by either the method of joint or the method of section. Theoretically, we can always solve this problem by $2j$ simultaneous equations

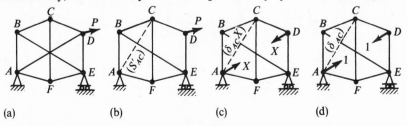

Fig. 4-17

of equilibrium for j joints of the system. The method is perfectly general, but it is difficult to do it by hand calculation. A better way to handle is to substitute for the bar AD a bar AC and thus obtain a stable simple truss, as in Fig. 4-17(b), which can be completely analyzed by the method of joint for the given loading. Next, let the same simple truss be loaded with two equal and opposite forces X at A and D along AD, as shown in Fig. 4-17(c). Again a complete analysis can be carried out by the method of joint such that the internal force for each member will be expressed in terms of the unknown X. Or for convenience sake, we put a pair of unit forces in place of the X's, as given in Fig. 4-17(d). It is apparent that the stresses obtained from (d) times X will give those of (c).

Now the analysis of (a) can be made equivalent to the superposing effects of (b) and (c) if we let the bar force of AC obtained from (b) and (c), or from (b) and (d) times X, be equal to zero. Thus if we let S_i' denote the force in any bar of (b) and δ_i the corresponding bar force of (d), then the corresponding internal force S_i in any bar of (a) is expressed by

$$S_i = S_i' + \delta_i X$$

in which X denotes the bar force in member AD solved by

$$S_{AC} = S_{AC}' + \delta_{AC} X = 0$$

Or
$$X = -\frac{S_{AC}'}{\delta_{AC}}$$

With the value of X determined, the force in any other bar of the given truss can be obtained without difficulty.

The following example will illustrate the application of the *substitute-member method* to the analysis of a complex truss.

Example 4-6. In analyzing the loaded complex truss shown in Fig 4-18(a),

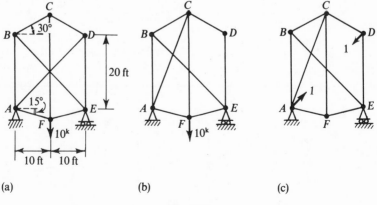

(a)   (b)   (c)

Fig. 4-18

we begin with the complete analyses of the corresponding simple trusses of Figs. 4-18(b) and (c). These can be done by the method of joints, and the results are recorded in columns (2) and (3), respectively, of Table 4-1. Next, setting

$$S_{AC} = S'_{AC} + \delta_{AC} X$$
$$= -10.89 + 0.465X = 0$$

we obtain $\qquad\qquad X = +23.42 \text{ kips}$

which is the value of the bar force in AD.

TABLE 4-1

(1) Member	(2) S'_i	(3) δ_i	(4) $\delta_i \cdot X$	(5) $S_i(= S'_i + \delta_i \cdot X)$
AB	$+6.22$	-1.382	-32.36	-26.1 (kips)
BC	$+4.55$	-1.010	-23.65	-19.1
CD	0	-0.816	-19.11	-19.1
DE	0	-1.115	-26.11	-26.1
EF	$+4.08$	-0.910	-21.31	-17.2
FA	$+4.08$	-0.910	-21.31	-17.2
BE	-5.58	$+1.240$	$+29.0$	$+23.4$
CF	$+7.89$	$+0.471$	$+11.03$	$+18.9$
AC	-10.89	$+0.465$	$+10.89$	0

Having found this, we may now fill in column (4) of Table 4-1 and then complete the stress analysis for the complex truss, as recorded in column (5) of Table 4-1. Note that in this particular case we obtain a special check by the condition of symmetry of the structure.

It should be noted that complex trusses may often be arranged so as to be geometrically unstable. However, it is not always possible to see a critical form just by inspection. For example, the complex truss shown in Fig. 4-19 has a critical form if $a = c$; otherwise it is a stable form. In such cases detection is based on the principle that if the analysis for the truss yields a unique solution then the truss is stable and determinate; on the other hand, if the analysis fails to yield a unique solution, then the truss has a critical form.

In this connection, one may resort to the denominator determinant, denoted by D, for solution by Cramer's Rule of the system of $2j$ simultaneous equations for j joints. Thus if

$$D \neq 0$$

the system has a unique solution, which indicates that the truss is stable and determinate. If

$$D = 0$$

then the truss has a critical form since many solutions are possible.

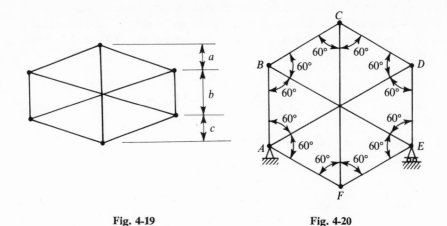

Fig. 4-19   **Fig. 4-20**

The above principle also suggests a convenient way of detecting critical form known as the *zero-load test*. For an unloaded truss, one possible solution satisfying equilibrium is that all bars of the truss have zero stress. Hence, if with no loads on a truss we can find a set of solutions for the bar stresses other than zero that will also satisfy the condition of equilibrium for the truss, then it has a critical form.

As an illustration, let us consider the complex truss shown in Fig. 4-20.

Without external forces acting on the truss, we assume a tension S in the member BC, i.e.,

$$S_{BC} = S$$

Fro m joint B $S_{AB} = S$ and $S_{BE} = -S$

and from joint C $S_{CD} = S$ and $S_{CF} = -S$

Because of the identity in each joint, we can easily see that the set of solution

$$S_{AB} = S_{BC} = S_{CD} = S_{DE} = S_{EF} = S_{FA} = S \qquad S_{AD} = S_{BE} = S_{CF} = -S$$

will satisfy the condition of equilibrium at all joints. Since S may be assigned any value different from zero, this indicates that the truss has a critical form. In such cases no matter how the truss is actually loaded, the truss will have an infinite number of solutions that can satisfy the equilibrium requirements of statics. Therefore it is statically indeterminate.

Although the given illustration is aimed at a complex truss, the principle and the method of detecting a critical form described above can be applied to other types of structures such as beams and rigid frames.

**4-5. GENERAL DESCRIPTION OF BRIDGE AND ROOF
TRUSS FRAMEWORKS**

Figure 4-21 shows a typical *through trussed bridge.* The word *through*
indicates that the trains (or vehicles) actually travel through the bridge. If the
bridge is installed under the floor or deck, then the bridge is called a *deck
bridge.* If the trains pass between trusses but the depth is insufficient to allow
the use of a top chord bracing system, the bridge is called *half-through.*

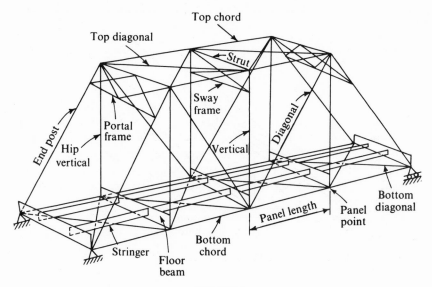

Fig. 4-21

Referring to Fig. 4-21, we place the road surface (or the rail and tie system
in railways) on the short longitudinal beams called *stringers,* assumed simply
supported on the *floor beams* which in turn are supported by the two *main
trusses.* The moving loads on bridge are transmitted to the main trusses
through the system of the connection of road surface (or rail and tie), stringer,
and floor beam.

The top series of truss members parallel to the stringer are called *top
chords;* while the corresponding bottom series of members are called *bottom
chords.* The members connecting the top and bottom chords form the web
system and are referred to as *diagonals* and *verticals.* The end diagonals are
called *end posts,* and the side verticals are called *hip verticals.* The point at
which web members connect to a chord is called a *panel point,* and the length
between two adjacent panel points on the same chord is called the *panel length.*

The cross struts at corresponding top-chord panel points, together with the top diagonals connecting the adjacent struts, make up the top-chord lateral system. The bottom-chord lateral system is composed of the floor beams and the bottom diagonals connecting the adjacent floor beams.

The two main trusses are also cross braced at each top-chord pannel point by *sway frames*. The frame in the plane of each pair of end posts is called a *portal frame*.

A typical roof truss framework supported by columns is shown in Fig. 4-22.

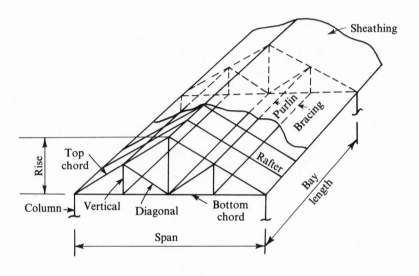

Fig. 4-22

A roof truss with its supporting columns is called a *bent*. The space between adjacent bents is called a *bay*. *Purlins* are longitudinal beams that rest on the top chord and preferably at the joints of the truss, in accordance with the definition of truss. The *roof covering* may be laid directly on the purlins for very short bay lengths but usually is laid on the wood *sheathings* that, in turn, rest either on the purlins or on the *rafters* (if provided). Rafters are the sloping beams extending from the *ridge* to the *eaves* and are supported by the purlins.

For a symmetrical roof truss the ratio of its *rise*, center height, to its *span*, the horizontal distance between the center lines of the supports, is called *pitch*.

The truss consists of top-chord, bottom-chord, and *web members*. Although the purlins act to strengthen the longitudinal stability, additional bracing is always necessary. The *bracing members* may run from truss to truss longitudinally or diagonally and may be installed in the plane of the bottom chord, the top chord, or both. Surface loads are transmitted from covering, sheathing, rafter, purlin and distributed to adjacent trusses.

4-6. CONVENTIONAL TYPES OF BRIDGE AND ROOF TRUSSES

The members of a main truss may be arranged in many different ways. However, the principal types of trusses encountered in bridges and buildings are shown in Figs. 4-23 and 4-24.

Among these types, the Pratt, Howe, and Warren trusses are more commonly used. We may note that in the Pratt truss the diagonals, except the end posts, are stressed in tension and that the verticles, except the hip verticals, are stressed in compression under dead load. On the other hand, in a Howe truss the diagonals are in compression and the verticals are in tension. Note also, of all the trusses shown in Fig. 4-23 under dead load, the upper chords are in compression and bottom chords in tension.

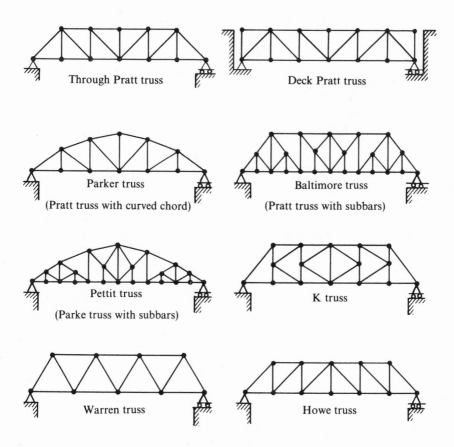

Fig. 4-23

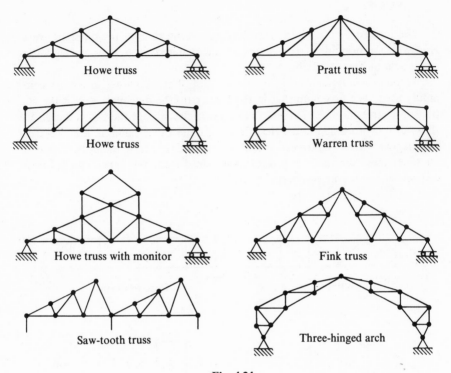

Fig. 4-24

PROBLEMS

4-1. Classify the trusses of Fig. 4-25 as being simple, compound, or complex.

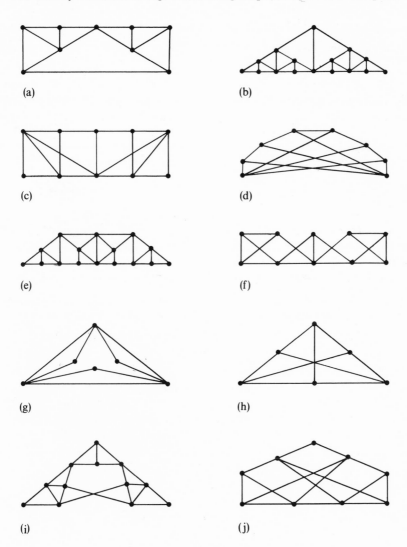

(a)

(b)

(c)

(d)

(e)

(f)

(g)

(h)

(i)

(j)

Fig. 4-25

4-2. Determine the bar force in each member of the trusses shown in Fig. 4-26 by the method of joint.

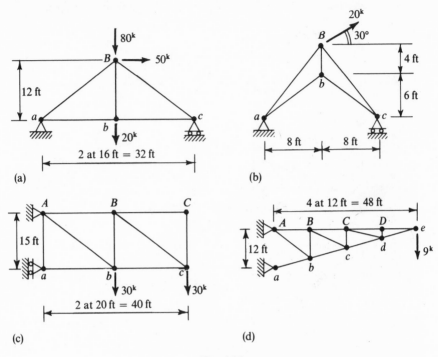

(a)

(b)

(c)

(d)

Fig. 4-26

4-3. By the method of section compute the bar forces in the lettered bars of the trusses shown in Fig. 4-27.

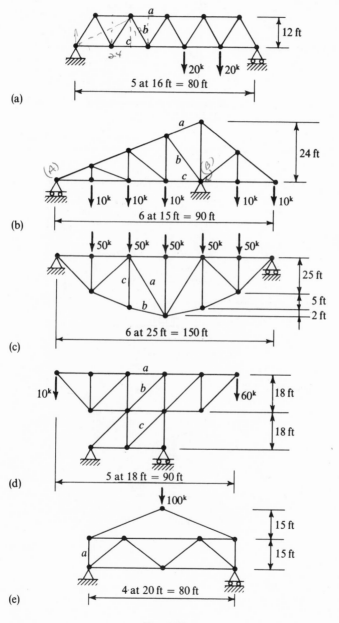

(a)

(b)

(c)

(d)

(e)

Fig. 4-27

4-4. By the mixed method of joint and section determine the bar forces in the lettered bars of the *K* truss shown in Fig. 4-28.

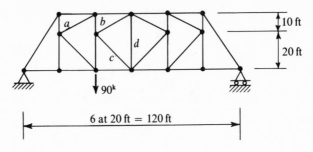

Fig. 4-28

4-5. Make a complete analysis of the compound truss shown in Fig. 4-29.

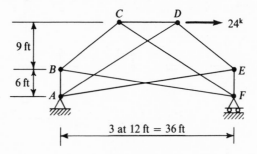

Fig. 4-29

4-6. Determine the bar forces in the lettered bars of the compound truss shown in Fig. 4-30.

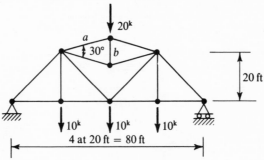

Fig. 4-30

4-7. Make a complete analysis of the compound truss shown in Fig. 4-31. Use the concept of secondary truss.

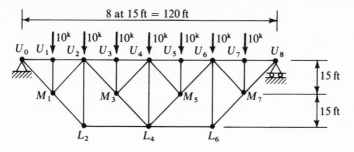

Fig. 4-31

4-8. Use the substitute-member method to make a complete analysis of the complex truss shown in Fig. 4-32.

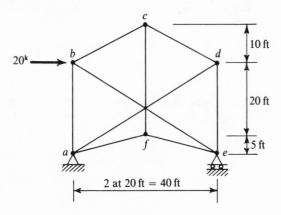

Fig. 4-32

5

STATICALLY DETERMINATE RIGID
FRAMES AND COMPOSITE STRUCTURES

5-1. RIGID JOINT

A *rigid frame* may be defined as a structure composed of a number of members connected together by joints some or all of which are rigid, i.e., capable of resisting both force and moment as distinguished from a pin-connected joint which offers no moment resistance. In steel structures, rigid joints may be formed by certain types of riveted or welded connections. In reinforced concrete structures, the materials in the joined members are mixed together in one unit so as to be substantially rigid. In the analysis of rigid frames, we assume that the centroidal axis of each member coincides with the line connecting the joint centers of the ends of the member. The so-called joint center is therefore the concurrent point of all centroidal axes of members meeting at the joint. With the joint rigid the ends of all connected members must not only translate but also rotate identical amounts at the joint.

5-2. ANALYSIS OF STATICALLY DETERMINATE RIGID FRAMES

Rigid frames are usually built to be highly statically indeterminate. The discussion of determinate rigid frames in this chapter is primarily of academic interest rather than of practical use and serves as a prelude to the analysis of indeterminate frames.

To analyze a statically determinate rigid frame, we start by finding the reaction components from statical equations for the entire structure. This done, we are able to determine the shear, moment, and axial force at any cross section of the frame by taking a free body cut through that section and by using the equilibrium equations. Based on the centroidal axis of each member, we can plot the shear, bending moment, and the direct force diagrams for the rigid frame. However, it is the bending-moment diagram with which we are mainly concerned in the analysis of a rigid frame.

The following numerical examples will serve to illustrate the above procedure.

Example 5-1. Analyze the simply supported rigid frame in Fig. 5-1(a). Let H_a and V_a, respectively, denote the horizontal and vertical reaction components at support a, and let V_d denote the vertical reaction at support d. From $\sum F_x = 0$, we find that $H_a = 10$ kips; from $\sum M_d = 0$, $V_a = 5$ kips; and from $\sum F_y = 0$, $V_d = 15$ kips, as indicated in Fig. 5-1(a). After all the external forces acting on the rigid frame are determined, the internal forces at each end of the members can easily be obtained, as shown in Fig. 5-1(b). Take member ab, for instance. At end b we find the shear force equal to zero by applying $\sum F_x = 0$ for the free body of member ab; also the axial force equal to 5 kips (down) from $\sum F_y = 0$, and the resisting moment equal to 50 ft-kips (counter-clockwise) from $\sum M_b = 0$.

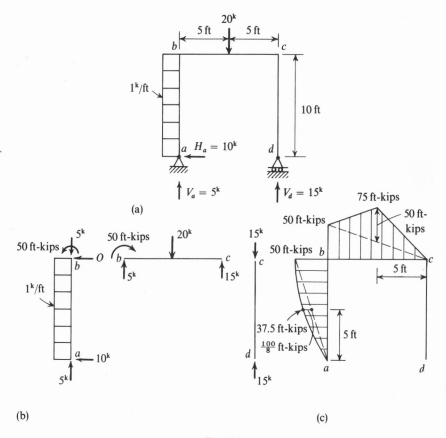

Fig. 5-1

The bending-moment diagram for each member can then be drawn, as shown in Fig. 5-1(c), by use of the method of superposition similar to that described for the simple beam (see Sec. 3-2). Note that the positive moment is drawn on the *compressive* side of the member.

Example 5-2. Analyze the simply supported gable frame, shown in Fig. 5-2(a), which is composed of two columns and two sloping members.

From $\sum F_x = 0$, $\sum M_e = 0$, and $\sum F_y = 0$ for the entire frame, the reaction elements are found to be

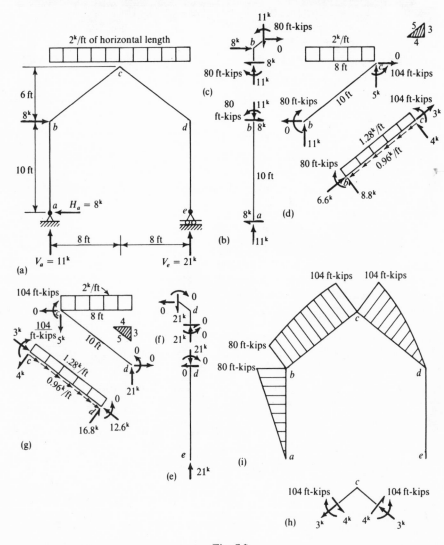

Fig. 5-2

$$H_a = 8 \text{ kips} \qquad V_a = 11 \text{ kips} \qquad V_e = 21 \text{ kips}$$

as shown in Fig. 5-2(a).

Next, we take member ab as a free body. With the end forces known at a, we can readily obtain those at the other end b from the equilibrium conditions, i.e.,

$$\text{shear} = 8 \text{ kips} \qquad \text{moment} = 80 \text{ ft-kips} \qquad \text{axial force} = 11 \text{ kips}$$

acting as indicated in Fig. 5-2(b).

Following this, we sketch the free body diagram for joint b, as shown in Fig. 5-2(c). Note that the joint is shown in an exaggerated manner, since theoretically it should be represented by a point and all forces acting on the joint should be concurrent at this point.

Next, let us take member bc as a free body subjected to the external load of 2 kips per horizontal unit length. With the internal forces known at end b, we can apply $\sum F_x = 0$, $\sum F_y = 0$, and $\sum M_c = 0$ to obtain the internal forces at end c as

$$\text{horizontal force} = 0 \qquad \text{vertical force} = 5 \text{ kips} \qquad \text{moment} = 104 \text{ ft-kips}$$

These act as indicated in the upper sketch of Fig. 5-2(d). To determine the stresses in each section of the member, we resolve all the indicated forces into components normal and tangential to the member section, as shown in the lower sketch of Fig. 5-2(d). For instance, at end b we have

$$\text{normal force (axial force)} = (11)(\tfrac{3}{5}) = 6.6 \text{ kips}$$

$$\text{tangential force (shear)} = (11)(\tfrac{4}{5}) = 8.8 \text{ kips}$$

Similarly, at end c we have

$$\text{normal force (axial force)} = (5)(\tfrac{3}{5}) = 3 \text{ kips}$$

$$\text{tangential force (shear)} = (5)(\tfrac{4}{5}) = 4 \text{ kips}$$

The total uniform load on member bc is 16 kips of which

$(16)(\tfrac{4}{5}) = 12.8$ kips acting transversely to the member axis

$(16)(\tfrac{3}{5}) = 9.6$ kips acting axially to the member axis

thus giving a uniform load of intensity:

$$\frac{12.8}{10} = 1.28 \text{ kips per ft acting transversely to the member axis}$$

$$\frac{9.6}{10} = 0.96 \text{ kip per ft acting axially to the member axis}$$

With these determined, the shear, bending moment, and direct force in any section of member bc can readily be obtained, as shown in Fig. 5-3.

In this manner we may proceed from member bc to joint c, then to member cd and joint d, and finally to member de. However, it seems more con-

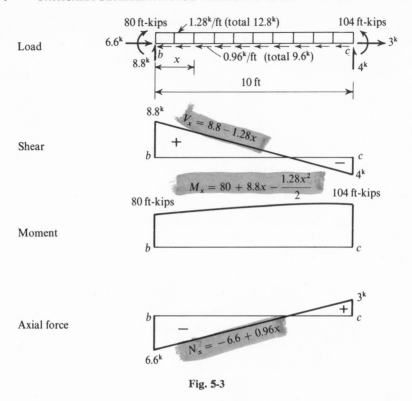

Fig. 5-3

venient to analyze de now and then to turn to joint d and member cd, and to leave the joint c as a final check, as shown in Figs. 5-2(e), (f), (g), and (h), respectively. The bending-moment diagram for the whole frame is plotted in Fig. 5-2(i).

Example 5-3. Consider the three-hinged frame loaded as in Fig. 5-4(a). The four reaction elements at supports a and e are first obtained by solving simultaneous equations, three from equilibrium and one from construction.

$$\sum F_x = 0 \qquad H_a - H_e = 0$$
$$\sum F_y = 0 \qquad V_a + V_e - 12 = 0$$
$$\sum M_e = 0 \qquad 12V_a - (12)(10) = 0$$
$$M_c = 0 \qquad 6V_e - 8H_e = 0$$

which give:

$$V_a = 10 \text{ kips} \qquad V_e = 2 \text{ kips} \qquad H_a = H_e = 1.5 \text{ kips}$$

The free-body diagrams for members ab, bd, de are then drawn as in Fig. 5-4(b). From these we plot the moment diagram for the frame, as shown in Fig. 5-4(c).

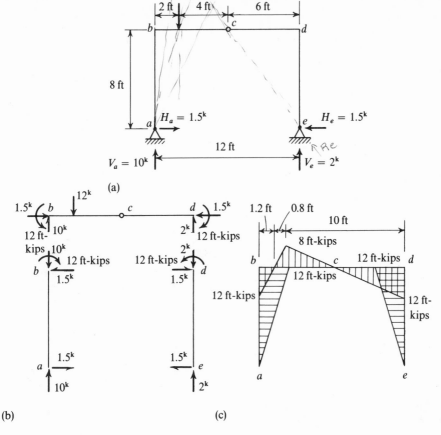

Fig. 5-4

It is interesting to note that in this particular case the portion to the right of hinge c (i.e., cde) carries no external load and is therefore a two-force member if isolated. The line of reaction at support e, called R_e, must be through points e and c and must meet the action line of the applied load at some point o, as shown in Fig. 5-5(a). Now if we take the whole frame as a free body, we see that the system constitutes a three-force member subjected to the applied load and support reactions. Thus the line of reaction at support a, called R_a, must be through points a and o so that the three forces are concurrent at point o as required by equilibrium. The vectors R_a and R_e can then easily be determined by the equilibrium triangle, as shown in Fig. 5-5(b).

In the case where loads are placed both to the left and to the right of the connecting hinge of a three-hinged frame, one way to analyze this is to use the method of superposition. This is illustrated in Fig. 5-6 in which the case of (a)

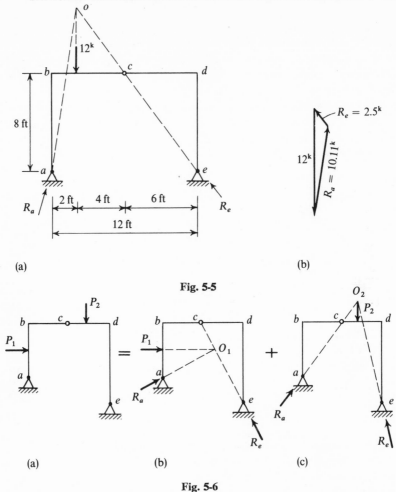

(a)

(b)

Fig. 5-5

(a)   (b)   (c)

Fig. 5-6

can be made equivalent to the sum of effects of (b) and (c), each analyzed by the method previously discussed.

5-3. ANALYSIS OF STATICALLY DETERMINATE COMPOSITE STRUCTURES

There are structures in which some members are two-force members subjected to axial forces only and others are not. Such structures may be called *beam trusses* or *truss beams* but are more frequently called *composite structures*.

In discussing a composite structure, we begin by investigating the *stability* and *determinacy* of the structure. The most fundamental approach is to separate the composite structure into a number of free bodies of beam and truss through connections or by cutting certain two-force members. After that, we may be in a position to see if the total number of equations of statics equals the total number of unknown elements as a *necessary* condition for statical determinacy. A final decision that the structure is both determinate and stable should be based on the criteria given in Ch.2.

For example, Fig. 5-7(a) shows a *mill bent* (bent of a mill building) which consists of a roof truss supported at two ends *a* and *g* by two columns hinged to the foundation at *A* and *G*. In this case all truss members are two-force members while the columns are not.

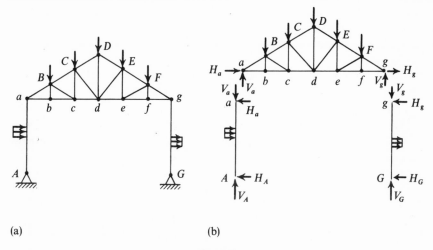

(a)   (b)

Fig. 5-7

To investigate the stability and determinacy of this composite structure, we separate it into three free bodies through connections at *a* and *g* as well as at *A* and *G*, as shown in Fig. 5-7(b). The total number of unknown elements involved is eight whereas the total number of equations of statics is nine, since each free body generally provides three equilibrium equations. Therefore, this is an unstable structure.

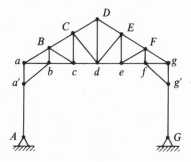

Fig. 5-8

However, the stability of the structure can be secured by adding two-force members $a'b$ and fg' called *knee braces*, as shown in Fig. 5-8. The introduction of two links adds two unknowns to the system and makes the structure statically indeterminate to the first degree.

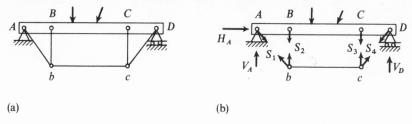

(a)   (b)

Fig. 5-9

As another example, consider the trussed beam shown in Fig. 5-9(a) in which the simple beam *AD* is strengthened by the addition of trussing. To investigate the determinacy of the structure, we separate the structure into two parts by cutting through the two-force members *Ab*, *Bb*, *Cc*, and *Dc*, as shown in Fig. 5-9(b). The total number of unknowns involved in two free bodies is seven, four from bar stresses and three from reactions, as indicated in Fig. 5-9(b); the total number of statical equations is six. Therefore, it is statically indeterminate to the first degree.

To illustrate the stress analysis of a statically determinate composite structure, let us consider the system shown in Fig. 5-10(a) in which members *ab* and *df* are two-force members since they are pin-connected at ends and carry

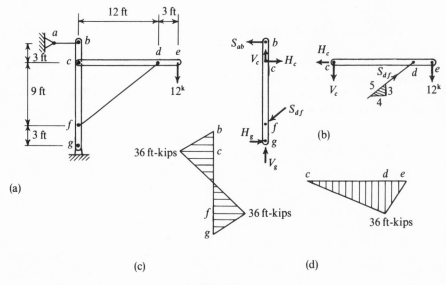

(a)

(b)

(c)   (d)

Fig. 5-10

no load. If we cut through these two members and take two free bodies, as shown in Fig. 5-10(b), we at once observe that the structure is statically determinate since there is a total of six unknowns involved in the two free bodies and six independent equations of statics to solve them.

Using the free body of beam ce and applying equilibrium equations, we obtain the unknown forces S_{df}, H_c, and V_c as follows:

$$\sum M_c = 0 \qquad (\tfrac{3}{5} S_{df})(12) - (12)(15) = 0 \qquad S_{df} = 25 \text{ kips}$$

$$\sum F_x = 0 \qquad \tfrac{4}{5} S_{df} - H_c = 0 \qquad H_c = 20 \text{ kips}$$

$$\sum F_y = 0 \qquad \tfrac{3}{5} S_{df} - 12 - V_c = 0 \qquad V_c = 3 \text{ kips}$$

Next, using the free body of member bg and applying equilibrium equations, we obtain all the remaining unknown forces as follows:

$$\sum M_g = 0 \qquad 15 S_{ab} - (20)(12) + (\tfrac{4}{5})(25)(3) = 0 \qquad S_{ab} = 12 \text{ kips}$$

$$\sum F_x = 0 \qquad H_g - S_{ab} + 20 - (\tfrac{4}{5})(25) = 0 \qquad H_g = 12 \text{ kips}$$

$$\sum F_y = 0 \qquad V_g + 3 - (\tfrac{3}{5})(25) = 0 \qquad V_g = 12 \text{ kips}$$

The results can be checked by taking the whole structure as a free body and applying $\sum M_g = 0$, $\sum F_x = 0$, and $\sum F_y = 0$.

After these unknowns are solved, we can determine the shear, the moment, and the axial force at any section of the structure without difficulty. The bending-moment diagrams for members bg and ce are plotted, as shown in Figs. 5-10(c) and (d) respectively.

5-4. APPROXIMATE ANALYSIS FOR STATICALLY INDETERMINATE STRUCTURES

As previously mentioned, the rigid frames of present-day construction are highly indeterminate. It will be seen in the later chapters of this book, which deal with statically indeterminate structures, that to obtain the solution for a building frame based on more exact analyses is often tedious and time consuming. In many cases, we cannot obtain the solution without the aid of modern electronic computers. For this reason empirical rules and approximate methods were often used in the past by structural and architectural engineers in designing various kinds of indeterminate structures. In order to do this, as many independent equations of statics as there are independent unknowns must be available. The additional equations of statics are worked out by reasonable assumptions based on experience and knowledge of the more exact analyses.

To illustrate, consider a frame subjected to uniform floor loads such as the one shown in Fig. 5-11(a). The frame is indeterminate to the 24th degree since eight cuts in the girders would render the frame into three stable and determinate parts and since each cut involves the removal of three elements of

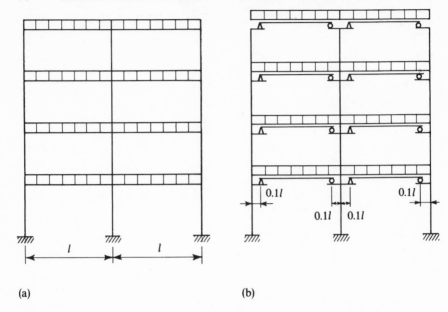

(a)   (b)

Fig. 5-11

restraint, i.e., bending moment, shear, and axial force. A preliminary survey of stresses may be performed by assuming the following so that the indeterminate frame can be solved by a determinate approach, i.e., by equations of statics alone:

1. The axial force in each girder is small and can be neglected.
2. A point of inflection (zero moment) occurs in each girder at a point one-tenth of the span length from the left end of the girder.
3. A point of inflection occurs in each girder at a point one-tenth span length from the right end of the girder.

This would render the frame equivalent to the one shown in Fig. 5-11(b), which is statically determinate.

Another case that may also be worth brief mention, without going into details, is the approximate analysis for wind stresses in building frames. Consider a frame subjected to lateral forces (equivalent wind) acting at the joints such as the one shown in Fig. 5-12(a). The frame is statically indeterminate to the 27th degree. There are several methods available for dealing with the problem. The method chosen to illustrate this is called the *cantilever method* and is based on the following assumptions:

1. A point of inflection exists at the center of each girder.
2. A point of inflection exists at the center of each column.

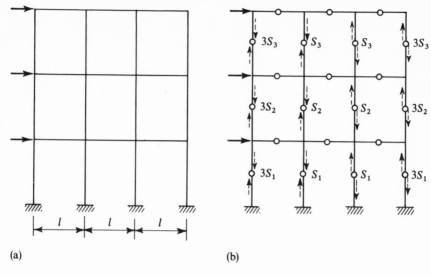

(a)   (b)

Fig. 5-12

3. The unit axial stresses in the columns of a story vary as the horizontal distances of the columns from the center of gravity of the bent. It is usually further assumed that all columns are identical in a story, so that the axial forces of the columns in a story will vary in proportion to the distances from the center gravity of the bent.

This would lead the frame to appear as the form shown in Fig. 5-12(b). Note that the last assumption virtually puts the column axial forces in one story in terms of a single unknown (see the dash arrows in Fig. 5-12(b)). It is therefore equivalent to making $(n - 1)$ additional assumptions for each story, n being the number of columns in one story. In this case, there are three for each story, or nine in total regarding column axial forces. As a result, the total number of additional equations is 30 (9 from column axial forces and 21 from inserting pins), which is three more than are necessary. However, it happens that a statical analysis can be carried out for the frame on the basis of the above assumptions without inconsistency.

Like building frames, statically indeterminate composite structures are often analyzed by a determinate approach. Take, for instance, the mill bent acted on by lateral loads, as shown in Fig. 5-13(a). It is statically indeterminate to the third degree. An approximate analysis of stresses may be carried out on the basis of the following assumptions:

1. The horizontal reactions at the bases of the columns are equal.
2. A point of inflection occurs midway between the base A and the end B of the knee brace of the left column.

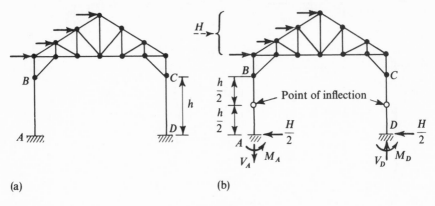

(a)   (b)

Fig. 5-13

3. A point of inflection occurs midway between the base D and the end C of the knee brace of the right column.

As indicated in Fig. 5-13(b), the six reaction elements can be determined by the three equilibrium equations together with the above three assumptions. After these are determined, the rest of analysis can be worked out easily.

PROBLEMS

5-1. Analyze each of the frames shown in Fig. 5-14, and draw the bending-moment diagram.

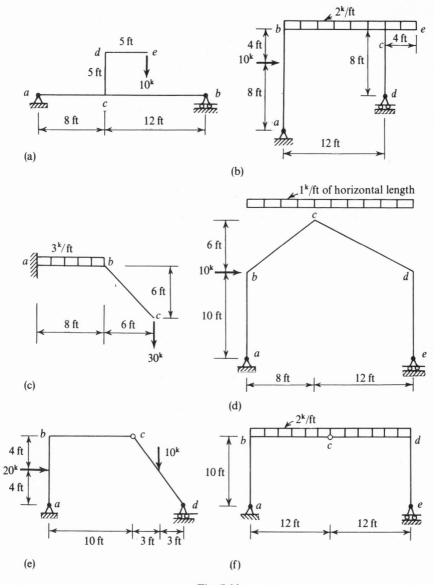

(a)

(b)

(c)

(d)

(e)

(f)

Fig. 5-14

5-2. Determine the forces acting on members *ad* and *ae* of the composite structure of Fig. 5-15.

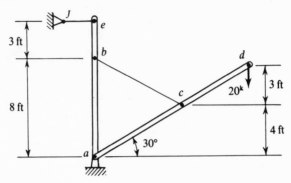

Fig. 5-15

5-3. Analyze the frame of Fig. 5-16 by the cantilever method. Assume constant *EI* for all members.

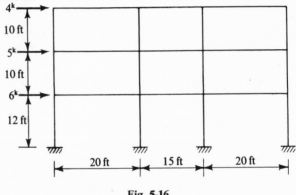

Fig. 5-16

5-4. Analyze the portal frame of Fig. 5-17 by assuming that the horizontal reactions are equal and that a point of inflection occurs halfway between the base and
the end of the knee brace for each column.

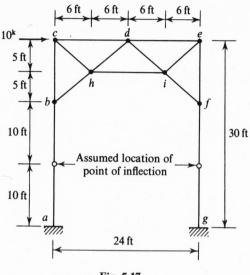

Fig. 5-17

6

INFLUENCE LINES FOR STATICALLY

DETERMINATE STRUCTURES

6-1. THE CONCEPT OF INFLUENCE LINE

In the design of a structure, as discussed in Sec. 1-2, the loading conditions for the structure must be established before the stress analysis can be made. For a static structure, we are mainly concerned with two kinds of load, dead load and live load (the impact load being a fraction of the live load). The former remains stationary with the structure; whereas the latter, either the moving or the movable load, may vary in position on the structure. When one is designing any specific part of a structure, particular attention should be paid to the placement of the live load so that it will cause the maximum live stresses for the part considered. The part of the structure and the type of stress may be the reaction of a support, the shear or moment of a beam section, or the bar force in a truss. In this connection we should note that the position of the load that causes the maximum bending moment at a section will not necessarily cause the maximum shear at the same section or that the condition of loading which causes the maximum axial force at one member may not cause the maximum axial force at some other member. This and other considerations with respect to the relationship between the stress and the corresponding critical position of live load lead to the construction of the influence line.

The concept of the influence line may be illustrated by considering a beam under a single concentrated load, as follows:

1. Suppose that a single concentrated load is placed transversely on a beam. Each part or each beam section is then affected by the load acting on the beam. For instance, each beam section will have certain shear and bending moment, and the supports will undergo certain reactions. When the load is fixed in position, a different section has different stresses. When the position of the load is varied, the corresponding value of stress at a fixed section varies.

2. As mentioned, we have touched on three distinct things—the value of the stress, the position of the beam section, and the position of the load. It should be noted that a plane curve is possible only when two variables are

related, not three. In this respect, we may express the value of the stress as a function of the position of the beam section, if we let the position of the load be fixed. Throughout the preceding chapters, involving the analysis of beams, the discussion was devoted to finding the value of the stresses on various sections, while the position of the applied load was assumed definite. The familiar shear or moment curve is nothing but the graphic representation of the relationship

$$y = f(x)$$

in which x is taken along the beam axis to indicate the variable position of beam section and y the corresponding value of the function (shear or moment) under fixed load.

3. On the other hand, we may express the value of stress in a fixed section as a function of variable load position. For simplicity and elegance, we use a single unit load to represent the moving live load. As the unit load travels across the span, we fix our attention on its effect upon a certain section of the beam and record the variation of stress (shear or moment) for that section by

$$y = g(x)$$

in which x is taken along the beam axis to indicate the variable position of the unit load and y the corresponding value of function (shear or moment) for the fixed section. The graphic representation of $y = g(x)$ is called the *influence line* for the value of the function under consideration.

4. A generalized definition of the influence line is:

An influence line is a curve the ordinate (y value) *of which gives the value of the function* (shear, moment, reaction, bar force, etc.) *in a fixed element* (member section, support, bar in truss, etc.) *when a unit load is at the ordinate.*

As an illustration, let us begin with the basic approach as to how to draw a bending moment influence line for the midspan section of a simple beam 10 ft long (Fig. 6-1 (a)). We may first divide the span into equal segments, say ten segments $AB, BC, \ldots, JK$, to indicate the position of load. As the unit load moves continuously from the left to the right, we focus our attention on the midspan section F and compute the bending moment at F for each 1-ft interval. The results are plotted in Fig. 6-1(b). This gives the bending moment influence line for section F. The abscissa coincides with the beam axis, indicating the position of the load, and the ordinate gives the corresponding moment at F due to the single unit load placed at the ordinate. For instance, the oridinate at D is 1.5, which is the value of the moment at F caused by a unit loat at D.

Of course we need not always plot the influence line in the above-indicated fashion, since it is time-consuming. In most cases we can find an equation $y = g(x)$ expressing the desired stress y at the given section in terms of the

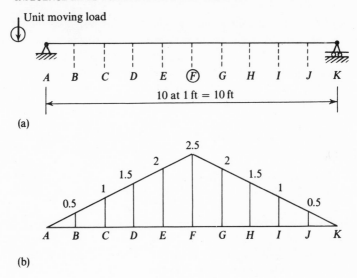

(a)

(b)

Fig. 6-1

load position x. The plane curve represented by the equation gives the desired influence line. To illustrate this technique, let us use the same problem previously mentioned but picture it in a different way, as shown in Fig. 6-2(a). The unit load is placed at a distance x from left end A. The reactions at ends A and K are expressed as functions of x,

$$R_A = \frac{(10-x)(1)}{10} = 1 - \frac{x}{10} \qquad R_K = \left(\frac{x}{10}\right)(1) = \frac{x}{10}$$

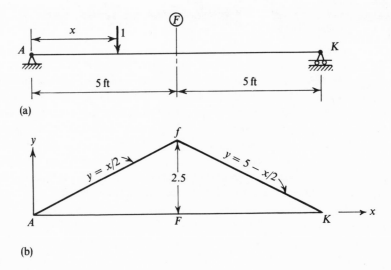

(a)

(b)

Fig. 6-2

respectively. When the moving load is confined to the left of section F (as shown), the bending moment at F may be found from R_K; i.e.,

$$M_F = 5R_K = \frac{x}{2}$$

or

$$y = \frac{x}{2}$$

where y denotes M_F. As the moving load is confined to the right of section F (not shown), the bending moment at F may be found from R_A; i.e.,

$$M_F = 5R_A = (5)\left(1 - \frac{x}{10}\right) = 5 - \frac{x}{2}$$

or

$$y = 5 - \frac{x}{2}$$

Selecting the coordinate axes as shown in Fig. 6-2(b), we plot $y = x/2$ and $y = 5 - (x/2)$ by two straight lines. The curve AfK given in Fig. 6-2(b) is the desired *bending-moment influence line* for section F, and the corresponding diagram $AFKf$ is called the *bending-moment influence diagram* for section F.

Although in this particular case the influence diagram of Fig. 6-2(b) is identical to the moment diagram for the same beam under a unit load at mid-span, we must not confuse the influence diagram with a bending-moment diagram for the beam. Whereas the latter shows by each ordinate the bending moment at the corresponding section due to a fixed load; the influence diagram shows by each ordinate the bending moment at a fixed section due to a unit load placed at the point where the ordinate stands.

6-2. USE OF THE INFLUENCE LINE

By definition an influence line indicates the effect of a unit load moving across a span. It should be pointed out that the unit load for which the influence line is drawn is usually vertical although it may be horizontal or inclined. The construction of influence lines is closely associated with the stress analysis of bridge structures, which are frequently subjected to the action of various systems of moving loads. However, the usefulness of influence lines is not limited to bridge structures. They are also important in the determination of the maximum stresses in other engineering structures subjected to live loads.

An influence line is a useful tool in stress analysis in two ways:

1. It serves as a criteria in determining the maximum stress—a guide to determine just what portion of the structure should be loaded in order to cause the maximum effect on the part under consideration.

2. It simplifies the computation.

To illustrate, consider a simple beam 10 ft long subjected to the passage of a moving uniform load of 1 kip per ft without limit in length and a movable concentrated load of 10 kips that may be placed at any point of the span (see Fig. 6-3(a)). Determine the maximum bending moment at the midspan section C.

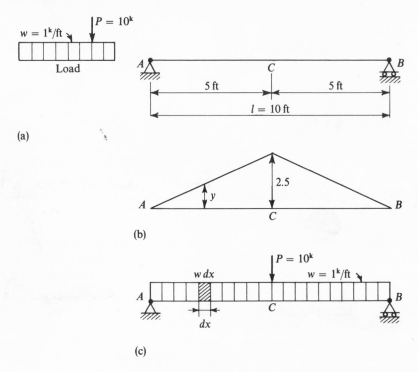

(a)

(b)

(c)

Fig. 6-3

We start by drawing the bending-moment influence line for section C, as in Fig. 6-3(b). It is apparent from the influence line that, to obtain the maximum M_C, the single concentrated load of 10 kips should be placed at the midpoint of the span, where the maximum ordinate of the influence line occurs, and the uniform load should be spread over the entire span.

Next, to compute the bending moment at C due to the live loads so placed, we simply multiply each load by the corresponding influence ordinate and add. Referring to Figs. 6-3(b) and (c), we obtain

$$M_C = P(2.5) + \sum (w \, dx) y$$

Note that the uniform load is treated as a series of infinitesimal concentrated loads each of magnitude $w \, dx$ and the total effect of the uniform load is the summation

$$\sum (w \, dx) y$$

Now, $\sum (w\,dx)y = \int_0^l wy\,dx = w\int_0^l y\,dx$

= (load intensity) × (area of influence diagram)

Therefore, the total bending moment at C is:

$$M_C = (10)(2.5) + (1)\frac{(2.5)(10)}{2} = 25 + 12.5 = 37.5 \text{ ft-kips}$$

This value may be checked by the conventional method of computing M_C:

$$M_C = \left(\frac{10}{2}\right)(5) + \frac{(1)(10)^2}{8} = 25 + 12.5 = 37.5 \text{ ft-kips}$$

In this simple case, such a conclusion may be drawn without the aid of the influence diagram; but for more complicated moving load systems, we find that the influence diagram can be of substantial help, as will be discussed in Ch. 7.

6-3. INFLUENCE LINES FOR STATICALLY DETERMINATE BEAMS

Example 6-1. In Fig. 6-4(a) is shown a simple beam. Draw the influence lines for

1. the reaction at $A(R_A)$ and the reaction at $B(R_B)$
2. the shear at any section $C(V_C)$
3. the moment at any section $C(M_C)$

To draw the influence line for R_A, we place a unit load distance x from A and express R_A as the function of x; i.e.,

$$R_A = \left(\frac{l - x}{l}\right)(1) = 1 - \frac{x}{l}$$

which represents a straight line with unit ordinate at A and zero at B as shown in Fig. 6-4(b). Similarly, we draw the influence line for R_B as shown in Fig. 6-4(c) based on

$$R_B = \frac{x}{l}$$

To draw the influence line for V_C, we note that as long as the unit load is applied at any position to the left of section C, $0 \leqslant x < a$, V_C is found to be equal to R_B but with the opposite sign. When the unit load is applied at any section to the right of C, $a < x \leqslant l$, V_C is found to be equal to R_A. Therefore, the influence line V_C in the portion of AC is the same as that of R_B but with negative sign, and the influence line V_C in the portion CB is the same as that of R_A. These two lines are parallel with equal slopes of $-1/l$, and there is an abrupt change of unity when the unit load passes from the left of C to the right of C (see Fig. 6-4(d) for the influence line of V_C).

To draw the influence line for M_C, we note that as long as the load is to the left of C, we can use segment CB as a free body and find that $M_C = R_B b$. When the load passes C to the right, we can use segment AC as a free body and find that $M_C = R_A a$. Both are straight lines but with different slopes, as shown in Fig. 6-4(e).

In connection with the influence lines of Fig. 6-4 we note the following:

①. The maximum reaction due to a single concentrated load occurs when the load is at the support.

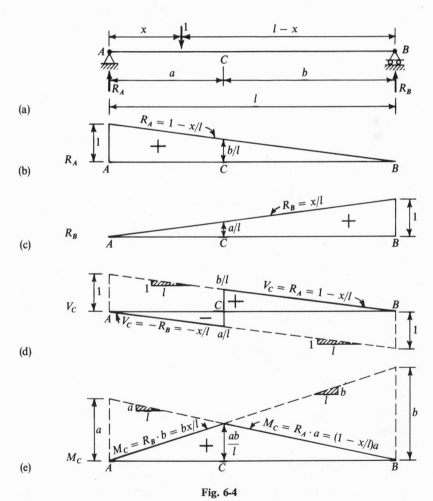

Fig. 6-4

2. The maximum reaction due to a uniform load occurs when the beam is fully loaded.

3. The maximum shear at any section C due to a single concentrated load occurs when the load is just to the right or left of the section and is on the longer of the two segments into which C divides the beam.

4. The maximum shear at any section C due to a uniform load occurs when the load extends from C to the more distant support. Referring to Fig. 6-4(d), if $b > a$, we find the load is distributed over BC to cause a maximum V_C.

5. The maximum moment at any section C due to a single concentrated load occurs when the load is at C.

6. The maximum moment at any section C due to a uniform load occurs when the beam is fully loaded.

By comparing the influence lines for the same function at different points, we note the following:

1. The maximum shear for a simple beam due to a single concentrated load occurs at the end and equals the load.

2. The maximum shear for a simple beam due to a uniform load occurs at the end when the beam is fully loaded and equals one-half the total load.

3. The maximum moment for a simple beam due to a single concentrated load Q occurs at the midspan section and equals $Ql/4$.

4. The maximum moment for a simple beam due to a uniform load of intensity q occurs at the midspan section when the beam is fully loaded and equals $ql^2/8$.

Example 6-2. Consider the simple beam with an overhang shown in Fig. 6-5(a). To construct the influence line for R_B, we place a unit load distance x from end A and apply $\sum M_C = 0$ to obtain

$$R_B = \frac{20 - x}{16}$$

The linear expression represents a straight line with the maximum ordinate $5/4$ at A and the minimum ordinate 0 at C, as shown in Fig. 6-5(b). Note that when the unit load is placed at B, the influence ordinate for R_B should be equal to unity.

The influence line for R_C may be found by applying $\sum F_y = 0$; i.e.,

$$R_C = 1 - R_B = 1 - \frac{20 - x}{16} = \frac{x - 4}{16}$$

which is also a linear function of x representing a straight line, as shown in Fig. 6-5(c). As a check, the ordinate at C should be equal to unity; and that at B zero.

The influence line for the shear at the section just to the left of B, called

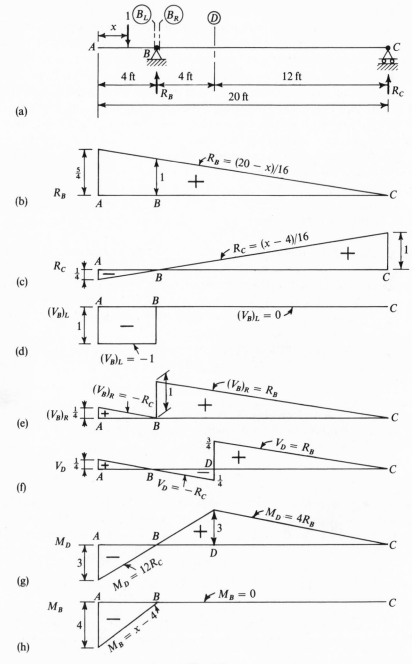

(a)

(b)

(c)

(d)

(e)

(f)

(g)

(h)

Fig. 6-5

$(V_B)_L$, is given in Fig. 6-5(d). As long as the unit load is on the overhanging portion of beam, $(V_B)_L = -1$; as load passes B to the right, $(V_B)_L = 0$.

The influence line for the shear at the section just to the right of B, called $(V_B)_R$, is shown in Fig. 6-5(e). As long as the unit load is on the overhanging portion of beam, $(V_B)_R$ equals R_C but with opposite sign. As the load is on the simple-beam portion, $(V_B)_R$ equals R_B.

By a similar approach, we construct the influence line for shear at section D, as shown in Fig. 6-5(f). As a check, when the unit load passes D from the left to the right, the shear at D increases suddenly from $-\frac{1}{4}$ to $+\frac{3}{4}$; i.e., there is an abrupt change of shear equal to unity at D.

The influence line for the moment at D is shown in Fig. 6-5(g). We note that as long as the unit load is confined to the portion AD, the moment at D may be found from R_C,

$$M_D = 12R_C = (12)\left(\frac{x - 4}{16}\right) = \frac{3x - 12}{4} \qquad (0 \leqslant x \leqslant 8)$$

which represents a straight line from A to D with the ordinates equal to -3 at A and $+3$ at D. When the load passes D to the right, the moment at D may be found from R_B; i.e.,

$$M_D = 4R_B = (4)\left(\frac{20 - x}{16}\right) = \frac{20 - x}{4} \qquad (8 \leqslant x \leqslant 20)$$

which represents a straight line from D to C with ordinates $+3$ at D and 0 at C.

Finally, we construct the influence line for the moment at B, as shown in Fig. 6-5(h). When the load is placed at A, M_B has its greatest negative value of 4. As the load travels from A to B, the moment varies linearly from -4 to 0. As the load enters the portion BC, there will be no moment at B.

Example 6-3. Consider the compound beam shown in Fig. 6-6(a). To construct the influence line for the reaction at A, we start by placing a unit load distance x from A. Note that, while it is traveling on the portion AC, the reaction at A can be determined by applying the condition equation $M_C = 0$; i.e.,

$$R_A = \frac{2a - x}{2a} = 1 - \frac{x}{2a} \qquad (0 \leqslant x \leqslant 2a)$$

which represents the straight line shown in Fig. 6-6(b). As the unit load passes C to the right, there will be no reaction at A.

The influence line for the shear at section D is shown in Fig. 6-6(c). Note that

$$V_D = R_A - 1 = -\frac{x}{2a} \qquad (0 \leqslant x < a)$$

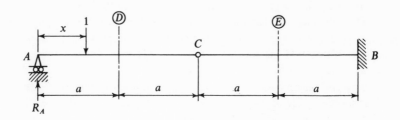

(a)

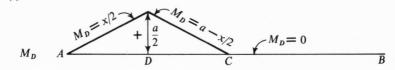

(b)

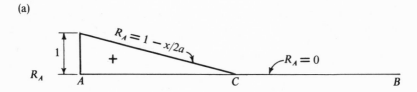

(c)

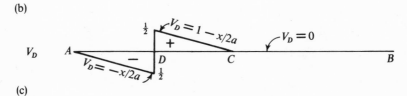

(d)

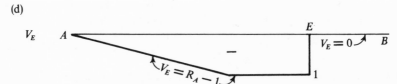

(e)

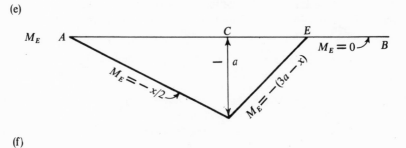

(f)

Fig. 6-6

$$V_D = R_A = 1 - \frac{x}{2a} \qquad (a < x \leqslant 2a)$$

There is an abrupt change in the influence ordinate equal to unity as the load passes from the left of D to the right of D, and there will be no shear at D as the load passes C to the right, since $R_A = 0$.

The moment influence line at section D is shown in Fig. 6-6(d) for which we note that

$$M_D = (R_A)(a) - (1)(a - x) = \left(1 - \frac{x}{2a}\right)(a) - (1)(a - x) = \frac{x}{2}$$
$$(0 \leqslant x \leqslant a)$$

$$M_D = (R_A)(a) = \left(1 - \frac{x}{2a}\right)(a) = a - \frac{x}{2} \qquad (a \leqslant x \leqslant 2a)$$

There will be no moment at D as the load passes C to the right, since $R_A = 0$.

The shear influence line at section E is shown in Fig. 6-6(e) for which we note that as long as the unit load is confined to the portion AE,

$$V_E = R_A - 1 \qquad (0 \leqslant x < 3a)$$

which means that each influence ordinate of V_E can be obtained by the corresponding influence ordinate of R_A minus one. As the load passes E to the right

$$V_E = R_A = 0 \qquad (x > 3a)$$

Finally, the moment influence line at section E is constructed as in Fig. 6-6(f). Note that as the load is confined to the portion AC,

$$M_E = (R_A)(3a) - (1)(3a - x) = \left(1 - \frac{x}{2a}\right)(3a) - (1)(3a - x) = -\frac{x}{2}$$
$$(0 \leqslant x \leqslant 2a)$$

As the load passes C to E, the value of R_A in the above equation becomes zero, so that

$$M_E = -(3a - x) \qquad (2a \leqslant x \leqslant 3a)$$

As the load passes E to B $\quad M_E = 0 \qquad (x \geqslant 3a)$

The influence line for beams, particularly for compound beams, can be more easily constructed by the method of virtual work, which we shall discuss in the next section.

6-4. BEAM INFLUENCE LINES BY VIRTUAL WORK

By the principle of virtual work we can derive a very simple and elegant method for constructing the influence lines for beams. It can be stated as follows:

1. To obtain an influence line for the reaction of any statically determinate beam, remove the support and make a positive unit displacement of its point of application. The deflected beam is the influence line for the reaction.

2. To obtain an influence line for the shear at a section of any statically determinate beam, cut the section and induce a unit relative transverse sliding displacement between the portion to the left of the section and the portion to the right of the section keeping all other constraints (both external and internal) intact. The deflected beam is the influence line for the shear at the section.

3. To obtain the influence line for the moment at a section of any statically determinate beam, cut the section and induce a unit rotation between the portion to the left of the section and the portion to the right of the section keeping all other constraints (both external and internal) intact. The deflected beam is the influence line for moment at the section.

To prove the theory, we take the case of a simple beam. The procedure of proof is generally applicable to more complicated beams. Figure 6-7(a) shows a simple beam subjected to a single unit moving load. To find the reaction at A by the method of virtual work, we remove the constraint at A, substitute R_A for it, and let A travel a small virtual displacement δs_A along R_A. We then have a deflected beam $A'B$, as shown in Fig. 6-7(b), where y indicates the transverse displacement at the point of unit load. Applying the virtual-work equation, we obtain

$$(R_A)(\delta s_A) - (1)(y) = 0$$

from which
$$R_A = \frac{y}{\delta s_A}$$

If we let
$$\delta s_A = 1$$

then
$$R_A = y$$

Since y is, on the one hand, the ordinate of the deflected beam at the point where the unit load stands and is, on the other hand, the value of function R_A due to the unit moving load (i.e., the influence ordinate at the point), we conclude that the deflected beam $A'B$ of Fig. 6-7(b) is the influence line for R_A if δs_A is set to be unity.

To determine the shearing force at any beam cross section C, we cut the beam at C and let the two portions AC and CB have a relative virtual transverse displacement δs_C at C without causing relative rotation between the two portions. This is equivalent to rotating AC and BC the same small angle about A and B respectively. Applying the virtual-work equation, we obtain

$$(V_C)(\delta s_C) - (1)(y) = 0$$

from which
$$V_C = \frac{y}{\delta s_C}$$

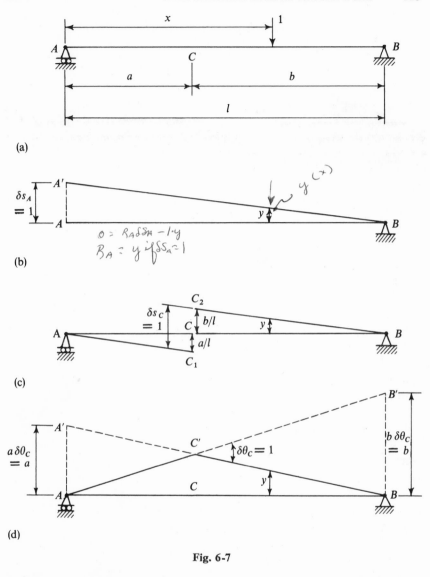

(a)

(b)

$0 = R_A \delta s_A - 1 \cdot y$

$B_A = y \ if \ \delta s_A = 1$

(c)

(d)

Fig. 6-7

If we let $\delta s_C = 1$

then $V_C = y$

This proves that the deflected beam AC_1C_2B of Fig. 6-7(c) is the influence line for V_C. It should be pointed out that the virtual displacement is supposed to be vanishingly small and that, when we say $\delta s_C = 1$, we do not mean that $\delta s_C = 1$ ft or 1 in. but one unit of very small distance for which the expressions

$$CC_1 = \frac{a}{l}$$

$$CC_2 = \frac{b}{l}$$

shown in Fig. 6-7(c) are justified.

To determine the moment at any beam cross section C by the method of virtual work, we cut the beam at C and induce a relative virtual rotation between the two portions AC and CB at C without producing relative transverse sliding between the two. Thus, by virtual work,

$$(M_C)(\delta\theta_C) - (1)(y) = 0$$

from which

$$M_C = \frac{y}{\delta\theta_C}$$

If we let

$$\delta\theta_C = 1$$

then

$$M_C = y$$

This proves that the deflected beam $AC'B$ of Fig. 6-7(d) is the influence line for M_C. Note that, when we say $\delta\theta_C = 1$, we do not mean that $\delta\theta_C = 1$ radian. One unit of $\delta\theta_C$ may be as small as $1/100$ radian for which it is justified to write

$$AA' = a \cdot \delta\theta_C = a \quad \text{units} \qquad BB' = b \cdot \delta\theta_C = b \quad \text{units}$$

as indicated in Fig. 6-7(d). Compare the influence lines of Fig. 6-7 with those obtained in Fig. 6-4; they are identical.

To illustrate the rules given above, we take a simple beam with an overhang and a compound beam so that we may clearly see that the method based on virtual work can be used to advantage in constructing influence lines for beams.

Example 6-4. Figure 6-8(a) shows a simple beam with an overhang the same as that in Fig. 6-5(a). We wish to obtain the influence lines for R_B, R_C, $(V_B)_L$, $(V_B)_R$, V_D, M_D, and M_B by the virtual-work method.

To construct the influence line for R_B, we remove the support at B and let it move a unit distance upward. The deflected beam $A'B'C$, shown in Fig. 6-8(b), is the desired influence line. The influence line for R_C is obtained in a similar manner, as shown in Fig. 6-8(c) by $A'BC'$.

To construct the influence line for $(V_B)_L$, we cut the section immediately to the left of B and let the left portion have a unit relative transverse displacement with respect to the right portion at the cut point without causing relative rotation between the two. Since the portion to the right of the cut section is a simple beam, it remains stable and rigid. Only the portion to the left of the cut section can go down a unit distance.[*] Thus the influence line for $(V_B)_L$ is represented by $A'B'BC$ in Fig. 6-8(d).

[*] Sign convent V_L ↓ V_R ↑

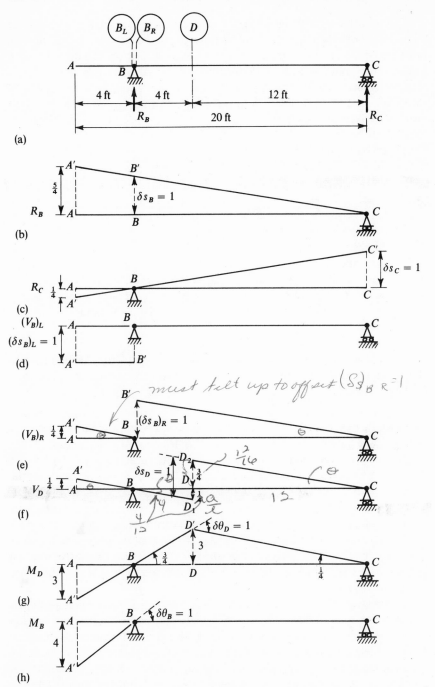

Fig. 6-8

To construct the influence line for $(V_B)_R$, we cut the section immediately to the right of B and let the two portions of the beam have a unit relative transverse displacement without producing relative rotation. To do this, end B of BA cannot move because of the presence of the support at B; however, it is allowed to rotate about the hinge at B. We, therefore, first raise the end B of BC a unit distance to B' and then turn BA to BA' parallel to $B'C$. The deflected beam so arrived at, as shown in Fig. 6-8(e) by $A'BB'C$, is the influence line for $(V_B)_R$.

To construct the influence line for V_D, we cut the beam at D and let the two portions DA and DC have a relative transverse displacement equal to unity. The two are kept parallel to each other so that no relative rotation is introduced. Then the deflected beam $A'D_1D_2C$ of Fig. 6-8(f) is the desired influence line. The slope of the deflected beam is found to be $-1/16$, with which the influence ordinates can easily be determined, as indicated in Fig. 6-8(f).

The influence line for M_D is shown in Fig. 6-8(g) by the deflected beam $A'D'C$. It results from cutting the beam at D and letting DA and DC have a unit rotation at D without allowing relative translation between the two. Referring to Fig. 6-8(g), we find that the rotation of DC about support C is $1/4$ and the rotation of DA about support B is $3/4$. Thus the influence ordinate at D is found to be $(12)(1/4)$ or 3.

Similarly, we construct the influence line for M_B, as shown in Fig. 6-8(h) by $A'BC$, by cutting the section just to the left of B and letting the left portion of the beam have a unit relative rotation with respect to the right portion of the beam at B.

Example 6-5. Figure 6-9(a) shows a compound beam, the same beam as in Fig. 6-6(a). Draw influence lines for R_A, V_D, M_D, V_E, M_E.

To construct the influence line for R_A, we remove support A and move end A up a unit distance. The deflected beam $A'CB$ shown in Fig. 6-9(b) is the influence line for R_A. Note that portion CB is a cantilever and will remain unmoved.

To construct the influence line for V_D, we cut the beam through D and let the left portion of beam have a relative transverse displacement equal to unity with respect to the right portion of beam at D without causing relative rotation between the two. The deflected beam AD_1D_2CB, shown in Fig. 6-9(c), is the influence line for V_D.

To construct the influence line for M_D, we cut the beam through D and let the left portion of beam rotate a unit angle with respect to the right portion at D. The deflected beam $AD'CB$ of Fig. 6-9(d) is the influence line for M_D.

The influence line for V_E is shown in Fig. 6-9(e) by $AC'E'EB$ which results from cutting the beam through E and moving the left portion of beam down a unit distance with respect to the right portion of beam at E, while keeping the deflected portion $C'E'$ parallel to BE.

The influence line for M_E is shown in Fig. 6-9(f) by $AC'EB$ which results

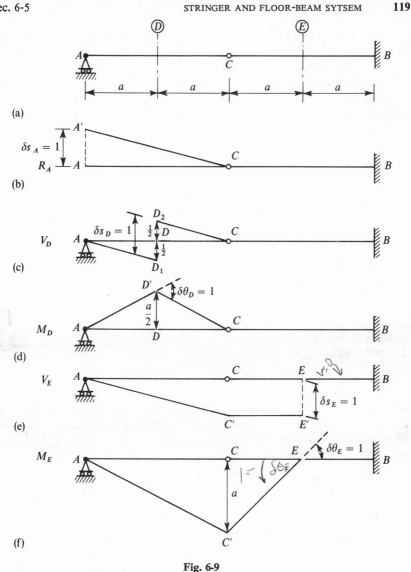

Fig. 6-9

from cutting the beam through E and rotating the left portion of beam a unit angle with respect to the right portion of beam at E. Point E is kept fixed in the original position.

6-5. STRINGER AND FLOOR-BEAM SYSTEM

In the case where a long-span girder is used in bridge construction, the live loads are usually not applied directly to the girder but are transmitted by a *stringer and floor-beam system* to the main girder. A similar arrangement is used in a trussed bridge where the moving wheels are carried by the bridge floor to the stringers, which are supported by the floor beams. The floor beams then carry the stringer reactions to the panel points of the main trusses. Regardless of actual details of construction, we may picture this system as in Fig. 6-10 where the main girder AB carries a number of floor beams that, in turn, support stringers running parallel to the girder. Those portions of the main

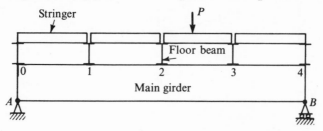

Fig. 6-10

girder between the floor beams, such as 0-1, 1-2, etc., are called *panels*, and the end points of the panels, such as 0, 1, 2, etc., are called *panel points*. It is assumed that the stringers are each simply supported on the floor beams. Consequently, a load P applied to any stringer will be transmitted to the main girder only at the two corresponding panel points. The girder (likewise the truss) is therefore *panel point loaded* for any condition of loading on the stringer. It is obvious that the division of the load between two panel points has no effect on the reactions at A and B. However, the internal shearing force in the main girder is made constant throughout any one panel by such a division. Thus, it is customary to say *panel shear* rather than shear at a particular section.

Note that it is not necessary for the stringers in every panel to be simply supported by the adjacent floor beams, as shown in Fig. 6-10. For instance, stringers may be arranged as in Fig. 6-11(a) or (b). However, the type in Fig. 6-10 is most commonly used and will be discussed.

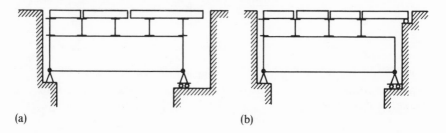

(a)   (b)

Fig. 6-11

To illustrate the analysis of the girder, let us consider the system shown in Fig. 6-12(a). When the loads are transmitted through stringers and floor beams,

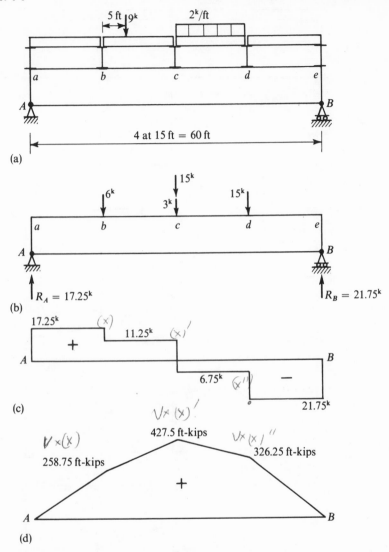

Fig. 6-12

we find that the girder is subjected to forces as shown in Fig. 6-12(b). The reaction at A is found to be 17.25 kips from $\sum M_B = 0$ by using either the loading shown in Fig. 6-12(a) or that shown in Fig. 6-12(b). Similarly, the reaction at B is found to be 21.75 kips.

Based on the load diagram of Fig. 6-12(b), the shear diagram is plotted as in Fig. 6-12(c) and the moment diagram as in Fig. 6-12(d). Since the girder is panel point loaded, the shear has the same value at any section in a given panel of the girder.

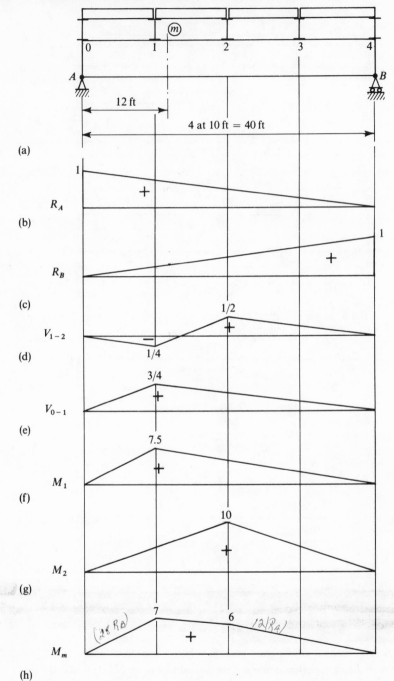

Fig. 6-13

6-6. INFLUENCE LINES FOR GIRDERS WITH FLOOR
SYSTEMS

Consider the girder shown in Fig. 6-13(a), which consists of 4 equal panels each 10 ft long.

The influence lines for R_A and R_B are readily drawn as shown in Figs. 6-13 (b) and (c) respectively. They are constructed in the same way as the influence lines for reactions of a simple beam 40 ft long, since end reactions of the girder are not affected by the presence of the floor system.

The influence line for the shear in panel 1-2, denoted by V_{1-2}, is shown in Fig. 6-13(d). When the load is confined to 0-1, we observe that V_{1-2} is equal to R_B but with a minus sign. When the load is confined to 2-4, V_{1-2} is equal to R_A. As the load travels from panel point 1 to panel point 2, the value of V_{1-2} has a linear variation. For, in general, in a structure where the loads are transmitted by members of a floor system to panel points, the influence line between consecutive panel points will be a straight line provided that the stringers act as simple beams spanning the adjacent floor beams. This can be proved as follows:

Consider a girder with a floor system, shown partially in Fig. 6-14(a). Let

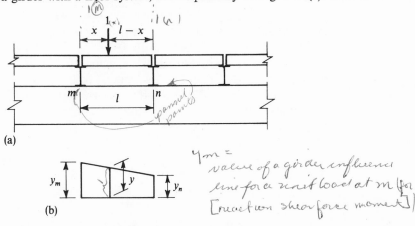

(a)

(b)

Fig. 6-14

a unit load travel along a panel m-n. When the unit load is at m, we let y_m be the corresponding influence ordinate for it; when the unit load is at n, we let y_n be the corresponding influence ordinate for it (see Fig. 6-14(b)). Now when the unit load is in any intermediate position, say distance x from m, it will be transmitted to the girder through the floor beams at m and n with amounts $(l - x)/l$ and x/l respectively. The effect of the load is found by multiplying each of these values by the corresponding value of the influence ordinate, and adding. This gives

$$(1)(y) = \left(\frac{l-x}{l}\right)(y_m) + \left(\frac{x}{l}\right)(y_n)$$

in which y denotes the influence ordinate where the unit load is located as indicated in Fig. 6-14(b). This, being a linear expression of x, specifies the influence ordinate for the general intermediate position in the panel.

Similarly, we draw the influence line for shear in panel 0-1 as shown in Fig. 6-13(e).

To construct the influence line for the moment at panel point 1, denoted by M_1, we note that when the unit load is at 0, $M_1 = 0$; when the unit load is located between panel points 1 and 4, $M_1 = 10R_A$, which represents a straight line with maximum ordinate 7.5 at panel point 1 and 0 at panel point 4. Connecting the influence ordinates at panel points 0 and 1 by a straight line, we complete the influence line for M_1, as shown in Fig. 6-13(f).

Similarly, we draw the influence line for moment at panel point 2 as shown in Fig. 6-13(g). *(arm length × ordinate) ($\frac{1}{2} \times 20$) = 10*

It should be pointed out that in this girder, in which each stringer has a length of one panel, the influence line for the moment at any panel point is the same as the influence line for the moment at the corresponding point in a simple beam.

To draw the moment influence line at any intermediate section, say section m in the second panel, we note that as the load is confined to 0-1, $M_m = 28R_B$, which represents a straight line with 0 ordinate at 0 and a maximum ordinate 7 at panel point 1. As the load is confined to 2-4, $M_m = 12R_A$, which also represents a straight line with 0 ordinate at panel point 4 and a maximum ordinate 6 at panel point 2. It remains for us to connect the influence ordinates at panel points 1 and 2 by a straight line to complete the influence line for M_m, as shown in Fig. 6-13(h).

6-7. INFLUENCE LINES FOR STATICALLY DETERMINATE BRIDGE TRUSSES

Influence lines for bridge trusses are drawn as a unit load travels across the loaded chord, i.e., the chord to which the floor systems are connected. Throughout our discussion we shall assume that the stringers act as simple beams between the adjacent floor beams so that the influence line will be a straight line between any two adjacent panel points as described in Sec. 6-6.

Example 6-6. Figure 6-15(a) shows a Pratt truss with parallel top and bottom chords. For convenience in further analysis, we may begin with the influence lines for the end reactions at a and g, which can readily be drawn as shown in Fig. 6-15(b). The dotted line gives the influence line for reaction at g.

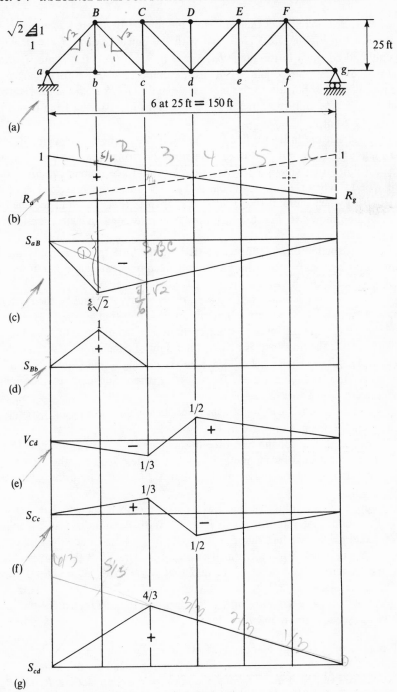

Fig. 6-15

From this, we construct the influence line for the bar force in the end post aB as shown in Fig. 6-15(c). Note that with the unit load at a, $S_{aB} = 0$; with the unit load confined in the range bg, $S_{aB} = (-\sqrt{2})(R_a)$, the minus sign indicates a compression (see joint a). Thus the influence diagram for S_{aB} from b to g is in direct proportion to the influence diagram for R_a. As the load travels from a to b, the influence line will be a straight line varying from 0 at a to $(-\sqrt{2})(5/6)$ at b.

To construct the influence line for the bar force in the hip verticle Bb, we note that, unless the unit load is on stringer ab or stringer bc, there will be no force transmitted to panel point b and therefore no force in member Bb. Now since $S_{Bb} = 0$ when the unit load is at a, $S_{Bb} = 1$ when the unit load is at b, and $S_{Bb} = 0$ when the unit load is at c, it is obvious that the influence line for S_{Bb} is the one shown in Fig. 6-15(d) where the plus sign indicates tension in the member.

The construction of an influence line for the bar force in a diagonal member is illustrated by Fig. 6-15(e) where the verticle component of the bar force in member Cd, denoted by V_{Cd}, is considered.

As the unit load is at, or to the left of, panel point c, we find by the method of shear (using the right portion of truss as a free body) that $V_{Cd} = -R_g$ (compression). Hence the influence line is a straight line from 0 at a to $-1/3$ at c. Similarly, when the unit load is at or to the right of panel point d, we find that $V_{Cd} = R_a$ (tension). Hence the influence line is a straight line from $1/2$ at d to 0 at g. As the unit load travels from c to d, the influence line is straight, varying from $-1/3$ to $1/2$.

If we wish to construct the influence line for the bar force in Cd, we simply multiply each influence ordinate of V_{Cd} by $\sqrt{2}$.

The influence line for the bar force in member Cc is shown in Fig. 6-15(f) and can be constructed by the method of shear as described for the truss diagonal. In this particular case, we observe that $S_{Cc} = -V_{Cd}$ by using joint C as a free body.

To construct the influence line for a chord member, such as bar cd, we apply the method of moment about joint C. With the unit load at, or to the left of, panel point c, we find that $S_{cd} = 100R_g/25 = 4R_g$. Hence the influence line is a straight line varying from 0 at a to $4/3$ at c. With the unit load at, or to the right of, panel point c, we find that $S_{cd} = 50R_a/25 = 2R_a$. Hence the influence line is a straight line varying from $4/3$ at c to 0 at g. Thus we complete the influence diagram for S_{cd}. As shown in Fig. 6-15(g), it is a triangle with its apex under the moment center C.

Example 6-7. In Fig. 6-16(a) is shown a Parker truss. Let us draw the influence lines for the bar forces (or components) in members BC and Bc.

As before, we first draw the influence lines for the end reactions R_a and R_g,

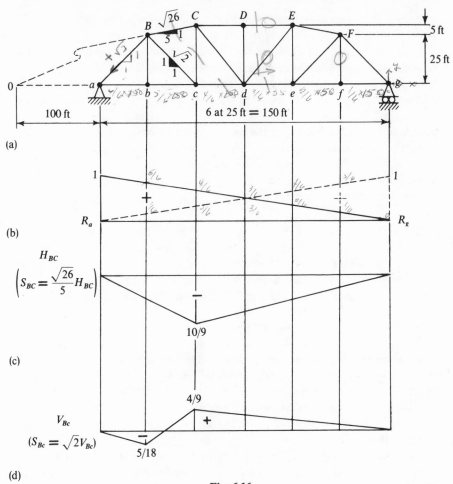

Fig. 6-16

as shown in Fig. 6-16(b), for reference, since the bar force can always be expressed in terms of support reactions.

To construct the influence line for the bar force in BC, or rather the influence line for its horizontal component H_{BC}, we use the method of moment about c by passing a diagonal section which cuts bars BC, Cc, and cd. As the unit load goes from a to c, it is evident, if the right portion of the structure is used as a free body, that $H_{BC} = -100R_g/30 = -10R_g/3$. The minus sign indicates a compressive force in H_{BC}. Hence the influence line is a straight line varying from 0 at a to $-10/9$ at c. As the unit load goes from c to g, we find, by using the left portion of the structure as a free body, that $H_{BC} = -50R_a/30 = -5R_a/3$. Hence the influence line is a straight line varying from $-10/9$ at c

to 0 at g. We thus complete the influence line for H_{BC}, as shown in Fig. 6-16(c).

To construct the influence line for bar force in Bc, or rather for its vertical component V_{Bc}, we use the method of moment about O (see Fig. 6-16(a)) by passing a vertical section which cuts bars BC, Bc, and bc. As the unit load travels from a to b, we find, by using the part of the structure to the right of the section, that $V_{Bc} = -250R_g/150 = -5R_g/3$. Hence it is a straight line from 0 at a to $-5/18$ at b. As the unit load travels from c to g, we find, by using the part of structure to the left of the section, that $V_{Bc} = 100R_a/150 = 2R_a/3$. Hence it is a straight line from $4/9$ at c to 0 at g. We need only connect the ordinates at b and c by a straight line to complete the influence line for V_{Bc}, as shown in Fig. 6-16(d).

Example 6-8. In Fig. 6-17(a) is shown a K truss. Let us draw the influence lines for the bar forces (or components) in members dp, dq, and Dp. To do this, we may either directly compute influence ordinates for each successive panel point or draw influence lines for members other than those under consideration and from these deduce the desired influence lines. The latter procedure will be used in this problem.

We begin with the influence diagram for the bar force of the lower chord cd, as shown in Fig. 6-17(b), the apex of which is at panel point c with the ordinate equal to $6/7$, found by taking section m-m and using C as the moment center. Similarly, we can draw the influence line for the bar force of the lower chord de, as shown in the same figure, by taking the section n-n and using D as the moment center. The influence diagram is also a triangle with its apex at panel point d and the ordinate equal to $15/14$. Both influence lines can be drawn without difficulty.

Now by use of the equation $\sum F_x = 0$ at joint d, we obtain the horizontal component of the bar force in dp equal to the difference between S_{de} and S_{cd}. That is

$$H_{dp} = S_{de} - S_{cd}$$

The influence diagram for H_{dp} can thus be obtained from the shaded portion of Fig. 6-17(b) rearranged as in Fig. 6-17(c).

To construct the influence line for the bar force in dq, we note, from $\sum F_y = 0$ at joint d, that

$$S_{dq} = -V_{dp} = -\frac{3}{4} H_{dp}$$

when the unit load is placed at any panel points other than d, and that

$$S_{dq} = 1 - V_{dp} = 1 - \frac{3}{4} H_{dp} = 1 - \left(\frac{3}{4}\right)\left(\frac{5}{14}\right) = \frac{41}{56}$$

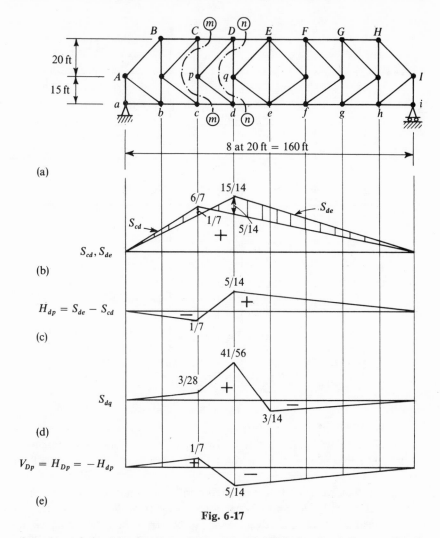

Fig. 6-17

when the unit load is placed at d. See Fig. 6-17(d) for the influence line for S_{dq}.

The influence line for the vertical component of the bar force in Dp is shown in Fig. 6-17(e), which is the same figure for H_{dp} but with opposite sign, since

$$V_{Dp} = H_{Dp} = -H_{dp}$$

by using $\sum F_x = 0$ at joint p.

Example 6-9. Let us consider the statically determinate long-span bridge truss of Fig. 6-18(a). The difficulty with the influence line for such a truss

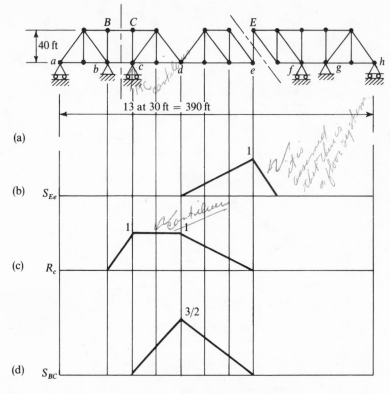

Fig. 6-18

disappears, sence we know that it is always possible to obtain the ordinate to the influence line of any element for a unit load at each panel point of the truss.

Suppose that we wish to construct the influence lines for the reaction at support c, denoted by R_c, and the bar force in member BC, denoted by S_{BC}, of the indicated truss. It is advantageous to construct first the influence line for the force in the hanger Ee, as shown in Fig. 6-18(b). This can easily be verified by cutting the hanger and taking the moments about d ($M_d = 0$) of the forces acting on the suspended span de.

Next, we construct the influence line for R_c, as shown in Fig. 6-18(c). Note that by taking a free body through the hanger Ee and the panel bc, we observe

$$R_c + S_{Ee} = 1$$

as long as the unit load is confined to portion ce.

The influence line for the bar force in member BC is then drawn, with the force computed by taking the moments about d ($M_d = 0$) of the forces acting on the part of structure between panel bc and d (see Fig. 6-18(a)). For instance, for a unit load at point d, $R_c = 1$; by taking $M_d = 0$, we obtain

$$S_{BC} = \frac{(1)(60)}{40} = \frac{3}{2}$$

The influence line for S_{BC} is shown in Fig. 6-18(d).

PROBLEMS

6-1. Given a simple beam 24 ft long, construct the influence lines for the shear and bending moment at a section 8 ft from the left end, and obtain the maximum shear and bending moment for the section resulting from a moving uniform load of 3 kips per ft and a movable concentrated load of 50 kips.

6-2. A cantilever beam 20 ft long is fixed at the right end. Construct the shear and moment influence lines for sections 5 ft, 10 ft, and 20 ft from the free end. Using the same loadings given in the previous problem, compute maximum shears and moments at these sections.

6-3. In Fig. 6-19 is shown a simple beam with an overhang. Draw the influence lines for R_B, R_C, V_D, M_B, and M_D.

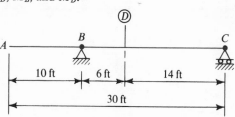

Fig. 6-19

6-4. Given a compound beam such as that shown in Fig. 6-20, construct the influence lines for R_A, R_C, R_E, V_B, M_B, and M_C. Compute the maximum value for each of them due to a moving uniform load of 2 kips per ft.

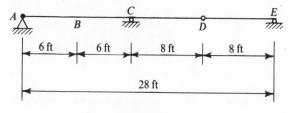

Fig. 6-20

6-5. Solve Prob. 6-3 by the method of virtual work.

6-6. Construct the influence lines for Prob. 6-4 by the method of virtual work.

6-7. For the girder of Fig. 6-21, construct influence lines for (a) the reaction at end A, (b) the reaction at end B, (c) the shear in panel 0-1, (d) the shear in panel 1-2,

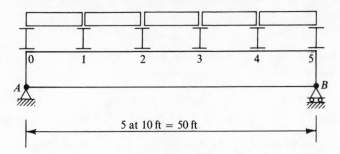

Fig. 6-21

(e) the moment at panel point 2, and (f) the moment at a point 15 ft from the left end.

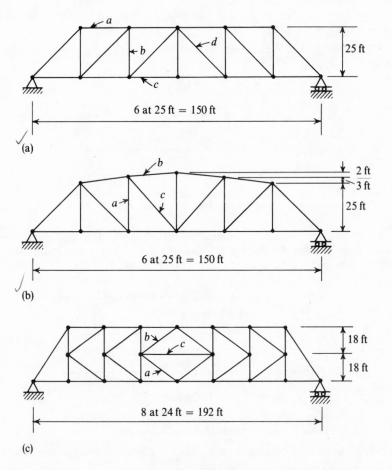

Fig. 6-22

$(a \ni b)$

6-8. For the trusses shown in Fig. 6-22, construct the influence lines for the bar force (or component) in each of the lettered bars.

6-9. For the subdivided Warren truss, shown in Fig. 6-23, draw the influence lines for the bar forces (or components) in members fF, ef, eF, FG, and gG.

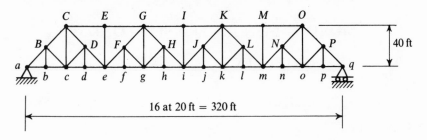

Fig. 6-23

6-10. For the long-span truss shown in Fig. 6-24, draw the influence lines for the bar forces in the lettered bars; also draw the influence lines for the reactions at a and c.

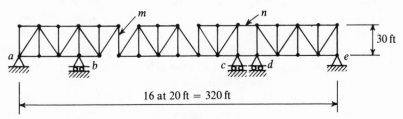

Fig. 6-24

7

MOVING CONCENTRATED LOADS:

CRITERIA FOR MAXIMA

7-1. GENERAL

As mentioned in Sec. 6-1, when designing any specific part of a structure, we should know where to place the live load in order to produce the maximum live stresses for the part considered. Under a single concentrated load or a uniform load the critical position causing certain maximum live stress can be spotted at once by inspection at the influence line. For more complicated conditions of loading, which are of various magnitude and spacing, such as a series of moving wheels on a locomotive, we cannot tell the critical position by just looking at the influence line. The method that should be followed for such cases is essentially one of trial aided by the use of criteria based on the influence line, in order to minimize computations.

For a different type of influence line, there will be a corresponding different criterion for maxima. In this chapter we shall discuss the criteria for maxima derived from the more common types of influence diagrams.

7-2. CRITERION FOR THE MAXIMUM VALUE OF A FUNCTION HAVING THE INFLUENCE DIAGRAM

This category includes the influence line for the end reaction in a simple beam and that for the end reaction of a simply supported girder or truss.

Consider a simple beam subjected to the passage of a series of wheel loads of fixed spacing and magnitude, the resultant of which is denoted by P located at a distance p from the right end B (see Fig. 7-1(a)).

To find the reaction at end A, we first construct the influence line for R_A, as shown in Fig. 7-1(b), from which

$$R_A = Py = P(p)\left(\frac{1}{l}\right)$$

where $1/l$ is the slope of the influence line, a constant. The reaction at A is, therefore, in direct proportion to P and p. If there is no change in the value

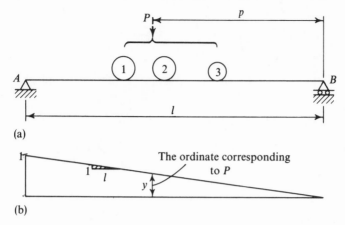

(a)

(b)

Fig. 7-1

of P (i.e., there is no load entering or leaving the span during the movement), then the maximum reaction will occur when wheel 1 is placed over the support in such a way that p will be the longest possible distance. If several systems of concentrated loads pass over a beam, the one which has the greatest value of Pp will cause the maximum reaction; and this reaction will always occur when one load is placed over the support.

Example 7-1. Find the maximum reaction that will occur in a girder of 60-ft span under the car loads shown in Fig. 7-2(a).

Using the criterion established in the foregoing discussion, we make three trials for the left-end reaction, as shown in Figs. 7-2(b), (c), and (d), respectively. Brief computations are also given in each figure. Note that in a simple case, such as this, it is not always necessary to try all three loadings. They serve, here, to illustrate the method of trial only.

By comparing the results, we conclude that the maximum reaction is 35 kips when wheel 3 is directly over the left-end support.

7-3. CRITERION FOR THE MAXIMUM VALUE OF A
FUNCTION HAVING THE INFLUENCE DIAGRAM

The influence line for the shear in a simple beam section is of this type. Consider the simple beam in Fig. 7-3(a) subjected to the passage of a group of wheel loads the resultant of which is P. We wish to find the maximum shear at section C, the influence line of which is shown in Fig. 7-3(b).

A glance at the influence line suggests that the first possible position for producing the maximum V_C is to place wheel 1 directly over C, as shown in Fig. 7-3(c). In this case, the shear at C, called $(V_C)_1$, equals the reaction at A; i.e.,

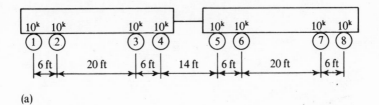

(a)

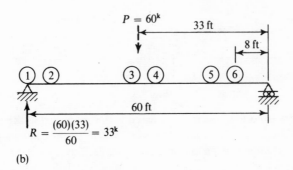

(b)

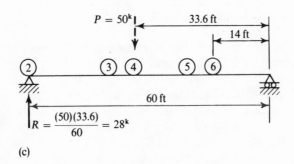

(c)

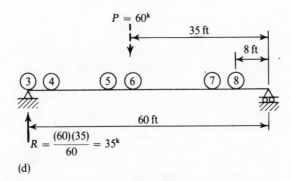

(d)

Fig. 7-2

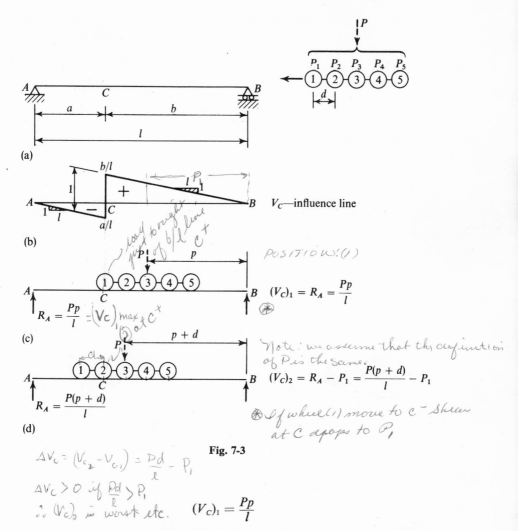

Fig. 7-3

If the group moves toward the left, we observe that as soon as wheel 1 passes C, there will be a sudden drop of V_C equal to P_1. Since the influence ordinate increases toward the left, another possible maximum shear will not be reached until wheel 2 reaches C (see Fig. 7-3(d)). In this case, the shear at C, called $(V_C)_2$, equals the reaction at A minus P_1; i.e.,

$$(V_C)_2 = \frac{P(p+d)}{l} - P_1$$

where d is the distance between wheel 1 and wheel 2—the distance the group moves in the second trial.

Now let ΔV_c be the difference between the second result and the first result:

$$\Delta V_c = (V_c)_2 - (V_c)_1 = \frac{P(p+d)}{l} - P_1 - \frac{Pp}{l} = \frac{Pd}{l} - P_1$$

Then if $\quad \dfrac{Pd}{l} > P_1 \left(\text{or } \dfrac{P}{l} > \dfrac{P_1}{d} \right) \quad\quad (V_c)_2 \quad$ will be greater

if $\quad\quad \dfrac{Pd}{l} < P_1 \left(\text{or } \dfrac{P}{l} < \dfrac{P_1}{d} \right) \quad\quad (V_c)_1 \quad$ will be greater

The above criteria will hold provided that there is no load entering or leaving the span. Note that the term Pd/l is the increase in V_c, whereas P_1 is the decrease in V_c during the movement. In general, if the increment in the value of the function is greater than the decrement, we let the system move; otherwise no progress should be made.

~ **Example 7-2.** Consider the group of wheels shown in Fig. 7-4(a), which will pass over a simple beam of 50-ft span in assumed direction. Let it be desired

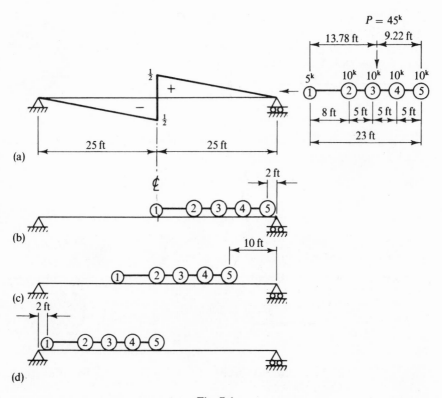

(a)

(b)

(c)

(d)

Fig. 7-4

to find the maximum shear $(+)$ and the minimum shear $(-)$ at the midspan section.

To do this, we construct the shear influence line for the midspan section as shown in Fig. 7-4(a).

1. Maximum shear $(+)$. The first trial position to find maximum positive shear at the midspan section is shown in Fig. 7-4(b). If the criterion

$$\frac{Pd}{l} \gtrless P_1$$

is applied to the first wheel, we find

$$\frac{(45)(8)}{50} = 7.2 > 5$$

Therefore, we let the group move forward 8 ft, as shown in Fig. 7-4(c). The second position produces the maximum shear since applying the foregoing criterion for the second wheel gives

$$\frac{(45)(5)}{50} = 4.5 < 10$$

which indicates that the loss in shear will be greater than the gain if the wheels move 5 ft farther to the left.

When the load is in the position in Fig. 7-4(c), we compute the maximum shear for the midspan section as

$$\frac{(45)(9.22 + 10)}{50} - 5 = 12.3 \text{ kips}$$

2. Maximum shear $(-)$. The position of loading for the minimum negative shear at the midspan section is obvious, as shown in Fig. 7-4(d), under which the shear is given by

$$-\frac{(45)(13.78 + 2)}{50} = -14.2 \text{ kips}$$

7-4. CRITERION FOR THE MAXIMUM VALUE OF A

FUNCTION HAVING THE INFLUENCE DIAGRAM

The influence line for the bending moment in a simple beam section, that for the bending moment at the panel point of a simple girder bridge, and that for the bar force in a chord member of certain bridge trusses are of this type.

Consider the simple beam with a system of concentrated loads entering from the right-hand side shown in Fig. 7-5(a). Ascertain the critical loading position for the maximum bending moment in section C.

We start by constructing the influence line for M_C, as shown in Fig. 7-5(b). The system of loads takes the position shown, with the loads on the span

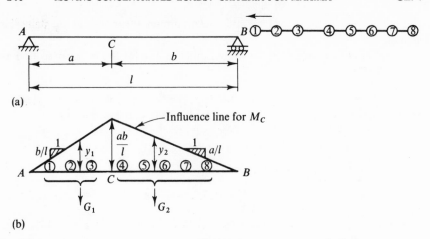

Fig. 7-5

divided into two groups, G_1 moving over the left segment length a; and G_2 over b. The bending moment at C under this loading is found to be

$$(M_C)_1 = G_1 y_1 + G_2 y_2$$

(see Fig. 7-5(b)). If the system moves a small distance Δx to the left, the bending moment at C will be

$$(M_C)_2 = G_1\left[y_1 - \left(\frac{b}{l}\right)\Delta x \right] + G_2\left[y_2 + \left(\frac{a}{l}\right)\Delta x \right]$$

As a result, the change in the bending moment is

$$\Delta M_C = (M_C)_2 - (M_C)_1 = (G_2)\left(\frac{a}{l}\right)\Delta x - (G_1)\left(\frac{b}{l}\right)\Delta x$$

Thus, the rate of change of M_C with respect to the the movement Δx is

$$\frac{\Delta M_C}{\Delta x} = G_2\left(\frac{a}{l}\right) - G_1\left(\frac{b}{l}\right)$$

for which we note the following:

1. As long as

$$G_2\left(\frac{a}{l}\right) > G_1\left(\frac{b}{l}\right) \quad \text{or} \quad \frac{G_1}{a} < \frac{G_2}{b}$$

the bending moment at C will increase as a result of the progressive movement of the loads to the left.

In order to obtain the maximum bending effect at C, the movement to the left must be continued until some one in group G_2 arrives at the peak ordinate.

Further movement to the left will change the value of G_1 and G_2. Unless the inequality still holds for the new arrangement, no further progress should be made.

2. Similarly, if

$$G_2\left(\frac{a}{l}\right) < G_1\left(\frac{b}{l}\right) \quad \text{or} \quad \frac{G_1}{a} > \frac{G_2}{b}$$

the bending moment at C will increase as a result of the movement to the right.

In order to obtain the maximum M_C, we should let the system move to the right until one of G_1 reaches the peak ordinate. Further movement to the right will depend on whether the above inequality is still valid for the new arrangement.

3. In both cases, the magnitude of the effect of the load system on M_C increases as the movement to the left, or right, is continued until some load is at the peak ordinate. Loads entering or leaving the span will increase the growth of M_C with this movement.

4. Thus the criterion for maximum bending moment at C is *to place some load at the peak ordinate* so that:

a. When this load is just to the right of the peak ordinate, i.e., included in G_2, we have

$$\frac{G_1}{a} < \frac{G_2}{b}$$

b. When this load is just to the left of the peak ordinate, i.e., included in G_1, we have

$$\frac{G_1}{a} > \frac{G_2}{b}$$

Example 7-3. For the beam and loading in Fig. 7-6(a), find the maximum bending moment at the midspan section C.

In this case, the influence diagram for M_C is an isosceles triangle with the peak ordinate at the midspan, as shown in Fig. 7-6(b). We then place wheel 3 at C and find that

$$\frac{G_1}{a} = \frac{15}{25} < \frac{G_2}{b} = \frac{30}{25}$$

if wheel 3 moves to the right of the midspan and that

$$\frac{G_1}{a} = \frac{25}{25} > \frac{G_2}{b} = \frac{20}{25}$$

if wheel 3 moves to the left of the midspan.

Therefore, the load position in Fig. 7-6(b) is the position that causes the maximum bending moment at C, which is found to be

$$M_C = \frac{(45)(24.22)}{50}(25) - (10)(5) - (5)(8+5) = 429.95 \text{ ft-kips}$$

430.00 ft-kips

wrong

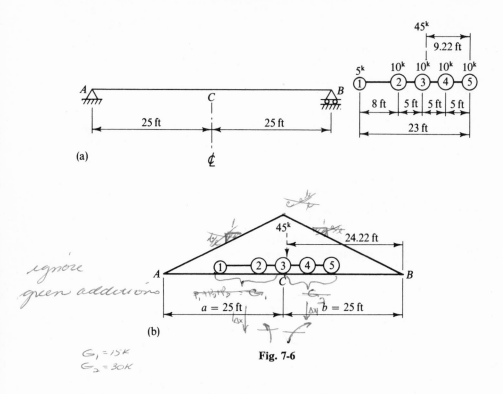

Fig. 7-6

7-5. CRITERION FOR THE MAXIMUM VALUE OF A
FUNCTION HAVING THE INFLUENCE DIAGRAM

This category includes the influence lines for the shear in an interior panel of a girder bridge and the bar force in the web member of certain bridge trusses.

Consider the girder bridge subjected to the passage of the group of wheel loads shown in Fig. 7-7(a). Find the maximum shear in the second panel from the left, whose influence line is shown in Fig. 7-7(b).

The best approach is to try several loading positions and to compare the changes in the value of the function because of the movement. Let us try the first loading position, shown in Fig. 7-7(c), with wheel 1 at the peak ordinate. Next, let the system move to the left until wheel 2 reaches the peak ordinate, as shown in Fig. 7-7(d). The computations to the right of Figs. 7-7(c) and (d) show that the movement results in an increase in the value of the function. Next, let this system move further to the left until wheel 3 reaches the peak ordinate, as shown in Fig. 7-7(e). The computations to the right of Figs. 7-7(d) and (e) show that this gives a decrease in the value of the function. Thus the second position of loading, shown in Fig. 7-7(d), produces the maximum shear in the second panel of the girder.

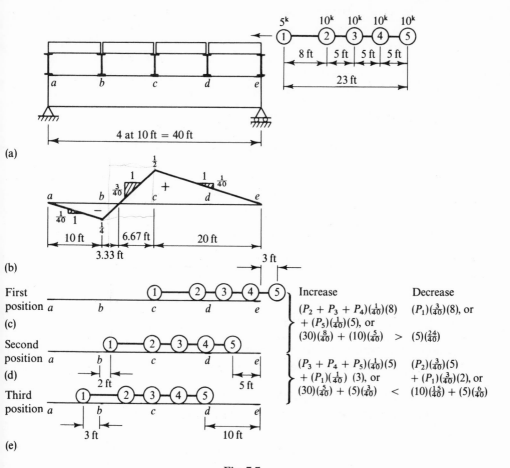

Fig. 7-7

By using the influence diagram, we find the maximum value of V_{b-c} to be

$$V_{b-c} = (10)\left(\frac{20 + 15 + 10 + 5}{40}\right) - (5)\left(\frac{1}{4}\right)\left(\frac{3.33 - 2}{3.33}\right)$$

$$= 12.5 - 0.5 = 12 \text{ kips}$$

Note that this method of locating a load system for the maximum effect is perfectly general and may always be employed in cases with more complicated influence diagrams.

7-6. ABSOLUTE MAXIMUM BENDING MOMENT

Under a single concentrated load or a uniform load the maximum bending moment in a simple beam occurs at the midspan section. However, when a simple beam is subjected to a group of concentrated loads, the maximum bending moment does not usually occur at midspan.

Consider a simple beam subjected to the passage of a series of wheel loads, as shown in Fig. 7-8. The maximum moment must occur under a certain load P_x, where the shear of the beam changes sign since the necessary condition for the maximum bending moment is $dM/dx = V = 0$.

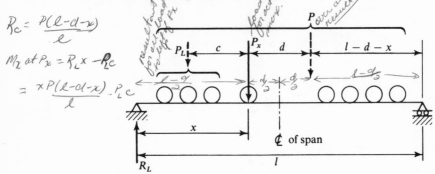

$R_c = \dfrac{P(l-d-x)}{l}$

$M_2 \text{ at } P_x = R_L x - R_c$

$= \dfrac{x P(l-d-x)}{l} - R_L c$

Fig. 7-8

Now let the resultant of the group be P; the resultant of the wheels to the left of P_x be P_L; the distance between P_x and P_L be c; and that between P_x and P be d. Also let the distance from the left end to P_x be x, as shown. Taking $\sum M = 0$ about the right end gives the left end reaction as

$$R_L = \frac{P(l - d - x)}{l}$$

The moment under P_x will be

$$M_x = (R_L)(x) - (P_L)(c)$$
$$= \frac{(P)(l - d - x)(x)}{l} - (P_L)(c)$$

Now setting

$$\frac{dM_x}{dx} = 0$$

we obtain the following expression

$$x = l - d - x$$

which means that in order to obtain the absolute maximum bending moment for a simple beam under a set of moving concentrated loads, the distance from one end of the beam to wheel load P_x should be equal to the distance from the other end of the beam to resultant P. In other words, the center line of the span bisects the distance between P_x and P.

Usually the absolute maximum bending moment occurs under the load which is closest to the resultant of all the wheel loads.

Example 7-4. Find the absolute bending moment for the beam and loading in Fig. 7-9(a).

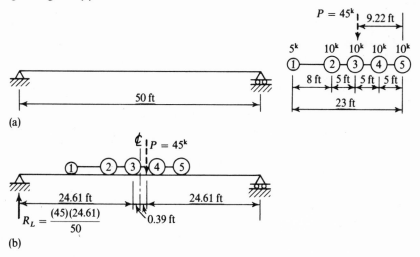

(a)

(b)

Fig. 7-9

To do this, we place the loads as shown in Fig. 7-9(b), in which the center line of the span bisects wheel 3 and the resultant P. The absolute maximum bending moment will occur under wheel 3 and is found to be

$$M = \frac{(45)(24.61)^2}{50} - (10)(5) - (5)(8 + 5) = 430.09 \text{ ft-kips}$$

It is interesting to compare the above results with the maximum bending moment at midspan due to the same loading. The latter has been found to be 429.95 ft-kips (see Example 7-3), which is very close to, but still smaller than, the absolute maximum bending moment.

PROBLEMS

7-1. A simple beam 45 ft long carries moving loads of 5 kips, 10 kips, and 10

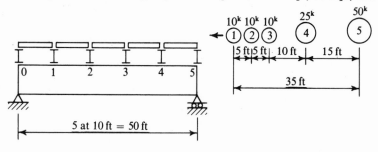

Fig. 7-10

kips spaced 5 ft apart. Calculate (a) the maximum left reaction and (b) the maximum shear and bending moment at a section 15 ft from the left end.

7-2. Find the maximum shear for panel 1-2 and the maximum bending moment at panel point 2 for the girder in Fig. 7-10 resulting from the passage of a set of moving wheels from the right.

7-3. For a simple beam of 60-ft span subjected to the loading given in Fig. 7-10 (let the set of wheels be from either of two directions), compute (a) the maximum end shear, (b) the maximum shear at the quarter point, (c) the maximum bending moment at the center, and (d) the absolute maximum bending moment in the beam.

7-4. For the truss and the loading shown in Fig. 7-11 compute the maximum forces in bars *a* and *b*. Consider both tension and compression in bar *a*.

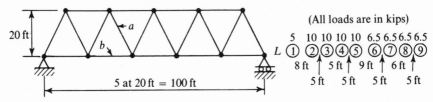

Fig. 7-11

8

ELASTIC DEFORMATIONS OF

STRUCTURES

8-1. GENERAL

The calculation of elastic deformations for structures, both the linear deformations of points and the rotational deformations of lines (slopes) from their original positions, is of great importance in the analysis, design, and construction of structures. For instance, in the erection of a bridge structure, especially when the cantilever method is used, the theoretical elevations of some or all joints must be computed for each stage of the work. In building design the sizes of beams and girders are sometimes governed by the allowable deflections. Most importantly the stress analysis for statically indeterminate structures is based largely upon an evaluation of their elastic deformations under load. By a statically indeterminate structure we mean a structure in which the number of unknown elements involved is greater than the number of equations of statics available for solution of the equations. If such is the case, there will be infinite number of solutions that can satisfy the statical equations. In order to reach a *unique* correct solution, the conditions of the *continuity* of structure, which are associated with the geometric and elastic properties of structure, are a necessary supplement.

Numerous methods of computing elastic deformations have been developed. Among them the following are considered the most general and significant in structural analysis and will, therefore, be discussed in this chapter:

1. The method of virtual work (unit-load method)
2. Castigliano's first theorem
3. The conjugate-beam method

8-2. CURVATURE OF ELASTIC LINE

The mathematical definition for curvature is *the rate at which a curve is changing direction.* To derive the expression for curvature, we shall consider a curve such as the one shown in Fig. 8-1. The average rate of change of

147

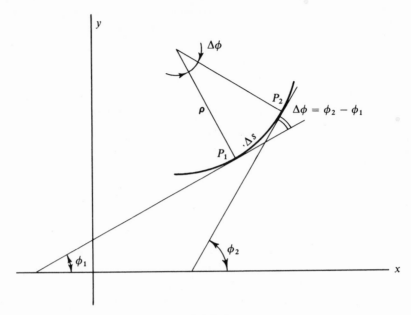

Fig. 8-1

direction between points P_1 and P_2 is $\Delta\phi/\Delta s$. The limiting value of this ratio as Δs approaches zero is called *curvature*, and the *radius of curvature* is the reciprocal of the curvature. Thus if we let κ denote the curvature and ρ the radius of curvature, we have

$$\kappa = \frac{1}{\rho} = \lim_{\Delta s \to 0} \frac{\Delta\phi}{\Delta s} = \frac{d\phi}{ds}$$

Now since $\tan\phi = dy/dx$,

$$\frac{d}{dx}\tan\phi = \frac{d^2y}{dx^2}$$

or

$$(1 + \tan^2\phi)\frac{d\phi}{dx} = \frac{d^2y}{dx^2}$$

This gives

$$\frac{d^2y}{dx^2} = \left[1 + \left(\frac{dy}{dx}\right)^2\right]\frac{d\phi}{dx}$$

whereby

$$\frac{d\phi}{dx} = \frac{d^2y/dx^2}{1 + (dy/dx)^2}$$

Also

$$\frac{dx}{ds} = \frac{1}{ds/dx} = \frac{1}{[(dx^2 + dy^2)/dx^2]^{1/2}} = \frac{1}{[1 + (dy/dx)^2]^{1/2}}$$

Hence

$$\kappa = \frac{d\phi}{ds} = \left(\frac{d\phi}{dx}\right)\left(\frac{dx}{ds}\right) = \frac{d^2y/dx^2}{[1 + (dy/dx)^2]^{3/2}} \tag{8-1}$$

For a loaded beam with its longitudinal axis taken as the x axis, we may set dy/dx in the above formula equal to zero if the deflection of beam is small. Thus we obtain:

$$\kappa = \frac{d\phi}{ds} \approx \frac{d^2 y}{dx^2} \qquad (8\text{-}2)$$

In general, except for very deep beams with a short span, the deflection due to the shearing force is negligible and only that due to the bending moment is considered. In order to develop a formula for the curvature due to elastic bending, let us consider a small element of a beam shown in Fig. 8-2. Owing

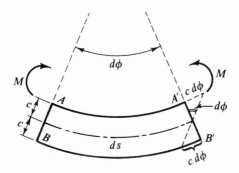

Fig. 8-2

to the action of bending moment M, the two originally parallel sections AB and $A'B'$ will change directions. This angle change is denoted by $d\phi$. If the length of the element is ds and the maximum bending stress, which occurs at the extreme fibers, is called f, the total elongation at the top or bottom fiber is $c\,d\phi$ (see Fig. 8-2), which equals $f\,ds/E$, E being the modulus of elasticity. Thus

$$c\,d\phi = \frac{f\,ds}{E}$$

or

$$\frac{d\phi}{ds} = \frac{f}{Ec}$$

Replacing f with Mc/I, I being the moment of inertia of the cross-sectional area of the beam about the axis of bending, gives

$$\frac{d\phi}{ds} = \frac{M}{EI} \qquad (8\text{-}3)$$

which expresses the relationship between the curvature and the bending moment. Now equating Eqs. 8-2 and 8-3, we obtain the approximate curvature for a loaded beam as

$$\frac{d^2 y}{dx^2} = \frac{M}{EI} \qquad (8\text{-}4)$$

Note that the above equation involves three major assumptions:

1. small deflection of beam
2. elastic material
3. only bending moment considered significant

The curvature, established in the coordinate axes of Fig. 8-1, clearly has the same sign as M, but the sign may be reversed if the direction of the y axis is reversed. In that case, we have

$$\frac{d^2y}{dx^2} = \frac{+M}{EI} \qquad (8\text{-}4a)$$

Small displacement (Bending only)

8-3. EXTERNAL WORK AND INTERNAL WORK

If a variable force F moves along its direction a distance ds, the work done is $F\,ds$. The total work done by F during a period of movement may be expressed by

$$W = \int_{s_1}^{s_2} F\,ds \qquad (8\text{-}5)$$

where s_1 and s_2 are the initial and final values of the position.

Consider a load gradually applied to a structure. Its point of application deflects and reaches a value Δ as the load increases from 0 to P. As long as the principle of superposition holds, a linear relationship exists between the load and the deflection as represented by the line oa in Fig. 8-3. The total work performed by the applied load during this period is given by

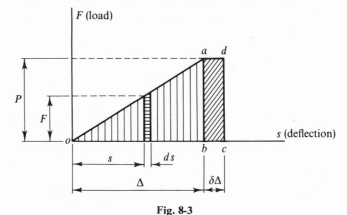

Fig. 8-3

$$W = \int_0^\Delta F\,ds = \int_0^\Delta \left(\frac{Ps}{\Delta}\right) ds = \frac{1}{2}\,P\Delta \qquad (8\text{-}6)$$

which equals the area of the shaded triangle oab in Fig. 8-3.

If further deflection $\delta\Delta$, caused by an agent other than P, occurs to the structure in the action line of P, then the additional amount of work done by the already existing load P will be $P\,\delta\Delta$, which equals the shaded rectangular area $abcd$ shown in Fig. 8-3.

Similarly, the work done by a couple M to turn an angular displacement $d\phi$ is $M\,d\phi$. The total work done by M is:

$$W = \int_{\phi_1}^{\phi_2} M\,d\phi \tag{8-7}$$

Also the work performed by a gradually applied couple C accompanied by a rotation increasing from 0 to Θ is given by

$$W = \tfrac{1}{2}\,C\,\Theta \tag{8-8}$$

Now consider a beam subjected to gradually applied forces. As long as the linear relationship between the load and the deflection maintains, all the external work will be converted into internal work or elastic strain energy. Let dW be the strain energy restored in an infinitesimal element of the beam (see Fig. 8-2). We have

$$dW = \tfrac{1}{2}\,M\,d\phi$$

if only the bending moment M produced by the forces on the element is considered significant. Using Eq. 8-3,

$$\frac{d\phi}{ds} = \frac{M}{EI}$$

$\dfrac{d\theta}{ds}$ = curvature $\dfrac{d\phi}{dx} \cdot \dfrac{d^2y}{dx^2}$

or

$$d\phi = \frac{M\,ds}{EI}$$

we have

$$dW = \frac{M^2\,ds}{2EI}$$

For a loaded beam with its longitudinal axis taken as the x axis, we let $ds \approx dx$ and obtain

$$dW = \frac{M^2\,dx}{2EI}$$

The total strain energy restored in the beam of length l is, therefore, given by

$$W = \int_0^l \frac{M^2\,dx}{2EI} \tag{8-9}$$

For a truss subjected to gradually applied loads, the internal work performed by a member with constant cross-sectional area A, length L, and internal axial force S is $S^2L/2AE$. The total internal work or elastic strain energy for the entire truss is

$$W = \sum \frac{S^2 L}{2AE} \tag{8-10}$$

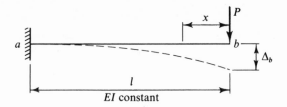

Fig. 8-4

In some special cases deformations of structures can be found by equating external work W_E and internal work (strain energy) W_I,

$$W_E = W_I \qquad (8\text{-}11)$$

For instance, to find the deflection at the free end of the loaded cantilever beam shown in Fig. 8-4, we have

$$W_E = \tfrac{1}{2} P\Delta_b$$

$$W_I = \int_0^l \frac{M^2 \, dx}{2EI}$$

$$= \int_0^l \frac{(-Px)^2 \, dx}{2EI} = \frac{P^2 l^3}{6EI}$$

Setting $W_E = W_I$ gives $\qquad \Delta_b = \dfrac{Pl^3}{3EI}$

Note that the method illustrated is quite limited in application since it is only applicable to deflection at a point of concentrated force. Furthermore, if more than one force is applied simultaneously to a structure, then more than one unknown deformation will appear in one equation, and a solution becomes impossible. Thus we do not consider this as a general method.

8-4. METHOD OF VIRTUAL WORK (UNIT-LOAD METHOD)

Consider the two cases in Figs. 8-5(a) and (b). Figure 8-5(a) illustrates a deformed elastic structure (be it a beam, a rigid frame, or a truss) subjected to the gradually applied loads $P_1, P_2, \ldots$ which move their points of application the distances $\Delta_1, \Delta_2, \ldots$ respectively. In order to find an expression for the deformation at any point of the structure, say the vertical deflection component at point C, we present the case of Fig. 8-5(b), which shows the same structure with all the actual loads removed but a virtual load of unity being gradually applied at point C along the desired deflection. Let δ denote the distance the unit load moves its point of application. Note that the virtual load is supposed to be vanishingly small and so are the corresponding virtual deformations.

Also shown in Fig. 8-5(a) is one of the typical deformed elements (be it a fiber in a beam or a rigid frame, or a bar in a truss) of length L subjected to

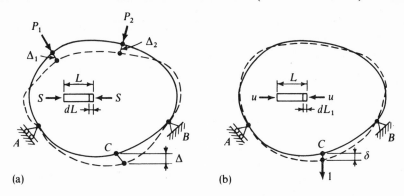

Fig. 8-5

internal forces, called S, with a corresponding change in length dL. In Fig. 8-5(b) the same element is subjected to internal forces, called u, with a corresponding change in length dL_1.

Since the external work done by the applied loads must equal the internal strain energy of all elements in the structure, we obtain for Fig. 8-5(a)

$$\tfrac{1}{2} P_1\Delta_1 + \tfrac{1}{2} P_2\Delta_2 = \tfrac{1}{2} \sum S \cdot dL \tag{8-12}$$

and for Fig. 8-5(b)

$$\tfrac{1}{2} (1)(\delta) = \tfrac{1}{2} \sum u \cdot dL_1 \tag{8-13}$$

Now imagine that the case in Fig. 8-5(b) exists first; the actual loads P_1 and P_2 are then gradually applied to it. Equating the total work done and the total strain energy restored during this period, we have

pre-existing internal forces

$$\tfrac{1}{2} (1)(\delta) + \tfrac{1}{2} P_1\Delta_1 + \tfrac{1}{2} P_2\Delta_2 + 1 \cdot \Delta$$
$$= \tfrac{1}{2} \sum u \cdot dL_1 + \tfrac{1}{2} \sum S \cdot dL + \sum u \cdot dL \tag{8-14}$$

Since the strain energy and work done must be same whether the loads are applied together or separately, we obtain from subtracting the sum of Eqs. 8-12 and 8-13 from Eq. 8-14.

$$\underset{\text{virtual}}{\overset{\text{actual}}{1 \cdot \Delta = \sum u \cdot dL}} \tag{8-15}$$

Note that Eq. 8-15 is the basic equation of the unit-load method. When the rotation of tangent at any point in the structure is desired, we need only replace the unit virtual force with a *unit virtual couple* in the above described procedure, and we obtain

$$\underset{\text{virtual}}{\overset{\text{actual}}{1 \cdot \theta = \sum u \cdot dL}} \tag{8-16}$$

where u is the internal force for a typical element caused by the unit couple and θ is the desired rotation angle.

To find a working formula for solving beam deformations, let us consider a statically determinate beam subjected to loads P_1 and P_2 shown in Fig. 8-6(a);

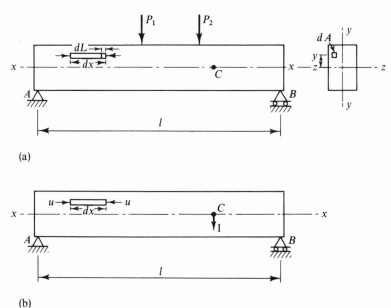

(a)

(b)

Fig. 8-6

the longitudinal axis of the beam is taken as the x axis. To find the vertical deflection Δ at an arbitrary point C, we place a unit vertical force at C, as shown in Fig. 8-6(b), and apply Eq. 8-12

$$1 \cdot \Delta = \sum u \cdot dL$$

$w = \frac{1}{2}(1)\delta + \frac{1}{2}\sum_{j=1}^{n} P_j \Delta_j + 1\Delta_j$

To interpret the terms dL and u involved in the above equation, let us first refer to Fig. 8-6(a) and observe that in the present case dL is the change of length of any fiber having length dx and cross-sectional area dA caused by the actual loads P_1 and P_2. dL equals unit elongation times dx and can, therefore, be expressed by $My\,dx/EI$ in which M is the bending moment at the section considered resulting from the actual loads, I the moment of inertia of the cross-sectional area of the beam about the axis of bending, y the distance from the fiber to the axis of bending, and E the modulus of elasticity. Next, refer to Fig. 8-6(b) and observe that u in this case is the internal force of the same fiber resulting from a fictitious unit load applied at C; u equals the bending stress of the fiber times dA, i.e., $u = my\,dA/I$ where m is the bending moment at the same section due to the unit load.

Substituting $dL = My\,dx/EI$ and $u = my\,dA/I$ in the basic equation gives

* 1 of case 2 (unit load) is only force pre-existing force

$$1 \cdot \Delta = \sum \left(\frac{my}{I}\, dA\right) \left(\frac{My}{EI}\, dx\right) \; \left(\frac{my}{EI}\, dx\right)$$

$$= \int_0^l \frac{Mm\, dx}{EI^2} \int_A y^2\, dA$$

Using $\int_A y^2\, dA = I$, we obtain

$$1 \cdot \Delta = \int_0^l \frac{Mm\, dx}{EI} \tag{8-17}$$

Equation 8-17 is the working formula for the determination of the deflection at any point of a beam. If the rotation of tangent at C is desired, we place a unit couple at C and apply the basic formula

$$1 \cdot \theta = \sum u \cdot dL$$

In a similar manner, we obtain

$$1 \cdot \theta = \int_0^l \frac{Mm\, dx}{EI} \tag{8-18}$$

where m is the bending moment at any section due to a unit couple at C.

Example 8-1. Find the deflection and slope at the free end of a cantilever beam subjected to a uniform load (Fig. 8-7(a)).

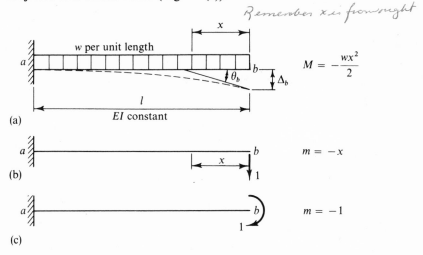

Remember x is from right

(a)

(b)

(c)

Fig. 8-7

To find Δ_b, we place a unit vertical downward load at b (Fig. 8-7(b)).

$$\Delta_b = \int_0^l \frac{Mm\, dx}{EI} = \int_0^l \frac{(-wx^2/2)(-x)\, dx}{EI} = \frac{wl^4}{8EI}$$

To find θ_b, we place a unit clockwise couple at b (Fig. 8-7(c)).

$$\theta_b = \int_0^l \frac{Mm\,dx}{EI} = \int_0^l \frac{(-wx^2/2)(-1)\,dx}{EI} = \frac{wl^3}{6EI}$$

The positive results indicate that Δ_b and θ_b are in the directions assumed.

Example 8-2. Find θ_A, θ_C, and Δ_C of the loaded beam in Fig. 8-8(a). Assume constant EI.

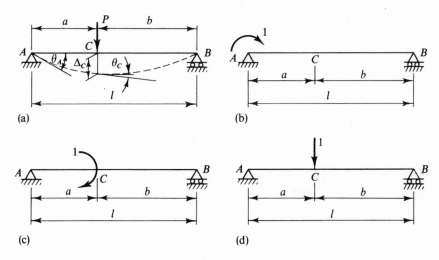

(a)

(b)

(c)

(d)

Fig. 8-8

To do this, we find it is advantageous to use double origins to perform the integration. That is,

$$\int_0^l \frac{Mm\,dx}{EI} = \int_0^a \frac{Mm\,dx}{EI} + \int_0^b \frac{Mm\,dx}{EI}$$

The terms of M and m in the above expression solving for θ_A, θ_C, and Δ_C are evaluated as in Table 8-1.

TABLE 8-1

Section	Origin	Limit	M	m for θ_A	m for θ_C	m for Δ_C
			Fig. 8-8(a)	Fig. 8-8(b)	Fig. 8-8(c)	Fig. 8-8(d)
AC	A	0 to a	$\dfrac{Pb}{l}x$	$1 - \dfrac{x}{l}$	$-\dfrac{x}{l}$	$\dfrac{b}{l}x$
BC	B	0 to b	$\dfrac{Pa}{l}x$	$\dfrac{x}{l}$	$\dfrac{x}{l}$	$\dfrac{a}{l}x$

$$\theta_A = \int_0^a \frac{(Pbx/l)[1 - (x/l)]\,dx}{EI} + \int_0^b \frac{(Pax/l)(x/l)\,dx}{EI}$$

$$= \frac{1}{EI}\left(\frac{Pa^2 b}{2l} - \frac{Pa^3 b}{3l^2} + \frac{Pab^3}{3l^2}\right) = \frac{Pab}{6EIl}\left(3a - \frac{2a^2}{l} + \frac{2b^2}{l}\right) = \frac{Pab(l + b)}{6EIl}$$

$$\theta_C = \int_0^a \frac{(Pbx/l)(-x/l)\,dx}{EI} + \int_0^b \frac{(Pax/l)(x/l)\,dx}{EI}$$

$$= \frac{1}{EI}\left(-\frac{Pa^3 b}{3l^2} + \frac{Pab^3}{3l^2}\right) = \frac{Pab(b - a)}{3EIl}$$

$$\Delta_C = \int_0^a \frac{(Pbx/l)(bx/l)\,dx}{EI} + \int_0^b \frac{(Pax/l)(ax/l)\,dx}{EI}$$

$$= \frac{1}{EI}\left(\frac{Pa^3 b^2}{3l^2} + \frac{Pa^2 b^3}{3l^2}\right) = \frac{Pa^2 b^2}{3EIl}$$

If $a = b = l/2$, then

$$\theta_C = 0 \qquad \theta_A = \frac{Pl^2}{16EI} \qquad \Delta_C = \frac{Pl^3}{48EI}$$

Example 8-3. Find the deflection at the center of the beam in Fig. 8-9. Use $E = 30,000$ kips per in.^{2}

Refer to Table 8-2 and obtain

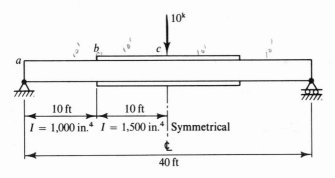

Fig. 8-9

TABLE 8-2

Section	Origin	Limit (ft)	M (ft-kips)	m (ft-kips)*	I (in.^{4})
ab	a	0 to 10	$5x$	$\dfrac{x}{2}$	1,000
bc	b	0 to 10	$5(10 + x)$	$\frac{1}{2}(10 + x)$	1,500

*We use a unit load of 1 kip.

$$E\Delta_c = \int_0^l \frac{Mm\,dx}{I}$$

$$= 2\left[\frac{1}{1{,}000}\int_0^{10}(5x)\left(\frac{x}{2}\right)dx + \frac{1}{1{,}500}\int_0^{10}\frac{5(10+x)^2}{2}dx\right]$$

$$= 2\left\{\frac{1}{1{,}000}\left[\frac{5}{6}x^3\right]_0^{10} + \frac{1}{1{,}500}\left(\frac{5}{2}\right)\left[100x + 10x^2 + \frac{x^3}{3}\right]_0^{10}\right\}$$

$$= 2\left(\frac{5}{6} + \frac{70}{18}\right) = 9.44$$

Now let us check the dimensions of both sides of the above expression. Note that a unit load of 1 kip must be included in the left side of the expression.

$$30{,}000\left(\frac{\text{kips}}{\text{in.}^2}\right)(1\text{ kip})(\Delta_c) = 9.44\,\frac{\text{ft-kips ft-kips ft}}{\text{in.}^4}$$

Thus
$$\Delta_c = \frac{9.44}{30{,}000}\frac{\text{ft}^3}{\text{in.}^2}$$

or
$$\Delta_c = \frac{(9.44)(1{,}728)\text{ in.}}{30{,}000} = 0.544\text{ in.}\qquad\text{(down)}$$

In a rigid frame the strain energy due to axial forces and shearing forces is usually much smaller as compared with that due to bending moment and can, therefore, be neglected. The formula

$$\int\frac{Mm\,dx}{EI}$$

previously derived for beam deformations is also good for finding the elastic deformations for a rigid frame, as illustrated in the following examples.

Example 8-4. Determine the horizontal, vertical, and rotational deflection components at end a of the rigid frame shown in Fig. 8-10(a). Assume that all members have the same value of EI.

To perform the integration for the entire frame denoted by

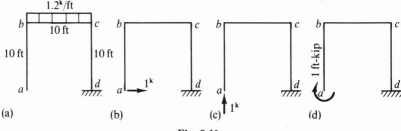

(a)   (b)   (c)   (d)

Fig. 8-10

$$\int_F ,$$

we must consider each member as a unit, the centroidal axis of the member being taken as the x axis. Thus,

$$\int_F \frac{Mm\,dx}{EI} = \int_{ab} \frac{Mm\,dx}{EI} + \int_{bc} \frac{Mm\,dx}{EI} + \int_{cd} \frac{Mm\,dx}{EI}$$

The terms of M and m in the above expression, when we solve for each of the deflection components at a, are listed in Table 8-3 in which we use m_1 to denote the bending moment at any section due to a unit horizontal force applied at a; m_2 that due to a unit vertical force at a; and m_3 that due to a unit couple at a. Note that the bending moment resulting in compression on the outside fibers of the frame is assumed to be positive.

TABLE 8-3

Member	Origin	Limit (ft)	M (ft-kips)	m_1 (ft-kips)	m_2 (ft-kips)	m_3 (ft-kips)
			Fig. 8-10(a)	Fig. 8-10(b)	Fig. 8-10(c)	Fig. 8-10(d)
ab	a	0 to 10	0	$-x$	0	1
bc	b	0 to 10	$-\dfrac{1.2x^2}{2}$	-10	x	1
cd	c	0 to 10	-60	$x-10$	10	1

To find the horizontal deflection at a, called Δ_1, we apply

$$\Delta_1 = \int_F \frac{Mm_1\,dx}{EI}$$

$$= \frac{1}{EI}\left[0 + \int_0^{10}\left(-\frac{1.2x^2}{2}\right)(-10)\,dx + \int_0^{10}(-60)(x-10)\,dx \right]$$

$$= 5{,}000\,\frac{\text{kips-ft}^3}{EI} \qquad \text{(right)}$$

Similarly, we have the vertical deflection at a, called Δ_2,

$$\Delta_2 = \int_F \frac{Mm_2\,dx}{EI}$$

$$= \frac{1}{EI}\left[0 + \int_0^{10}\left(-\frac{1.2x^2}{2}\right)(x)\,dx + \int_0^{10}(-60)(10)\,dx \right]$$

$$= -7{,}500\,\frac{\text{kips-ft}^3}{EI} \qquad \text{(down)}$$

and the rotational displacement at a, called Δ_3,

$$\Delta_3 = \int_F \frac{M m_3 \, dx}{EI}$$

$$= \frac{1}{EI} \left[0 + \int_0^{10} \left(-\frac{1.2 x^2}{2} \right) (1) \, dx + \int_0^{10} (-60)(1) \, dx \right]$$

$$= -800 \frac{\text{kips-ft}^2}{EI} \qquad \text{(counter-clockwise)}$$

Example 8-5. Find the deflection components at a of the same frame for each of three loading cases shown in Figs. 8-10(b), (c) and (d). Let

δ_{11} = the horizontal deflection at a due to a unit horizontal force at a;

δ_{21} = the vertical deflection at a due to a unit horizontal force at a;

δ_{31} = the rotational displacement at a due to a unit horizontal force at a;

Then $\qquad \delta_{11} = \int_F \frac{(m_1)^2 \, dx}{EI} \qquad \delta_{21} = \int_F \frac{m_1 m_2 \, dx}{EI} \qquad \delta_{31} = \int_F \frac{m_1 m_3 \, dx}{EI}$

since in this case (see Fig. 8-10(b)) $M = m_1$.

Likewise, if the frame is subjected only to a unit vertical force at a (see Fig. 8-10(c)), the three deflection components at a are found to be

$$\delta_{12} = \int_F \frac{m_2 m_1 \, dx}{EI} \qquad \delta_{22} = \int_F \frac{(m_2)^2 \, dx}{EI} \qquad \delta_{32} = \int_F \frac{m_2 m_3 \, dx}{EI}$$

And if the frame is subjected only to a unit couple at a (see Fig. 8-10(d)), the three deflection components at a are found to be

$$\delta_{13} = \int_F \frac{m_3 m_1 \, dx}{EI} \qquad \delta_{23} = \int_F \frac{m_3 m_2 \, dx}{EI} \qquad \delta_{33} = \int_F \frac{(m_3)^2 \, dx}{EI}$$

Taking numerical values from Table 8-3 and substituting in each of the above expressions, we find

$$\delta_{11} = \frac{1}{EI} \left[\int_0^{10} x^2 \, dx + \int_0^{10} (-10)(-10) \, dx + \int_0^{10} (x - 10)^2 \, dx \right]$$

$$= 1{,}667 \frac{\text{kips-ft}^3}{EI} \qquad \text{(right)}$$

$$\delta_{21} = \frac{1}{EI} \left[0 + \int_0^{10} (-10)(x) \, dx + \int_0^{10} (x - 10)(10) \, dx \right]$$

$$= -1{,}000 \frac{\text{kips-ft}^3}{EI} \qquad \text{(down)}$$

$$\delta_{31} = \frac{1}{EI} \left[\int_0^{10} (-x)(1) \, dx + \int_0^{10} (-10)(1) \, dx + \int_0^{10} (x - 10)(1) \, dx \right]$$

$$= -200 \frac{\text{kips-ft}^2}{EI} \qquad \text{(counter-clockwise)}$$

$$\delta_{12} = \delta_{21} = -1,000 \frac{\text{kips-ft}^3}{EI} \qquad \text{(left)}$$

$$\delta_{22} = \frac{1}{EI} \left[0 + \int_0^{10} x^2 \, dx + \int_0^{10} (10)^2 \, dx \right] = 1,333 \frac{\text{kips-ft}^3}{EI} \qquad \text{(up)}$$

$$\delta_{32} = \frac{1}{EI} \left[0 + \int_0^{10} (x)(1) \, dx + \int_0^{10} (10)(1) \, dx \right] = 150 \frac{\text{kips-ft}^2}{EI} \qquad \text{(clockwise)}$$

$$\delta_{13} = -200 \frac{\text{kips-ft}^3}{EI} \qquad \text{(left)}$$

$$\delta_{23} = 150 \frac{\text{kips-ft}^3}{EI} \qquad \text{(up)}$$

$$\delta_{33} = \frac{1}{EI} \left[\int_0^{10} dx + \int_0^{10} dx + \int_0^{10} dx \right] = 30 \frac{\text{kips-ft}^2}{EI} \qquad \text{(clockwise)}$$

Note that δ_{13} has the same value as δ_{31}, but they differ one dimension of length. The same is true for δ_{23} and δ_{32}.

The working formula for deflection of any joint of a loaded truss can be evaluated from the basic equation, Eq. 8-15,

$$1 \cdot \Delta = \sum u \cdot dL$$

by considering each member of the truss as an element. Thus the term dL is the shortening or lengthening of a bar due to applied loads and can be expressed by SL/AE. The above equation becomes

$$1 \cdot \Delta = \sum \frac{SuL}{AE} \qquad (8\text{-}19)$$

where

  $S =$ the internal force in any member due to actual loads
  $u =$ the internal force in the same member due to a fictitious unit load at the point where the deflection is sought, acting along the desired direction
  $L =$ the length of the member
  $A =$ the cross-sectional area of the member
  $E =$ the modulus of elasticity of the member

Sometimes the change of bar length dL is not caused by any external force but is due to the effect of temperature. If this is the case, we let $dL = \alpha t L$ and the working formula for finding deflection due to temperature change is given by

$$1 \cdot \Delta = \sum u \cdot \alpha t L \qquad (8\text{-}20)$$

where

  $\alpha =$ the coefficient of linear thermal expansion
  $t =$ the temperature rise in degrees

Example 8-6. Find the vertical deflection of joint b of the loaded truss shown in Fig. 8-11(a). Assume that L (ft)$/A$ (in.^{2}) = 1 and that E = 30,000 kips per in.^{2} for all members.

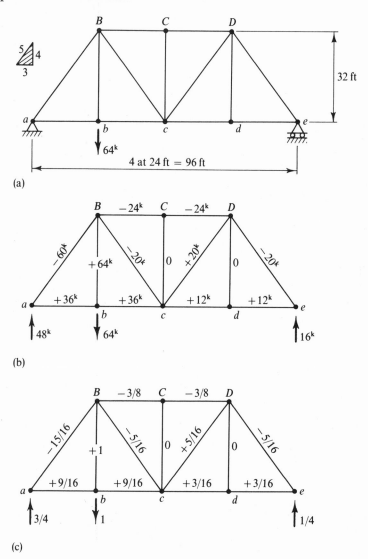

Fig. 8-11

We begin with the evaluation of S and u, and the answer diagrams for them are shown in Figs. 8-11(b) and (c) respectively. Next we apply Eq. 8-19. The complete solution is given in Table 8-4.

TABLE 8-4

Member	$\dfrac{L}{A}\left(\dfrac{\text{ft}}{\text{in.}^2}\right)$	S (kips)	$u*$	$\dfrac{SuL}{A}\left(\dfrac{\text{ft-kips}}{\text{in.}^2}\right)$
ab	1	$+36$	$+\ 9/16$	$+\ 20.25$
bc	1	$+36$	$+\ 9/16$	$+\ 20.25$
cd	1	$+12$	$+\ 3/16$	$+\ 2.25$
de	1	$+12$	$+\ 3/16$	$+\ 2.25$
BC	1	-24	$-\ 3/8$	$+\ 9.0$
CD	1	-24	$-\ 3/8$	$+\ 9.0$
aB	1	-60	$-15/16$	$+\ 56.25$
Bb	1	$+64$	$+1$	$+\ 64.0$
Bc	1	-20	$-\ 5/16$	$+\ 6.25$
Cc	1	0	0	0
cD	1	$+20$	$+\ 5/16$	$+\ 6.25$
Dd	1	0	0	0
De	1	-20	$-\ 5/16$	$+\ 6.25$
			Σ	$+202.0$

*We use a fictitious load of 1 (not 1 kip) for determining u values.

$$\Delta_v = \sum \frac{SuL}{AE} = \frac{+202}{30,000} = +0.00673 \text{ ft} \quad \text{(down)}$$

Example 8-7. For the loaded structure in the preceding example, find the absolute deflection of joint b.

To do this, we have to obtain the horizontal deflection of joint b in addition to the vertical deflection of that joint already found. The vector sum of these two displacement components is the solution.

When a unit horizontal load is applied at joint b to the right, only the member ab is under the stress of tension (i.e., $u = 1$); all other members are unstressed. The horizontal movement, called Δ_h, at joint b is thus given by

$$\Delta_h = \left(\frac{SuL}{AE}\right)_{ab} = \frac{(36)(1)}{30,000} = +0.0012 \text{ ft} \quad \text{(right)}$$

And the absolute deflection of joint b is given by

$$\Delta = \sqrt{(\Delta_v)^2 + (\Delta_h)^2} = \sqrt{(0.00673)^2 + (0.0012)^2} = 0.00684 \text{ ft}$$

moving down to the right and making an angle ϕ with the horizontal direction,

$$\phi = \tan^{-1}\frac{0.00673}{0.00120} = \tan^{-1} 5.72 \approx 80°$$

Example 8-8. For the same loaded truss (Fig. 8-11(a)) find the relative displacement between the two truss joints b and D.

To do this, we place opposing loads of unity at the joints along the line joining them, obtaining the u value in each member, as indicated in Fig. 8-12.

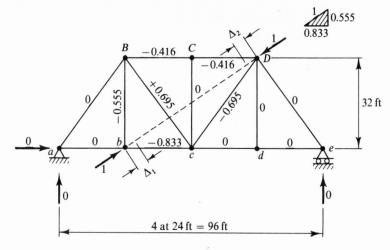

Fig. 8-12

TABLE 8-5

Member	$\dfrac{L}{A}\left(\dfrac{\text{ft}}{\text{in.}^2}\right)$	S (kips)	u	$\dfrac{SuL}{A}\left(\dfrac{\text{ft-kips}}{\text{in.}^2}\right)$
bc	1	$+36$	-0.833	-30
BC	1	-24	-0.416	$+10$
CD	1	-24	-0.416	$+10$
Bb	1	$+64$	-0.555	-35.5
Bc	1	-20	$+0.695$	-13.9
cD	1	$+20$	-0.695	-13.9
			Σ	-73.3

The solution is contained in Table 8-5. The deflection thus obtained is the algebraic sum of Δ_1 and Δ_2 (Fig. 8-12), i.e., the relative displacement between joints b and D.

$$\Delta = -\frac{73.3}{30,000} = -0.00244 \text{ ft}$$

Negative sign indicates that the two joints are moved away from each other.

Example 8-9. For the same loaded truss (Fig. 8-11(a)), find the rotation of member bc.

Finding the rotation of member bc is equivalent to finding the relative displacement between ends b and c (in the direction perpendicular to bc) divided by the length of bc. Assume counter-clockwise rotation. We then apply a pair of unit fictitious loads to joints b and c and evaluate the u value for each

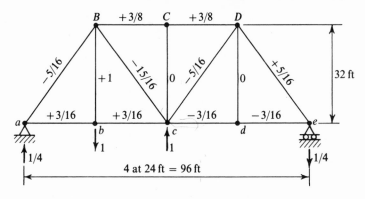

Fig. 8-13

member, as shown in Fig. 8-13. The computation leading to the solution of the relative displacement between joints b and c perpendicular to the original line of bc is contained in Table 8-6. The rotation of the member, denoted by

TABLE 8-6

Member	$\dfrac{L}{A}\left(\dfrac{\text{ft}}{\text{in.}^2}\right)$	S (kips)	u	$\dfrac{SuL}{A}\left(\dfrac{\text{ft-kips}}{\text{in.}^2}\right)$
ab	1	$+36$	$+\ 3/16$	$+27/4$
bc	1	$+36$	$+\ 3/16$	$+27/4$
cd	1	$+12$	$-\ 3/16$	$-\ 9/4$
de	1	$+12$	$-\ 3/16$	$-\ 9/4$
BC	1	-24	$+\ 3/\ 8$	-9
CD	1	-24	$+\ 3/\ 8$	-9
aB	1	-60	$-\ 5/16$	$+75/4$
Bb	1	$+64$	$+1$	$+64$
Bc	1	-20	$-15/16$	$+75/4$
Cc	1	0	0	0
cD	1	$+20$	$-\ 5/16$	$-25/4$
Dd	1	0	0	0
De	1	-20	$+\ 5/16$	$-25/4$
			Σ	$+80$

θ, is then determined;

$$\theta = \frac{80}{24E} = \frac{80}{(24)(30{,}000)} = \frac{1}{9{,}000} \text{ radian}$$

The positive value of the angle indicates a counter-clockwise rotation.

Example 8-10. Find the vertical deflection at joint b resulting from a rise in temperature of $50°F$ in the top chords BC and CD (Fig. 8-14). $\alpha = 0.0000065$ in. per in. per $1°F$.

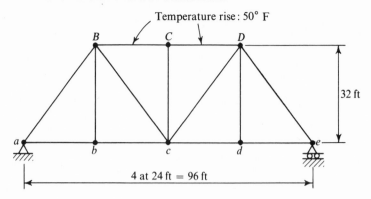

Fig. 8-14

On a statically determinate truss, no reactions or internal forces can be developed because of a temperature rise or drop in truss members. However, certain changes of bar length will take place if the temperature rises or drops in a bar. This in turn will cause the distortion of the whole truss.

To find the vertical deflection of joint b, we apply Eq. 8-20,

$$\Delta_b = \sum u \cdot \alpha t L$$

Note that, in this problem, only bars BC and CD are involved in computation since the rest of the members undergo no change of length. Now $u = -\frac{3}{8}$ (see Fig. 8-11(c)), and $\alpha t L = (0.0000065)(50)(24) = 0.0078$ for BC and CD. Thus

$$\Delta_b = 2(-\tfrac{3}{8})(0.0078) = -0.00585 \text{ ft}$$

The negative sign indicates an upward movement of joint b.

8-5. CASTIGLIANO'S FIRST THEOREM

In 1876, Alberto Castigliano published a notable paper in which he presented two important theorems. Castigliano's first theorem, which provides a general method for determining the elastic deformations of structures, can be stated as follows:

The first partial derivative of the total strain energy of the structure with respect to one of the applied actions gives the displacement along that action.

If we use P to denote the action (force or couple), Δ_P the corresponding displacement (deflection or rotation) along P, and W the total strain energy, the statement can be expressed by

$$\Delta_P = \frac{\partial W}{\partial P} \tag{8-21}$$

To demonstrate the theorem, consider the loaded beam in Fig. 8-15. The deflected position is represented by the dashed line.

If we consider only the internal work resulting from the bending moment, we have the total strain energy of the beam (see Eq. 8-9);

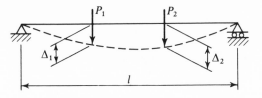

Fig. 8-15

$$W = \int_0^l \frac{M^2 \, dx}{2EI}$$

Now let M_1 be the bending moment at any section due to the gradually applied load P_1, and let M_2 be the bending moment at the same section due to the gradually applied load P_2. We have the total bending moment at any section given by

$$M = M_1 + M_2 = m_1 P_1 + m_2 P_2$$

where

$m_1 =$ bending moment at any section due to a unit load in place of P_1

$m_2 =$ bending moment of the same section due to a unit load in place of P_2

Thus $\quad \dfrac{\partial W}{\partial P_1} = \dfrac{\partial}{\partial P_1} \int_0^l \dfrac{M^2 \, dx}{2EI} = \int_0^l \dfrac{M \, (\partial M/\partial P_1) \, dx}{EI} = \int_0^l \dfrac{M m_1 \, dx}{EI} = \Delta_1$

and $\quad \dfrac{\partial W}{\partial P_2} = \dfrac{\partial}{\partial P_2} \int_0^l \dfrac{M^2 \, dx}{2EI} = \int_0^l \dfrac{M \, (\partial M/\partial P_2) \, dx}{EI} = \int_0^l \dfrac{M m_2 \, dx}{EI} = \Delta_2$

The last step of equality in each of the above two expressions is based on Eq. 8-17 from virtual work.

Let us now turn to the loaded truss in. Fig. 8-16. The total strain energy (see Eq. 8-10) is

$$W = \sum \frac{S^2 L}{2AE}$$

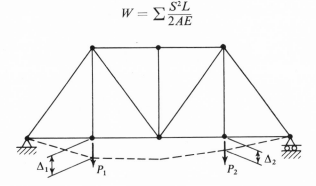

Fig. 8-16

If we let

S_1 = the internal force in any bar due to the gradually applied load P_1

S_2 = the internal force in the same bar due to the gradually applied load P_2

we have the total internal force in any bar given by

$$S = S_1 + S_2 = P_1 u_1 + P_2 u_2$$

where

u_1 = the internal force in any bar due to a unit load in place of P_1

u_2 = the internal force in the same bar due to a unit load in place of P_2

Thus
$$\frac{\partial W}{\partial P_1} = \frac{\partial}{\partial P_1} \sum \frac{S^2 L}{2AE} = \sum \frac{S(\partial S/\partial P_1)L}{AE} = \sum \frac{S u_1 L}{AE} = \Delta_1$$

and
$$\frac{\partial W}{\partial P_2} = \frac{\partial}{\partial P_2} \sum \frac{S^2 L}{2AE} = \sum \frac{S(\partial S/\partial P_2)L}{AE} = \sum \frac{S u_2 L}{AE} = \Delta_2$$

The last step of equality in each of the above expressions is based on Eq. 8-19 from virtual work.

It is interesting to point out that the Castigliano's first theorem basically does not differ from the method of virtual work for the analysis of linear structures subjected to external forces. The difference is only a matter of the arrangement of calculation. Using the method of virtual work, we have from Eqs. 8-17 and 8-19

$$\Delta = \int \frac{Mm \, dx}{EI} \qquad \text{for a beam or rigid frame}$$

and
$$\Delta = \sum \frac{SuL}{AE} \qquad \text{for a truss}$$

while in the application of Castigliano's first theorem we have

$$\Delta = \int \frac{M(\partial M/\partial P) \, dx}{EI} \qquad \text{for a beam or rigid frame} \qquad (8\text{-}22)$$

and
$$\Delta = \sum \frac{S(\partial S/\partial P)L}{AE} \qquad \text{for a truss} \qquad (8\text{-}23)$$

Example 8-11. Find the vertical deflection at the free end b of a cantilever beam ab subjected to a concentrated load P at b (see Fig. 8-4).

$$\Delta_b = \frac{\partial W}{\partial P} = \int_0^l \frac{M(\partial M/\partial P) \, dx}{EI} = \frac{1}{EI} \int_0^l M \frac{\partial M}{\partial P} \, dx$$

Now
$$M = -Px \qquad \frac{\partial M}{\partial P} = -x$$

Therefore
$$\Delta_b = \frac{1}{EI} \int_0^l (-Px)(-x) \, dx = \frac{Pl^3}{3EI} \qquad \text{(down)}$$

Example 8-12. Find the vertical deflection and the slope at the free end b

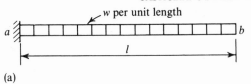

(a)

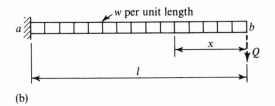

(b)

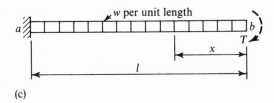

(c)

Fig. 8-17

of the cantilever beam ab subjected to a uniform load w (see Fig. 8-17(a)). Assume constant EI.

To find the vertical deflection at the free end, we note that in this case no concentrated force actually acts at the free end and thus Castigliano's first theorem cannot be directly applied. We must assume an imaginary force Q acting at the free end, as shown in Fig. 8-17(b), in order to carry out the partial derivative and to reduce it to zero in the final operation.

Thus the deflection along Q is given by

$$\Delta_b = \frac{\partial W}{\partial Q} = \int_0^l \frac{M(\partial M/\partial Q)\,dx}{EI} = \frac{1}{EI}\int_0^l M\frac{\partial M}{\partial Q}\,dx$$

Now since

$$M = -\frac{wx^2}{2} - Qx \qquad \frac{\partial M}{\partial Q} = -x$$

we have

$$\Delta_b = \frac{1}{EI}\int_0^l \left(-\frac{wx^2}{2} - Qx\right)(-x)\,dx$$

Setting $Q = 0$ in the above expression yields

$$\Delta_b = \frac{1}{EI}\int_0^l \left(-\frac{wx^2}{2}\right)(-x)\,dx = \frac{wl^4}{8EI} \qquad \text{(down)}$$

Similarly, to find the slope at b, we must apply a fictitious moment T at b (Fig. 8-17(c)) in order to use the theorem. Thus

$$\theta_b = \frac{\partial W}{\partial T} = \int_0^l \frac{M(\partial M/\partial T)\,dx}{EI} = \frac{1}{EI}\int_0^l M\frac{\partial M}{\partial T}\,dx$$

Since $\qquad M = -\dfrac{wx^2}{2} - T \qquad \dfrac{\partial M}{\partial T} = -1$

we have $\qquad \theta_b = \dfrac{1}{EI}\displaystyle\int_0^l \left(-\dfrac{wx^2}{2} - T\right)(-1)\, dx$

Setting $T = 0$ gives $\qquad \theta_b = \dfrac{1}{EI}\displaystyle\int_0^l \dfrac{wx^2}{2}\, dx = \dfrac{wl^3}{6EI} \qquad$ (clockwise)

Example 8-13. Find, by Castigliano's first theorem, the horizontal, vertical, and rotational displacements at end a of the loaded rigid frame in Fig. 8-18(a). Assume constant EI.

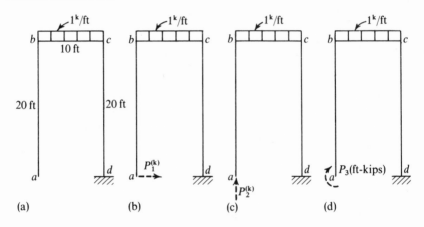

Fig. 8-18

To do this, we must in each case place an imaginary load at end a, as shown dotted in Figs. 8-18(b), (c), and (d). The solutions are completely shown in Table 8-7 in which Δ_1, Δ_2, and Δ_3 denote, respectively, the required horizontal, vertical, and rotational displacements at end a.

Example 8-14. Find, by Castigliano's first theorem, the horizontal, vertical, and rotational displacements at end a for the loaded frame in Fig. 8-19. Assume constant EI.

The necessary computations leading to evaluating the integral

$$\int_F \frac{M(\partial M/\partial P)\, dx}{EI}$$

are shown in Table 8-8.

TABLE 8-8

Member	Origin	Limit (ft)	M (ft-kips)	$\dfrac{\partial M}{\partial F_1}$ (ft)	$\dfrac{\partial M}{\partial F_2}$ (ft)	$\dfrac{\partial M}{\partial F_3}$
ab	a	0 to 20	$-F_1 x + 0 + F_3$	$-x$	0	1
bc	b	0 to 10	$-20F_1 + F_2 x + F_3$	-20	x	1
cd	c	0 to 20	$(x - 20)F_1 + 10F_2 + F_3$	$x - 20$	10	1

TABLE 8-7

Deflection	Member	Origin	Limit	M	$\frac{\partial M}{\partial P}$	$\frac{1}{EI}\int M\frac{\partial M}{\partial P}dx$	Answer
	ab	a	0 to 20	$-P_1 x$	$-x$	0	
Δ_1 Fig. 8-18(b)	bc	b	0 to 10	$-20P_1 - \frac{x^2}{2}$	-20	$\frac{1}{EI}\int_0^{10}\left(-\frac{x^2}{2}\right)(-20)\,dx = \frac{3,333}{EI}$	$13,333\,\dfrac{\text{kips-ft}^3}{EI}$ (right)
	cd	c	0 to 20	$-P_1(20-x)-50$	$x-20$	$\frac{1}{EI}\int_0^{20}(-50)(x-20)\,dx = \frac{10,000}{EI}$	
	ab	a	0 to 20	0	0	0	
Δ_2 Fig. 8-18(c)	bc	b	0 to 10	$P_2 x - \frac{x^2}{2}$	x	$\frac{1}{EI}\int_0^{10}\left(-\frac{x^2}{2}\right)(x)\,dx = \frac{-1,250}{EI}$	$-11,250\,\dfrac{\text{kips-ft}^3}{EI}$ (down)
	cd	c	0 to 20	$10P_2 - 50$	10	$\frac{1}{EI}\int_0^{20}(-50)(10)\,dx = \frac{-10,000}{EI}$	
	ab	a	0 to 20	P_3	1	0	
Δ_3 Fig. 8-18(d)	bc	b	0 to 10	$P_3 - \frac{x^2}{2}$	1	$\frac{1}{EI}\int_0^{10}\left(-\frac{x^2}{2}\right)(1)\,dx = \frac{-167}{EI}$	$-1,167\,\dfrac{\text{kips-ft}^2}{EI}$ (counter-clockwise)
	cd	c	0 to 20	$P_3 - 50$	1	$\frac{1}{EI}\int_0^{20}(-50)(1)\,dx = \frac{-1,000}{EI}$	

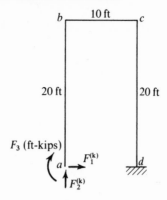

Fig. 8-19

$$\Delta_1 = \int_F \frac{M(\partial M/\partial F_1)\,dx}{EI} = \frac{1}{EI}\left[\int_0^{20} (-F_1 x + 0 + F_3)(-x)\,dx\right.$$

$$+ \int_0^{10} (-20F_1 + F_2 x + F_3)(-20)\,dx$$

$$\left.+ \int_0^{20} (x - 20)F_1 + 10F_2 + F_3)(x - 20)\,dx\right]$$

$$= (9{,}333F_1 - 3{,}000F_2 - 600F_3)\frac{\text{kips-ft}^3}{EI}$$

$$\Delta_2 = \int_F \frac{M(\partial M/\partial F_2)\,dx}{EI} = \frac{1}{EI}\left[0 + \int_0^{10} (-20F_1 + F_2 x + F_3)(x)\,dx\right.$$

$$\left.+ \int_0^{20} (x - 20)F_1 + 10F_2 + F_3)(10)\,dx\right]$$

$$= (-3{,}000F_1 + 2{,}333F_2 + 250F_3)\frac{\text{kips-ft}^3}{EI}$$

$$\Delta_3 = \int_F \frac{M(\partial M/\partial F_3)\,dx}{EI} = \frac{1}{EI}\left[\int_0^{20} (-F_1 x + 0 + F_3)(1)\,dx\right.$$

$$+ \int_0^{10} (-20F_1 + F_2 x + F_3)(1)\,dx$$

$$\left.+ \int_0^{20} (x - 20)F_1 + 10F_2 + F_3)(1)\,dx\right]$$

$$= (-600F_1 + 250F_2 + 50F_3)\frac{\text{kips-ft}^2}{EI}$$

Example 8-15. Given the loaded truss in Fig. 8-11(a), find the horizontal deflection at D.

Assume that joint D will move to the right. To apply the theorem we place an imaginary horizontal force Q acting at D, as shown in Fig. 8-20(a). The bar forces thus obtained are shown in Fig. 8-20(b). The complete solution is shown in Table 8-9.

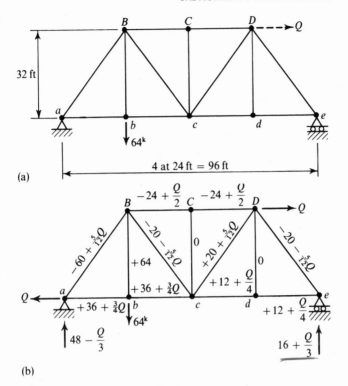

Fig. 8-20

TABLE 8-9

Member	$\dfrac{L}{A}$ (ft/in.²)	S (kips)	$\dfrac{\partial S}{\partial Q}$	$\dfrac{S(\partial S/\partial Q)L}{A}$ (ft-kips/in.²)
ab	1	$+36 + 3/4\,Q$	$+3/4$	$+27$
bc	1	$+36 + 3/4\,Q$	$+3/4$	$+27$
cd	1	$+12 + 1/4\,Q$	$+1/4$	$+\ 3$
de	1	$+12 + 1/4\,Q$	$+1/4$	$+\ 3$
BC	1	$-24 + 1/2\,Q$	$+1/2$	-12
CD	1	$-24 + 1/2\,Q$	$+1/2$	-12
aB	1	$-60 + 5/12\,Q$	$+5/12$	-25
Bb	1	$+64$	0	0
Bc	1	$-20 - 5/12\,Q$	$-5/12$	$+\ 8.33$
Cc	1	0	0	0
cD	1	$+20 + 5/12\,Q$	$+5/12$	$+\ 8.33$
Dd	1	0	0	0
De	1	$-20 - 5/12\,Q$	$-5/12$	$+\ 8.33$

$$\Sigma \quad +36.0$$

$$\Delta = \Sigma \frac{S(\partial S/\partial Q)L}{AE} = +\frac{36}{30,000} = +0.0012 \text{ ft} \qquad \text{(right)}$$

8-6. CONJUGATE-BEAM METHOD

The purpose of this method is to transform the problem of solving the slopes and deflections of a structure resulting from the applied loads (actual loads) to a problem of solving the shears and moments of a conjugate beam due to the elastic load derived from the angle changes of structural elements. The advantage of this method over the method of virtual work or Castigliano's first theorem are:

1. Unlike the previous methods, which are used to find one item of deformation at one point of the structure in an operation, this method enables us to find deformations at many points of the structure in a single set-up.

2. It is generally acknowledged that structural engineers prefer to deal with shear and bending moment rather than to do tedious integral calculus.

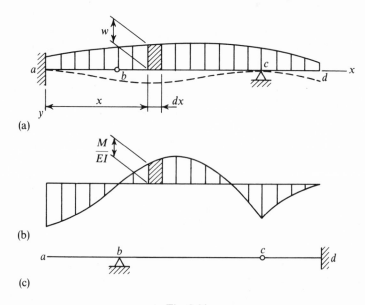

Fig. 8-21

Consider a typical beam subjected to load of intensity w, such as the one shown in Fig. 8-21(a), for which we may plot the moment diagram and, therefore, the M/EI diagram as in Fig. 8-21(b).

We recall that the curvature at any point of the beam of Fig. 8-21(a) is given by Eq. 8-4a; i.e.,

$$\frac{d^2y}{dx^2} = -\frac{M}{EI}$$

Now since the slope at any point of the beam is expressed by

$$\frac{dy}{dx} = \tan \theta \approx \theta$$

for small deformation, we have

$$\frac{d\theta}{dx} = -\frac{M}{EI}$$

or

$$d\theta = -\frac{M\,dx}{EI}$$

Integrate.

$$\theta = -\int \frac{M}{EI} dx \tag{8-24}$$

Substituting for θ by dy/dx and integrating again gives

$$y = \int \theta\,dx = -\iint \frac{M}{EI} dx\,dx \tag{8-25}$$

Next, consider an element dx of the beam (Fig. 8-21(a)) for which we may write the relationships between the load, the shear, and the bending moment (see Sec. 3-4).

$$\frac{dV}{dx} = -w$$

and

$$\frac{dM}{dx} = V$$

Thus over a portion of the beam

$$V = -\int w\,dx \tag{8-26}$$

and

$$M = \int V\,dx = -\iint w\,dx\,dx \tag{8-27}$$

Now suppose that we have a beam, called a *conjugate beam*, whose length equals that of the actual beam in Fig. 8-21(a). Let this beam be subjected to the so-called *elastic load* of intensity M/EI given in Fig. 8-21(b). (Elastic load is sometimes referred to as the *angle load*, a term obviously associated with $d\theta = M\,(dx/EI)$). The integral expressions for the shear and moment over a portion of the conjugate beam, denoted by $\bar{V}$ and $\bar{M}$ respectively, can be obtained by replacing w in Eqs. 8-26 and 8-27 with M/EI; i.e.,

$$\bar{V} = -\int \frac{M}{EI} dx \tag{8-28}$$

and

$$\bar{M} = -\iint \frac{M}{EI} dx\,dx \tag{8-29}$$

When we compare Eqs. 8-24 and 8-25 with Eqs. 8-28 and 8-29, it follows logically that, with properly prescribed boundary conditions for the conjugate beam, we may reach the following results:

1. The slope at a given section of a loaded beam (actual beam) equals the shear in the corresponding section of the conjugate beam subjected to the elastic load.

2. The deflection at a given section of a loaded beam equals the bending moment in the corresponding section of the conjugate beam subjected to the elastic load.

Thus far we have only stated that the conjugate beam is identical to the actual beam with regard to the length of the beam. In order that the above-stated identities be possible, the set-up of the support and connection of the conjugate beam must be such as to induce shear and moment in the conjugate beam in conformity to the slope and deflection induced by the counterparts in the actual beam. This is given in Table 8-10 and can be briefly summarized as

$$\text{Fixed end} \longleftrightarrow \text{Free end}$$
$$\text{Simple end} \longleftrightarrow \text{Simple end}$$
$$\text{Interior connection} \longleftrightarrow \text{Interior support}$$

The symbols between the two groups indicate conjugation to each other.

TABLE 8-10

Actual Beam Subjected to Applied Load			Conjugate Beam Subjected to Elastic Load
Fixed end	$\begin{cases} \theta = 0 \\ y = 0 \end{cases}$	$\left. \begin{array}{l} \bar{V} = 0 \\ \bar{M} = 0 \end{array} \right\}$	Free end
Free end	$\begin{cases} \theta \neq 0 \\ y \neq 0 \end{cases}$	$\left. \begin{array}{l} \bar{V} \neq 0 \\ \bar{M} \neq 0 \end{array} \right\}$	Fixed end
Simple end (hinge or roller)	$\begin{cases} \theta \neq 0 \\ y = 0 \end{cases}$	$\left. \begin{array}{l} \bar{V} \neq 0 \\ \bar{M} = 0 \end{array} \right\}$	Simple end (hinge or roller)
Interior support (hinge or roller)	$\begin{cases} \theta \neq 0 \\ y = 0 \end{cases}$	$\left. \begin{array}{l} \bar{V} \neq 0 \\ \bar{M} = 0 \end{array} \right\}$	Interior connection (hinge or roller)
Interior connection (hinge or roller)	$\begin{cases} \theta \neq 0 \\ y \neq 0 \end{cases}$	$\left. \begin{array}{l} \bar{V} \neq 0 \\ \bar{M} \neq 0 \end{array} \right\}$	Interior support (hinge or roller)

Thus, the conjugate beam for the beam in Fig. 8-21(a) is the one as shown in Fig. 8-21(c). If we use the M/EI diagram of Fig. 8-21(b) as the load to put

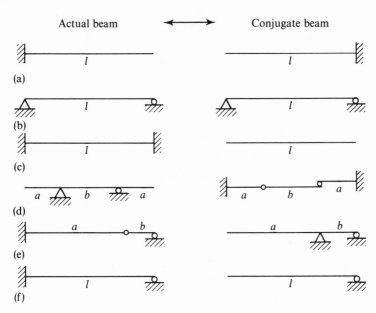

Fig. 8-22

on the beam in Fig. 8-21(c), the resulting shear and bending moment for any section of this beam will give the slope and deflection for the corresponding section of the original beam. Other examples of conjugate beams are shown in Fig. 8-22.

Unlike the actual beams, the conjugate beams may be unstable in themselves. For instance, the conjugate beams shown in Figs. 8-22(c) and (f) are unstable beams. However, they will maintain an unstable equilibrium under the action of an elastic load.

The same figure indicates that the conjugate-beam method is not limited to the analysis of statically determinate beams; in fact, the conjugate-beam method is also applicable to statically indeterminate beams.

The sign convention we use may be stated as follows:

The origin of the loaded beam is taken at the left end of the beam with y positive downward and x positive to the right. As a result, a positive deflection means a downward deflection and a positive slope means a clockwise rotation of the beam section. Recall that the derivation from the relationships among load, shear, and bending moment is based on taking the downward load as positive. Therefore, a positive M/EI should be taken as a downward elastic load.

Having defined these, we readily see that a positive shear at a section of the conjugate beam corresponds to a clockwise rotation at the section of the actual beam. A positive moment at a section of the conjugate beam corresponds to a downward deflection at the section of the actual beam.

Example 8-16. Find, by the conjugate-beam method, the vertical deflection at the free end c of the cantilever beam shown in Fig. 8-23(a). Assume constant EI.

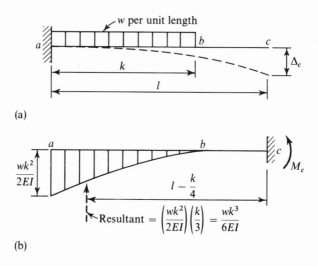

(a)

(b)

Fig. 8-23

To do this, we place an elastic load on the conjugate beam, as shown in Fig. 8-23(b). The vertical deflection at c of the actual beam is the moment at c of the conjugate beam. Thus,

$$\Delta_c = \left(\frac{wk^3}{6EI}\right)\left(l - \frac{1}{4}k\right) \qquad \text{(down)}$$

Example 8-17. Find θ_A, θ_C, and Δ_C for the loaded beam shown in Fig. 8-24(a) by the conjugate-beam method. Assume constant EI. Note that the deformations were solved by the method of virtual work in Example 8-2.

The conjugate beam together with the elastic load is shown in Fig. 8-24(b) in which the resultant of the loading is found to be $Pab/2EI$ acting at a distance $(l + b)/3$ from the right end as indicated. Thus

$$\theta_A = \frac{Pab(l + b)}{6EIl} \qquad \text{(clockwise)}$$

$$\theta_C = \frac{Pab(l + b)}{6EIl} - \left(\frac{Pab}{EIl}\right)\left(\frac{a}{2}\right) = \frac{Pab(b - a)}{3EIl} \qquad \begin{array}{l}\text{(clockwise,}\\ \text{if } b > a)\end{array}$$

$$\Delta_C = \frac{Pa^2b(l + b)}{6EIl} - \left(\frac{Pab}{EIl}\right)\left(\frac{a}{2}\right)\left(\frac{a}{3}\right) = \frac{Pa^2b^2}{3EIl} \qquad \text{(down)}$$

Example 8-18. Find, by the conjugate-beam method, the deflection at the center of the beam shown in Fig. 8-25(a). Assume $E = 30,000$ kips per in.^{2}

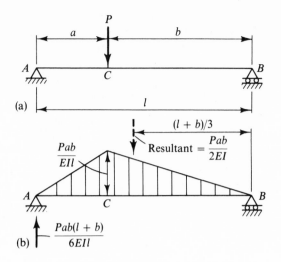

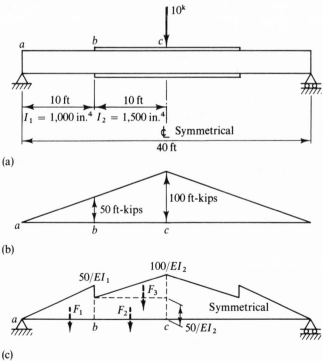

Fig. 8-24

Note that the deflection was solved by the method of virtual work in Example 8-3.

Fig. 8-25

To do this, we first plot the moment diagrams, as shown in Fig. 8-25(b). The diagram for the elastic load on the conjugate beam is given in Fig. 8-25(c) in which a sudden drop of M/EI occurs at b because of the enlargement of the cross-sectional area of the original beam at that section.

For convenience, we may divide the elastic load of the left half span into three parts denoted by F_1, F_2, and F_3, as indicated in Fig. 8-25(c). Thus the reaction at end a of the conjugate beam is given by

$$F_a = F_1 + F_2 + F_3 = \frac{250}{EI_1} + \frac{500}{EI_2} + \frac{250}{EI_2} = 250\frac{\text{kips-ft}^2}{EI_1} + 750\frac{\text{kips-ft}^2}{EI_2}$$

The deflection at c of the original beam is the bending moment about c of the conjugate beam subjected to the elastic load. This is given by

$$\Delta_c = \left(\frac{250}{EI_1} + \frac{750}{EI_2}\right)(20) - \left(\frac{250}{EI_1}\right)\left(10 + \frac{10}{3}\right) - \left(\frac{500}{EI_2}\right)(5) - \left(\frac{250}{EI_2}\right)\left(\frac{5}{3}\right)$$

$$= 1{,}667\frac{\text{kips-ft}^3}{EI_1} + 11{,}667\frac{\text{kips-ft}^3}{EI_2}$$

Using $E = 30{,}000$ kips per in.^{2}, $I_1 = 1{,}000$ in.^{4}, $I_2 = 1{,}500$ in.^{4}, we obtain

$$\Delta_c = \frac{(1{,}667)(1{,}728)}{(30{,}000)(1{,}000)} + \frac{(11{,}667)(1{,}728)}{(30{,}000)(1{,}500)} = 0.096 + 0.448 = 0.544 \text{ in.}$$

(down)

PROBLEMS

8-1. Use the method of virtual work to determine the vertical deflection at center and the slope at left end for a simply supported beam subjected to a uniform load over the entire span. Assume constant EI.

8-2. Use the method of virtual work to determine the vertical deflections at the load point and at the center of the beam in Fig. 8-26. E and I are both constant.

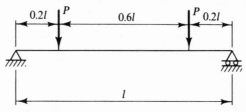

Fig. 8-26

8-3. Use the method of virtual work to determine the slope and deflection at the load point of the beam in Fig. 8-27. Use $E = 30{,}000$ kips per in.^{2}

8-4. By the method of virtual work, find the horizontal, vertical, and rotational displacement components at point a of the frame shown in Fig. 8-28. Use $E = 30{,}000$ kips per in.^{2} and $I = 500$ in.^{4}

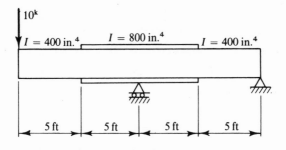

Fig. 8-27

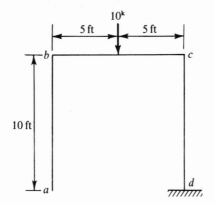

Fig. 8-28

8-5. For the truss in Fig. 8-29, the area of each bar in square inches equals one-half its length in feet. $E = 30,000$ kips per in.^{2} Use the method of virtual work to compute: (a) the vertical deflection at point B; (b) the horizontal deflection at point C; (c) the relative deflection between points b and C along the line joining them; (d) the rotation of member bc.

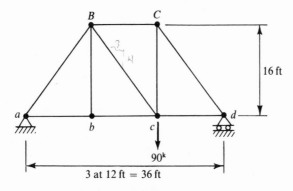

Fig. 8-29

8-6. Solve Prob. 8-1 by Castigliano's first theorem.

8-7. Solve Prob. 8-2 by Castigliano's first theorem.

8-8. Solve Prob. 8-3 by Castigliano's first theorem.

8-9. Solve Prob. 8-4 by Castigliano's first theorem.

8-10. Solve Prob. 8-5 by Castigliano's first theorem.

8-11. Solve Prob. 8-1 by the conjugate-beam method.

8-12. Solve Prob. 8-2 by the conjugate-beam method.

8-13. Solve Prob. 8-3 by the conjugate-beam method.

8-14. Use the conjugate-beam method to determine the deflection and slope at point b in Fig. 8-30. $E = 30,000$ kips per in.^{2}

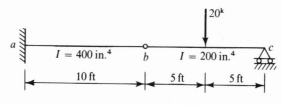

Fig. 8-30

9

ANALYSIS OF STATICALLY
INDETERMINATE STRUCTURES BY THE
METHOD OF CONSISTENT DEFORMATIONS

9-1. GENERAL

Statically indeterminate structures can be analyzed by direct use of the theory of elastic deformations developed in the preceding chapter. Any statically indeterminate structure can be made statically determinate and stable by removing the extra restraints called *redundant forces* or simply *redundants*, i.e., the force elements which are more than the minimum necessary for the static equilibrium of the structure. The statically determinate and stable structure that remains after removal of the extra restraints is called the *primary structure*. We may now consider the primary structure as being subjected to the combined action of the original loads plus the unknown redundants. The conditional equations for geometric consistence of the original structure, often referred to as the *compatibility equations*, are then obtained from the primary structure by superposition of the deformations caused by the original loads and redundants. We can have as many compatibility equations as the number of unknown redundants so that the redundants can be determined by solving these simultaneous equations. This method known as *consistent deformations* is generally applicable to the analysis of any structure, whether it is being analyzed for the effect of loads, support settlement, temperature change, or any other case. However, there is only one restriction on the use of this method, that is the principle of superposition must hold.

As an illustration, consider the loaded continuous beam with nonyielding supports shown in Fig. 9-1(a). It is statically indeterminate to the second degree, that is to say, with two redundants. The first step in the application of the method is to remove, say, the two interior supports and to introduce in these releases the redundant actions called X_1 and X_2, respectively, and by so doing to reduce or cut back the structure to a condition of determinateness and stability. Thus, the primary structure is a simple beam subjected to the combined action of a number of external forces and two redundants X_1 and X_2, as shown in Fig. 9-1(b).

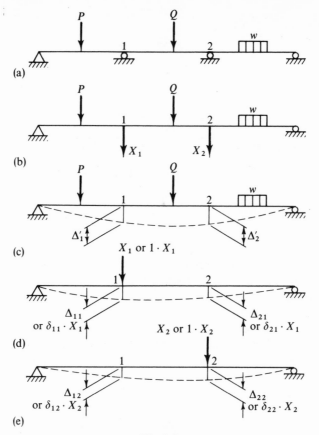

Fig. 9-1

The resulting structure in Fig. 9-1(b) can be regarded as the superposition of those shown in Figs. 9-1(c), (d), and (e). Consequently, any deformation of the structure can be obtained by the superposition of these effects.

Referring to Fig. 9-1(b), for unyielding supports, we find that compatibility requires

$$\Delta_1 = 0 \qquad\qquad (9\text{-}1)$$

$$\Delta_2 = 0 \qquad\qquad (9\text{-}2)$$

where

$\Delta_1 =$ the deflection at redundant point 1 (in the line of redundant force X_1);

$\Delta_2 =$ the deflection at redundant point 2 (in the line of redundant force X_2).

By the principle of superposition we may expand the above equations as

$$\Delta_1' + \Delta_{11} + \Delta_{12} = 0 \tag{9-3}$$

$$\Delta_2' + \Delta_{21} + \Delta_{22} = 0 \tag{9-4}$$

where

$\Delta_1' = $ the deflection at redundant point 1 due to external loads (see Fig. 9-1(c));

$\Delta_{11} = $ the deflection at redundant point 1 due to redundant force X_1 (see Fig. 9-1(d));

$\Delta_{12} = $ the deflection at redundant point 1 due to redundant force X_2 (see Fig. 9-1(e)).

The rest are similar.

The above equations may be expressed in terms of the *flexibility coefficients*. A typical flexibility coefficient δ_{ij} is defined by

$\delta_{ij} = $ the displacement at point i due to a unit action at j, all other points being assumed unloaded.

Thus Eqs. 9-3 and 9-4 may be written as

$$\Delta_1' + \delta_{11} X_1 + \delta_{12} X_2 = 0 \tag{9-5}$$

$$\Delta_2' + \delta_{21} X_1 + \delta_{22} X_2 = 0 \tag{9-6}$$

Apparently,

$\delta_{11} = $ the deflection at point 1 due to a unit force at point 1 (see Fig. 9-1(d))

$\delta_{12} = $ the deflection at point 1 due to a unit force at point 2 (see Fig. 9-1(e))

and so on.

Both the deflections resulting from the original external loads and the flexibility coefficients for the primary structure can be obtained by any method described in the preceding chapter. The remaining redundant unknowns are then solved by simultaneous equations. This process can be generalized. Thus for a structure with n redundants, we have

$$\begin{aligned}
\Delta_1' + \delta_{11} X_1 + \delta_{12} X_2 + \cdots + \delta_{1n} X_n &= 0 \\
\Delta_2' + \delta_{21} X_1 + \delta_{22} X_2 + \cdots + \delta_{2n} X_n &= 0 \\
&\vdots \\
\Delta_n' + \delta_{n1} X_1 + \delta_{n2} X_2 + \cdots + \delta_{nn} X_n &= 0
\end{aligned} \tag{9-7}$$

9-2. ANALYSIS OF STATICALLY INDETERMINATE BEAMS BY THE METHOD OF CONSISTENT DEFORMATIONS

The method of consistent deformations is quite easy to understand and can be most effectively demonstrated by a series of illustrations. In all the following examples we assume that only the bending distortion is significant.

Example 9-1. Analyze the propped beam shown in Fig. 9-2(a), which is statically indeterminate to the first degree. Assume constant EI.

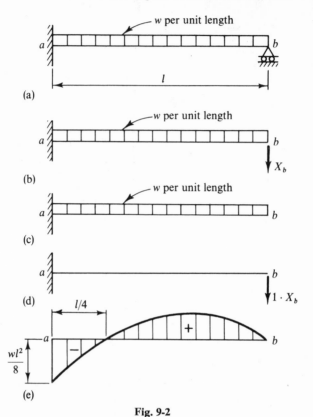

Fig. 9-2

Solution (1). One of the reactions may be considered as being extra. In this case let us first choose the vertical reaction at b as the redundant assumed to be acting downward, as shown in Fig. 9-2(b). By the principle of superposition we may consider the beam as being subjected to the sum of the effects of the original uniform loading and the unknown redundant X_b, as shown in Figs. 9-2(c) and (d) respectively.

Next, we find that the vertical deflection at b resulting from the uniform load (Fig. 9-2(c)) is given by

$$\Delta'_b = \frac{wl^4}{8EI}$$

and that the vertical deflection at b because of a unit load applied at b in place of X_b (Fig. 9-2(d)) is given by

$$\delta_{bb} = \frac{l^3}{3EI} \qquad \begin{cases} P = 1 \\ \dfrac{Pl^3}{3EI} \end{cases}$$

Note that Δ_b' and the flexibility coefficient δ_{bb} may be found by any method described in Ch. 8.

Applying compatibility equation

$$\Delta_b = \Delta_b' + \delta_{bb} X_b = 0$$

we obtain

$$\frac{wl^4}{8EI} + \left(\frac{l^3}{3EI}\right) X_b = 0$$

from which

$$X_b = -\frac{3wl}{8}$$

The minus sign indicates an upward reaction.

With reaction at b determined, we find the beam reduces to a statically determinate one. We can readily obtain reaction components at a from the equilibrium equations:

$$\sum F_y = 0 \qquad V_a = wl - \tfrac{3}{8} wl = \tfrac{5}{8} wl \qquad \text{(upward)}$$

$$\sum M_a = 0 \qquad M_a = \tfrac{1}{2} wl^2 - \tfrac{3}{8} wl^2 = \tfrac{1}{8} wl^2 \qquad \text{(counter-clockwise)}$$

The moment diagram for the beam is shown in Fig. 9-2(e).

Solution (2). The beam in Fig. 9-2(a) can be rendered statically determinate by removing the fixed support and replacing it with a hinged support. In addition to the original uniform loading, a redundant moment M_a is then applied to the primary structure, a simple beam, as shown in Fig. 9-3(a). The un-

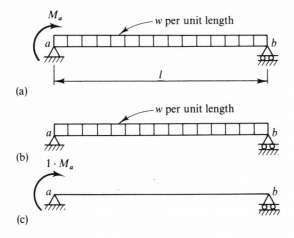

Fig. 9-3

known M_a can be solved by the condition of compatibility that the rotation at end a must be zero.

The rotation at end a for the primary structure due to the uniform loading alone (Fig. 9-3(b)) is given by

$$\theta'_a = \frac{wl^3}{24EI}$$

and that due to a unit couple applied at end a (Fig. 9-3(c)) is given by

$$\delta_{aa} = \frac{l}{3EI}$$

Using the compatibility equation

$$\theta_a = \theta'_a + \delta_{aa}M_a = \frac{wl^3}{24EI} + \frac{M_a l}{3EI} = 0$$

we solve for $\qquad\qquad M_a = -\frac{1}{8}wl^2$

The minus sign indicates a counter-clockwise moment. After M_a is determined, the rest of the analysis can be carried out without difficulty.

Solution (3). From the previous solutions we come to recognize that we are free to select redundants in analyzing a statically indeterminate structure, the only restriction being that the redundants should be so selected that a *stable* cut structure remains. Figure 9-4 will serve as an illustration. Let us cut the

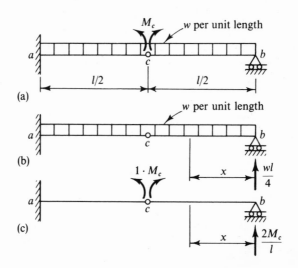

Fig. 9-4

beam at midspan section c and introduce in its place a hinge, so that the beam is stable and determinate. A pair of redundant couples, called M_c, together with the original loading are then applied to the primary structure, as shown in Fig. 9-4(a).

The redundant M_c is solved by the condition of compatibility that the rotation of the left side relative to the right side at section c must be zero.

Using the method of virtual work we evaluate the relative rotation at c due to the external loading alone (Fig. 9-4(b)) as

$$\theta'_c = \int_0^l \frac{Mm\,dx}{EI} = \int_0^l \frac{[(wlx/4) - (wx^2/2)](2x/l)\,dx}{EI} = -\frac{wl^3}{12EI}$$

and that due to a pair of unit couples acting at c (Fig. 9-4(c)) as

$$\delta_{cc} = \int_0^l \frac{m^2\,dx}{EI} = \int_0^l \frac{(2x/l)^2\,dx}{EI} = \frac{4l}{3EI}$$

Setting the total relative angular displacement at c equal to zero, we have

$$-\frac{wl^3}{12EI} + M_c\left(\frac{4l}{3EI}\right) = 0$$

from which
$$M_c = +\frac{wl^2}{16}$$

After M_c is determined, the rest of the analysis can easily be carried out.

Example 9-2. Suppose that the support at b of the previous problem is elastic and yields at k, per unit force, as shown in Fig. 9-5(a). Determine reaction at b, denoted by X_b.

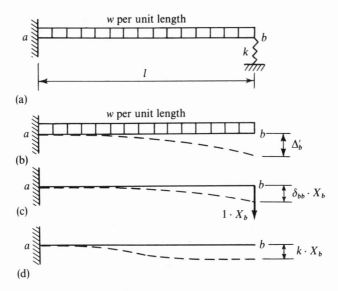

Fig. 9-5

To do this we apply the compatibility condition

$$\Delta'_b + \delta_{bb} X_b + k_b X = 0$$

This is explained in Figs. 9-5(b), (c), and (d). The above equation indicates that the total deflection caused by the uniform load plus redundant X_b (assumed to be acting downward) is not 0 but is equal to $-kX_b$. Since X_b is found to be an upward force opposite to the assumed direction, it gives $-kX_b$ a positive value, i.e., a downward deflection consistent with the contraction in an elastic support due to compressive force.

By substituting $\Delta_b' = wl^4/8EI, \delta_{bb} = l^3/3EI$ in the above equation, we obtain

$$\frac{wl^4}{8EI} + \frac{X_b l^3}{3EI} + kX_b = 0$$

from which
$$X_b = -\frac{wl/8}{1/3 + kEI/l^3}$$

The minus sign indicates an upward reaction.

If k equals zero, i.e., the support is nonyielding, then the above equation gives

$$X_b = -\tfrac{3}{8} wl$$

as previously found.

Example 9-3. Find the reactions for the beam with two sections shown in Fig. 9-6(a).

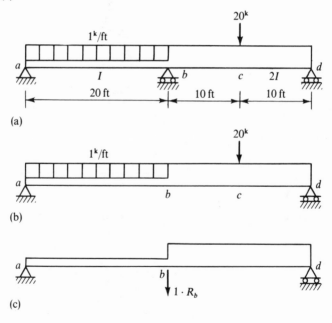

(a)

(b)

(c)

Fig. 9-6

In this problem it may be convenient to select the vertical reaction at support b as redundant. The beam is then considered as a simple beam subject to the original loading and the redundant R_b as shown in Figs. 9-6(b) and (c) respectively.

The compatibility requires

$$\Delta_b = \Delta'_b + \delta_{bb} R_b = 0$$

Using the method of virtual work, we have

$$\int \frac{Mm_b \, dx}{EI} + R_b \int \frac{(m_b)^2 \, dx}{EI} = 0$$

from which

$$R_b = -\frac{\int Mm_b \, dx/EI}{\int (m_b)^2 \, dx/EI}$$

where

$M =$ bending moment at any section of the primary beam caused by the original loading (Fig. 9-6(b));

$m_b =$ bending moment at the same section of the primary beam caused by a unit load in place of the redundant R_b (Fig. 9-6(c)).

The solution is completely shown in Table 9-1.

$$R_b = -\frac{\int_0^{20} \frac{\left(20x - \frac{x^2}{2}\right)\left(\frac{x}{2}\right) dx}{EI} + \int_0^{10} \frac{(20x)\left(\frac{x}{2}\right) dx}{2EI} + \int_{10}^{20} \frac{(200)\left(\frac{x}{2}\right) dx}{2EI}}{\int_0^{20} \frac{\left(\frac{x}{2}\right)^2 dx}{EI} + \int_0^{10} \frac{\left(\frac{x}{2}\right)^2 dx}{2EI} + \int_{10}^{20} \frac{\left(\frac{x}{2}\right)^2 dx}{2EI}}$$

$$= -25.84 \text{ kips}$$

The negative sign indicates an upward reaction at support b.

TABLE 9-1

Section	Origin	Limit (ft)	M (ft-kips)	m_b (ft-kips)	I
ab	a	0 to 20	$20x - \frac{(1)(x)^2}{2}$	$\frac{x}{2}$	I
dc	d	0 to 10	$20x$	$\frac{x}{2}$	$2I$
cb	d	10 to 20	$20x - 20(x-10)$ or 200	$\frac{x}{2}$	$2I$

After R_b is obtained, we can readily find the reactions at the other two supports by statics. That is:

$$R_a = R_d = 20 - (\tfrac{1}{2})(25.84) = 7.08 \text{ kips}$$

acting upward.

The end moments for a fixed-end beam, called *fixed-end moments*, are important in the methods of slope deflection and of moment distribution, which will be discussed in later chapters. The following examples are attempts to solve fixed-end moments due to common types of loading by the method of consistent deformations.

Example 9-4. The fixed-end beam of uniform cross section subjected to a single concentrated load shown in Fig. 9-7(a) is statically indeterminate to the

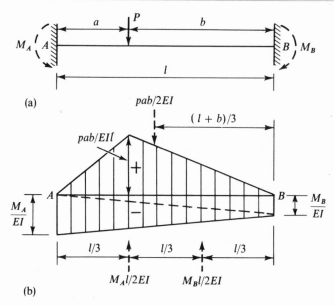

Fig. 9-7

second degree since the horizontal force does not exist. End moments M_A and M_B are selected as redundants. The original beam is then considered as equivalent to a simple beam (not shown) under the combined action of a concentrated force P and redundant moments M_A and M_B. It is convenient to apply the conjugate-beam method to determine M_A and M_B based on the condition that the slope and deflection at either end of the fixed-end beam must be zero. In other words, there will be no support reactions for the conjugate beam, and the positive and negative M/EI diagrams (elastic loads) given in Fig. 9-7(b) must form a balanced system. Thus from $\sum F_y = 0$,

$$\frac{Pab}{2EI} - \frac{M_A l}{2EI} - \frac{M_B l}{2EI} = 0$$

or,

$$M_A + M_B = \frac{Pab}{l} \qquad (9\text{-}8)$$

$\sum M_B = 0$,

$$\left(\frac{Pab}{2EI}\right)\left(\frac{l+b}{3}\right) - \left(\frac{M_A l}{2EI}\right)\left(\frac{2l}{3}\right) - \left(\frac{M_B l}{2EI}\right)\left(\frac{l}{3}\right) = 0$$

or,

$$2M_A + M_B = \frac{Pab}{l} + \frac{Pab^2}{l^2} \tag{9-9}$$

Solving Eqs. (9-8) and (9-9) simultaneously, we obtain

$$M_A = \frac{Pab^2}{l^2} \qquad M_B = \frac{Pa^2 b}{l^2}$$

Example 9-5. Find the end moments of a fixed-end beam of constant EI caused by a uniform load, as shown in Fig. 9-8(a).

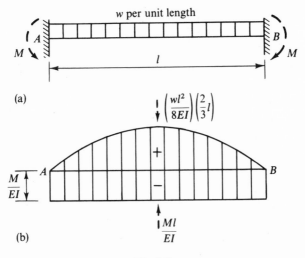

Fig. 9-8

Because of symmetry, the beam is statically indeterminate to the first degree, since $M_A = M_B = M$ as indicated in Fig. 9-8(a). By the method of conjugate beam (Fig. 9-8(b)); $\sum F_y = 0$,

$$\left(\frac{wl^2}{8EI}\right)\left(\frac{2l}{3}\right) - \frac{Ml}{EI} = 0$$

from which $\qquad\qquad M = \tfrac{1}{12} wl^2$

Example 9-6. If the fixed-end beam is loaded with an external couple M as shown in Fig. 9-9(a), the deflected elastic shape will be somewhat like that shown by the dotted line, which gives the sense of the end moments as indicated.

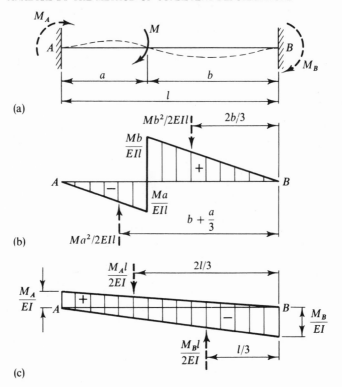

Fig. 9-9

As before, end moments M_A and M_B are chosen as redundants. The elastic loads based on the moment diagrams divided by EI plotted for external moment M and redundants M_A and M_B, as given in Figs. 9-9(b) and (c), must be in equilibrium themselves. From $\sum F_y = 0$,

$$\frac{M_A l}{2EI} + \frac{Mb^2}{2EIl} - \frac{M_B l}{2EI} - \frac{Ma^2}{2EIl} = 0$$

or,

$$M_A - M_B = \frac{M(a^2 - b^2)}{l^2} \tag{9-10}$$

From $\sum M_B = 0$,

$$\left(\frac{M_A l}{2EI}\right)\left(\frac{2l}{3}\right) + \left(\frac{Mb^2}{2EIl}\right)\left(\frac{2b}{3}\right) - \left(\frac{M_B l}{2EI}\right)\left(\frac{l}{3}\right) - \left(\frac{Ma^2}{2EIl}\right)\left(b + \frac{a}{3}\right) = 0$$

or,

$$2M_A - M_B = \frac{M[a^2 + 2b(a - b)]}{l^2} \tag{9-11}$$

Solving Eqs. (9-10) and (9-11) simultaneously, we obtain

$$M_A = \frac{Mb}{l^2}(2a - b) \qquad M_B = \frac{Ma}{l^2}(2b - a)$$

Note that M_A and M_B bear the same sense as the externally applied M, as indicated in Fig. 9-9(a), if $a > l/3$ and $b > l/3$.

9-3. ANALYSIS OF STATICALLY INDETERMINATE RIGID FRAMES BY THE METHOD OF CONSISTENT DEFORMATIONS

The general procedure illustrated in the previous section in solving statically indeterminate beams can be applied equally well to the analysis of statically indeterminate rigid frames, as in the following example.

Example 9-7. For the loaded rigid frame shown in Fig. 9-10(a), find the reaction components at the fixed end a, and plot the moment diagram for the entire frame. Assume the same EI for all members.

To do this, we start by removing support a and introducing in its place three redundant reaction components X_1, X_2, and X_3, as shown in Fig. 9-10(b). These can be taken as the superposition of four basic cases, as shown in Figs. 9-10(c), (d), (e), and (f) respectively. Since end a is fixed, compatibility requires

$$\Delta_1' + \delta_{11}X_1 + \delta_{12}X_2 + \delta_{13}X_3 = 0 \qquad (9\text{-}12)$$

$$\Delta_2' + \delta_{21}X_1 + \delta_{22}X_2 + \delta_{23}X_3 = 0 \qquad (9\text{-}13)$$

$$\Delta_3' + \delta_{31}X_1 + \delta_{32}X_2 + \delta_{33}X_3 = 0 \qquad (9\text{-}14)$$

Taking advantages of Examples 8-4 and 8-5, we note that

$$\Delta_1' = \frac{5,000}{EI} \qquad \delta_{11} = \frac{1,667}{EI} \qquad \delta_{12} = -\frac{1,000}{EI} \qquad \delta_{13} = -\frac{200}{EI}$$

$$\Delta_2' = -\frac{7,500}{EI} \qquad \delta_{21} = -\frac{1,000}{EI} \qquad \delta_{22} = \frac{1,333}{EI} \qquad \delta_{23} = \frac{150}{EI}$$

$$\Delta_3' = -\frac{800}{EI} \qquad \delta_{31} = -\frac{200}{EI} \qquad \delta_{32} = \frac{150}{EI} \qquad \delta_{33} = \frac{30}{EI}$$

Substituting these values in Eqs. 9-12, 9-13, and 9-14, we obtain

$$5,000 + 1,667X_1 - 1,000X_2 - 200X_3 = 0 \qquad (9\text{-}15)$$

$$-7,500 - 1,000X_1 + 1,333X_2 + 150X_3 = 0 \qquad (9\text{-}16)$$

$$-800 - 200X_1 + 150X_2 + 30X_3 = 0 \qquad (9\text{-}17)$$

solving for $X_1 = 1.0$ kip $X_2 = 6$ kips $X_3 = 3.33$ ft-kips

Note that the solution of this problem could be simplified by setting $X_2 = 6$ kips in Eqs. 9-15, 9-16, and 9-17, since we know this value beforehand because of the symmetry of the loaded frame.

The final results are shown in Fig. 9-10(g); the moment diagram for the

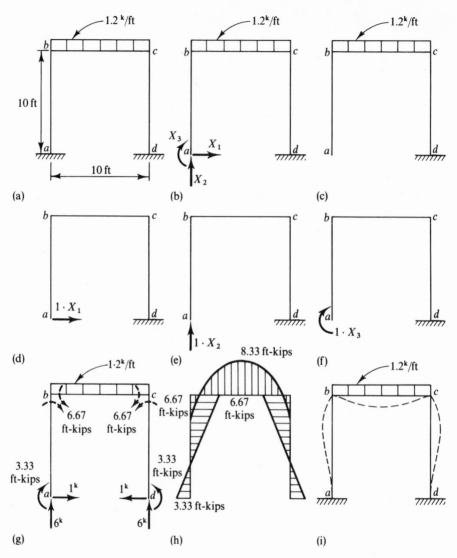

Fig. 9-10

whole frame is shown in Fig. 9-10(h). A sketch of the elastic deformation of the frame due to bending distortion is shown by the dotted line in Fig. 9-10(i). Note that in this case there is one point of inflection in each column and two points of inflection in the beam.

By referring to the foregoing example, we see that by using the method of consistent deformations in analyzing a rigid frame, we encounter tedious calculations of the flexibility coefficients. The work, if done by hand, will become intolerable if the problem involves as many redundants as a rigid frame usually does. As matter of fact, the method of consistent deformations is seldom used for analysis of rigid frames by hand calculation, since a solution can be much more easily obtained by the method of slope deflection or of moment distribution. However, with the development of high-speed *electronic computers*, this method has regained considerable strength in the scope of structural analysis, as will be seen in Ch. 17.

9-4. ANALYSIS OF STATICALLY INDETERMINATE TRUSSES BY THE METHOD OF CONSISTENT DEFORMATIONS

The indeterminateness of a truss may be due to redundant supports or redundant bars or both. If it results from redundant supports, the procedure for attack is the same as that described for a continuous beam. If the superfluous element is a bar, then the bar is considered to be cut at a section and replaced by two equal and opposite axial redundant forces representing the internal action for that bar. The condition equation is such that the relative displacement between the two sides at the cut section caused by the combined effect of the original loading and the redundants should be zero.

Example 9-8. Analyze the continuous truss in Fig. 9-11(a). Assume $E =$ 30,000 kips per in.² and $L(\text{ft})/A(\text{in.}^2) = 1$ for all members.

In this problem, it is convenient to select the central support as the redundant element. We begin by removing support c and introducing in its place

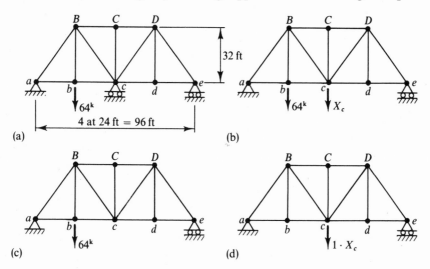

Fig. 9-11

a redundant reaction X_c, as shown in Fig. 9-11(b). The primary structure is then a simply supported truss subjected to an external load of 64 kips at joint b and a redundant X_c. The effects can be separated, respectively, as shown in Figs. 9-11(c) and (d).

Since support c is on a rigid foundation, the compatibility equation can be expressed by

$$\Delta_c = \Delta_c' + \delta_{cc} X_c = 0$$

Using virtual work gives

$$\sum \frac{S' u_c L}{AE} + X_c \sum \frac{u_c^2 L}{AE} = 0$$

from which

$$X_c = -\frac{\sum (S' u_c L / AE)}{\sum (u_c^2 L / AE)}$$

where

$S' =$ the internal force in any member of the primary truss due to the original loading (Fig. 9-11(c));

$u_c =$ the internal force in the same member of the primary truss due to a unit force at c (Fig. 9-11(d)).

The solution is completely shown in Table 9-2.

$$X_c = -\frac{122}{3.25} = -37.5 \text{ kips}$$

TABLE 9-2

Member	$\dfrac{L}{A}$ (ft/in.²)	S' (kips)	u_c	$\dfrac{S' u_c L}{A}$ (ft-kips/in.²)	$\dfrac{u_c^2 L}{A}$ (ft/in.²)	$S = S' + u_c X_c$ (kips)
ab	1	$+36$	$+3/8$	$+13.5$	$+9/64$	$36 - 14.1 = +21.9$
bc	1	$+36$	$+3/8$	$+13.5$	$+9/64$	$36 - 14.1 = +21.9$
cd	1	$+12$	$+3/8$	$+4.5$	$+9/64$	$12 - 14.1 = -2.1$
de	1	$+12$	$+3/8$	$+4.5$	$+9/64$	$12 - 14.1 = -2.1$
BC	1	-24	$-3/4$	$+18$	$+36/64$	$-24 + 28.2 = +4.2$
CD	1	-24	$-3/4$	$+18$	$+36/64$	$-24 + 28.2 = +4.2$
aB	1	-60	$-5/8$	$+37.5$	$+25/64$	$-60 + 23.4 = -36.6$
Bb	1	$+64$	0	0	0	$+64 + \ 0 \ = +64$
Bc	1	-20	$+5/8$	-12.5	$+25/64$	$-20 - 23.4 = -43.4$
Cc	1	0	0	0	0	0
cD	1	$+20$	$+5/8$	$+12.5$	$+25/64$	$+20 - 23.4 = -3.4$
Dd	1	0	0	0	0	0
De	1	-20	$-5/8$	$+12.5$	$+25/64$	$-20 + 23.4 = +3.4$
		Σ		$+122$	$+3.25$	

The negative sign indicates an upward reaction at support c.

After X_c is determined, we can readily obtain each bar force S from

$$S = S' + u_c X_c$$

as given in the last column of Table 9-2.

Example 9-9. Analyze the truss in Fig. 9-12(a). Assume $E = 30,000$ kips per in.^{2} and $L(\text{ft})/A(\text{in.}^2) = 1$ for all members.

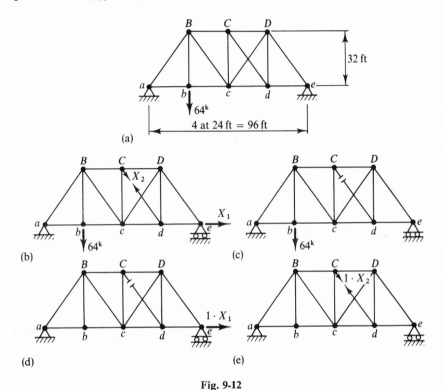

Fig. 9-12

The truss in Fig. 9-12(a) has two redundant elements, one in the reaction component and the other in the bar. Let us select the horizontal component of reaction at the right end hinge and the internal force in bar Cd as redundants. We then have a primary truss loaded as shown in Fig. 9-12(b) in which the original hinged support e is replaced by a roller acted on by a redundant horizontal reaction X_1 and the bar Cd is cut and a pair of redundant forces X_2 applied to it. This may again be replaced by the three basic cases shown in Figs. 9-12(c), (d), and (e). Since both the horizontal movement at support e and the relative displacement between the cut ends of bar Cd are zero, we have

$$\Delta_1 = \Delta_1' + \delta_{11} X_1 + \delta_{12} X_2 = 0$$

or

$$\left[\ \sum\frac{S'u_1L}{AE}+X_1\sum\frac{u_1^2L}{AE}+X_2\sum\frac{u_1u_2L}{AE}=0\ \right]\qquad(9\text{-}18)$$

and

$$\Delta_2=\Delta_2'+\delta_{21}X_1+\delta_{22}X_2=0$$

or

$$\left[\ \sum\frac{S'u_2L}{AE}+X_1\sum\frac{u_1u_2L}{AE}+X_2\sum\frac{u_2^2L}{AE}=0\ \right]\qquad(9\text{-}19)$$

Note that

S' = the internal force in any bar of the primary truss due to the original loading (Fig. 9-12(c));

u_1 = the internal force in the same bar of the primary truss due to a unit horizontal force acting at e (Fig. 9-12(d));

u_2 = the internal force in the same bar of the primary truss due to a pair of unit axial forces acting at the cut ends of bar Cd (Fig. 9-12(e)).

The redundants X_1 and X_2 can be determined by solving Eqs. 9-18 and 9-19 simultaneously, and the rest are obtained by

$$S=S'+u_1X_1+u_2X_2$$

The complete solution is shown in Table 9-3.

$$96+4X_1-\tfrac{3}{5}X_2=0$$
$$27.2-\tfrac{3}{5}X_1+4X_2=0$$

Solving simultaneously gives

$$X_1=-25.6\text{ kips}\qquad X_2=-10.6\text{ kips}$$

The negative signs indicate that the horizontal reaction at hinge e acts to the left and that the axial force in member Cd is compressive.

Example 9-10. Analyze the truss in Fig. 9-13(a) subject to rise of 50°F at the top chords BC and CD. Assume $\alpha = 0.0000065$ in. per in. per 1°F; $E = 30,000$ kips per in.^{2}; and $L(\text{ft})/A(\text{in.}^2) = 1$ for all members.

The truss is statically indeterminate to the first degree. Cut bar Cd and

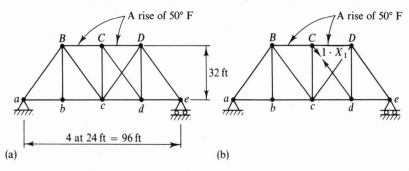

Fig. 9-13

TABLE 9-3

Member	$\dfrac{L}{A}$ (ft/in.²)	S' (kips)	u_1	u_2	$\dfrac{S'u_1L}{A}$ (ft-kips/in.²)	$\dfrac{S'u_2L}{A}$ (ft-kips/in.²)	$\dfrac{u_1^2L}{A}$ (ft/in.²)	$\dfrac{u_2^2L}{A}$ (ft/in.²)	$\dfrac{u_1u_2L}{A}$ (ft/in.²)	$S = S' + u_1X_1 + u_2X_2$ (kips)
ab	1	+36	+1	0	+36	0	+1	0	0	$36 - 25.6 + 0 = +10.4$
bc	1	+36	+1	0	+36	0	+1	0	0	$36 - 25.6 + 0 = +10.4$
cd	1	+12	+1	−3/5	+12	−7.2	+1	+9/25	−3/5	$12 - 25.6 + 6.4 = -7.0$
de	1	+12	+1	0	+12	0	+1	0	0	$12 - 25.6 + 0 = -13.6$
BC	1	−24	0	0	0	0	0	0	0	$-24 + 0 + 0 = -24$
CD	1	−24	0	−3/5	0	+14.4	0	+9/25	0	$-24 + 0 + 6.4 = -17.6$
aB	1	−60	0	0	0	0	0	0	0	$-60 + 0 + 0 = -60$
Bb	1	+64	0	0	0	0	0	0	0	$+64 + 0 + 0 = +64$
Bc	1	−20	0	0	0	0	0	0	0	$-20 + 0 + 0 = -20$
Cc	1	0	0	−4/5	0	0	0	+16/25	0	$0 + 0 + 8.5 = +8.5$
Cd	1	0	0	+1	0	0	0	+1	0	$0 + 0 - 10.6 = -10.6$
cD	1	+20	0	+1	0	+20	0	+1	0	$20 + 0 - 10.6 = +9.4$
Dd	1	0	0	−4/5	0	0	0	+16/25	0	$0 + 0 + 8.5 = +8.5$
De	1	−20	0	0	0	0	0	0	0	$-20 + 0 + 0 = -20$
				Σ	+96	+27.2	+4	+4	−3/5	

201

select its bar force X_1 as the redundant as shown in Fig. 9-13(b). The primary truss is then a simply supported truss subjected to the temperature rise in the top chords and the redundant axial force X_1. Since the relative displacement between the cut ends due to the combined effect of temperature rise and X_1 must be zero, we have

$$\Delta_1 = \Delta_1' + \delta_{11} X_1 = 0$$

or

$$\sum u_1(\alpha t° L) + X_1 \sum \frac{u_1^2 L}{AE} = 0$$

where

Δ_1' = the relative displacement between the cut ends of the primary truss due to the temperature rise = $\sum u_1(\alpha t° L)$ (see Eq. 8-20);

u_1 = the internal force in any member of the primary truss due to a pair of unit axial forces acting at the ends of the cut section.

The solution is shown in Table 9-4.

TABLE 9-4

Member	$\dfrac{L}{A}$ (ft/in.²)	u_1	$\alpha t° L$ (ft)	$u_1 \alpha t° L$ (ft)	$\dfrac{u_1^2 L}{A}$ (ft/in.²)	$S = u_1 X_1$ (kips)
ab	1	0	0	0	0	0
bc	1	0	0	0	0	0
cd	1	−3/5	0	0	+9/25	−21.1
de	1	0	0	0	0	0
BC	1	0	+0.0078	0	0	0
CD	1	−3/5	+0.0078	−0.00468	+9/25	−21.1
aB	1	0	0	0	0	0
Bb	1	0	0	0	0	0
Bc	1	0	0	0	0	0
Cc	1	−4/5	0	0	+16/25	−28.1
Cd	1	+1	0	0	+1	+35.1
cD	1	+1	0	0	+1	+35.1
Dd	1	−4/5	0	0	+16/25	−28.1
De	1	0	0	0	0	0
			Σ	−0.00468	+4	

$$\frac{4X_1}{30,000} - 0.00468 = 0$$

$$X_1 = 35.1 \text{ kips} \qquad \text{(tension)}$$

Although these illustrations are aimed at statically indeterminate trusses with one or two redundants, the procedure described can be extended to trusses with many degrees of redundancy.

PROBLEMS

9-1. Analyze the beam in Fig. 9-14 by the method of consistent deformations: (a) Use the reaction at center support b as redundant; (b) Use the internal moment at b as redundant. Assume constant EI.

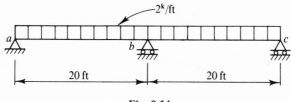

Fig. 9-14

9-2. Determine the reaction at b in Fig. 9-15 by the method of consistent deformations. Assume constant EI.

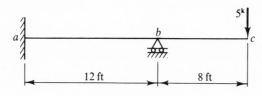

Fig. 9-15

9-3. Find the reaction at b in Fig. 9-16 by the method of consistent deformations. Assume constant E.

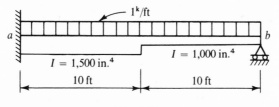

Fig. 9-16

9-4. Find the fixed-end moments for the beam in Fig. 9-17 by the method of consistent deformations. Assume constant EI.

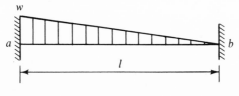

Fig. 9-17

9-5. Analyze the rigid frame shown in Fig. 9-18 by the method of consistent deformations. All members have the same value of *EI*.

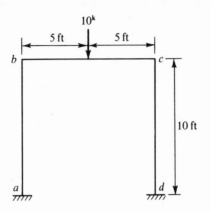

Fig. 9-18

9-6. Analyze each of the trusses in Fig. 9-19 by the method of consistent deformations. Assume $E = 30,000$ kips per in.^{2} and $L(\text{ft})/A(\text{in.}^2) = 2$ for all members.

9-7. Analyze the truss in Fig. 9-19(a) (without the external load) subject to a rise in temperature of $50°$F for member *BC*. Assume $\alpha = 0.0000065$ in. per in. per $1°$F.

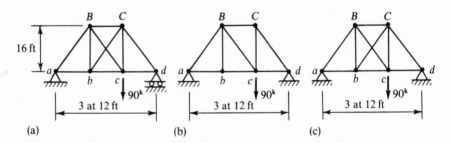

Fig. 9-19

10

ANALYSIS OF STATICALLY

INDETERMINATE STRUCTURES BY

THE METHOD OF LEAST WORK

10-1. GENERAL

The method of consistent deformations, discussed in the preceding chapter, involves superposition equations for the elastic deformations of the primary structure at the points of application of the redundants $X_1, X_2, \ldots, X_n$, the primary structure being stable and determinate and subjected to external actions, together with n redundant forces. The expressions that the displacement at each of n redundants equals zero for a loaded structure with nonyielding supports may be set up by the use of Castigliano's first theorem as

$$\Delta_1 = \frac{\partial W}{\partial X_1} = 0$$

$$\Delta_2 = \frac{\partial W}{\partial X_2} = 0 \qquad\qquad (10\text{-}1)$$

$$\vdots$$

$$\Delta_n = \frac{\partial W}{\partial X_n} = 0$$

where W is the *total strain energy* of the primary structure and is therefore a function of the external loads and the unknown redundant forces $X_1, X_2, \ldots, X_n$. There are as many simultaneous equations as the number of unknown redundants involved. Equation 10-1,

$$\frac{\partial W}{\partial X_1} = \frac{\partial W}{\partial X_2} = \cdots = \frac{\partial W}{\partial X_n} = 0$$

is known as Castigliano's second theorem, and it may be stated as *the redundants must have such value that the total strain energy of the structure is a minimum consistent with equilibrium.* For this reason it is sometimes referred to as the *theorem of least work.*

It is apparent from the foregoing proof that the method of least work and the method of consistent deformations are, in reality, identical. The choice between the two methods is a matter of personal preference. Because of the simpler set-up, the method of least work is more widely used by structural engineers. However, this method is limited to the computation of the internal forces produced only by external loads on a structure mounted on unyielding supports. It cannot be used to determine stresses caused by temperature change, support movements, fabrication errors, etc., and thus is not so general as the method of consistent deformations.

10-2. ANALYSIS OF STATICALLY INDETERMINATE
BEAMS BY THE METHOD OF LEAST WORK

The procedure involves expressing the total strain energy in the primary structure in terms of the original loads and the one or more redundant elements and placing each of the derivatives of this expression with respect to the unknown redundants equal to zero, thus solving for the unknowns.

As before, we consider bending moment to be the only significant factor contributing to the internal energy of the beams or rigid frames. Therefore, the total strain energy can be expressed by

$$W = \int \frac{M^2\, dx}{2EI}$$

Setting the derivative of this expression with respect to any redundant X_i equal to zero gives the equation of least work as

$$\int \frac{M(\partial M/\partial X_i)\, dx}{EI} = 0$$

Therefore, for a statically indeterminate beam (or rigid frame) with n redundants, we can write a set of n simultaneous least-work equations:

$$\frac{\partial W}{\partial X_1} = \int \frac{M(\partial M/\partial X_1)\, dx}{EI} = 0$$

$$\frac{\partial W}{\partial X_2} = \int \frac{M(\partial M/\partial X_2)\, dx}{EI} = 0 \qquad (10\text{-}2)$$

$$\vdots$$

$$\frac{\partial W}{\partial X_n} = \int \frac{M(\partial M/\partial X_n)\, dx}{EI} = 0$$

to solve all the unknown redundants.

Example 10-1. The beam in Fig. 10-1(a) was solved by consistent deformations in Example 9-1. Let us re-solve it by the method of least work.

Solution (1). Select the reaction at b, called R_b, as redundant, and take the

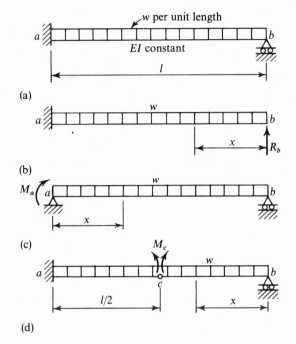

Fig. 10-1

end b as the origin (see Fig. 10-1(b)). Thus,

$$M = R_b x - \frac{wx^2}{2} \quad \text{and} \quad \frac{\partial M}{\partial R_b} = x$$

Substituting these in the expression

$$\frac{\partial W}{\partial R_b} = \int_0^l \frac{M(\partial M/\partial R_b)\, dx}{EI} = 0$$

we obtain

$$\int_0^l \frac{[R_b x - (wx^2/2)]x\, dx}{EI} = 0$$

or

$$\frac{R_b l^3}{3} - \frac{wl^4}{8} = 0$$

from which

$$R_b = \frac{3wl}{8}$$

acting in the direction shown.

Solution (2). Let us choose the end moment at a, called M_a, as redundant and take the origin at a as shown in Fig. 10-1(c).

$$M = M_a + \left(\frac{wl}{2} - \frac{M_a}{l}\right) x - \frac{wx^2}{2} \quad \text{and} \quad \frac{\partial M}{\partial M_a} = 1 - \frac{x}{l}$$

Setting

$$\frac{\partial W}{\partial M_a} = \int_0^l \frac{M(\partial M/\partial M_a)\,dx}{EI}$$

$$= \int_0^l \frac{\{M_a + [(wl/2) - (M_a/l)]x - (wx^2/2)\}\{1 - (x/l)\}\,dx}{EI} = 0$$

we solve for $$M_a = -\frac{1}{8}wl^2$$

The negative sign indicates a counter-clockwise moment.

Solution (3). Suppose that we select the internal moment at midspan section *c*, called M_c, as the redundant. We will then have the loaded primary structure shown in Fig. 10-1(d).

By statics, the reaction at end *b* is $(wl/4 + 2M_c/l)$ acting upward. Using *b* as the origin, we have

$$M = \left(\frac{wl}{4} + \frac{2M_c}{l}\right)x - \frac{wx^2}{2} \quad \text{and} \quad \frac{\partial M}{\partial M_c} = \frac{2x}{l}$$

Setting $$\frac{\partial W}{\partial M_c} = \frac{1}{EI}\int_0^l \left[\left(\frac{wl}{4} + \frac{2M_c}{l}\right)x - \frac{wx^2}{2}\right]\frac{2x}{l}\,dx = 0$$

we obtain $$M_c = \frac{wl^2}{16}$$

acting in the assumed direction.

Example 10-2. The reaction at support *b* in Fig. 10-2(a) was solved by the method of consistent deformations in Example 9-3 and will now be re-solved by the method of least work.

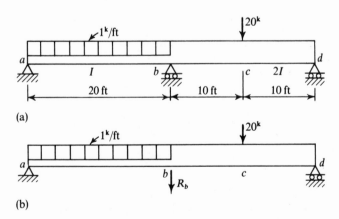

(a)

(b)

Fig. 10-2

If we choose the reaction at support *b*, called R_b, as redundant, we will have the loaded primary structure shown in Fig. 10-2(b).

The solution for the equation

$$\frac{\partial W}{\partial R_b} = \int \frac{M(\partial M/\partial R_b)\,dx}{EI} = 0$$

is contained in Table 10-1.

TABLE 10-1

Section	Origin	Limit (ft)	M (ft-kips)	$\dfrac{\partial M}{\partial R_b}$ (ft)	I
ab	a	0 to 20	$\left(20 + \dfrac{R_b}{2}\right)x - \dfrac{(1)x^2}{2}$	$\dfrac{x}{2}$	I
dc	d	0 to 10	$\left(20 + \dfrac{R_b}{2}\right)x$	$\dfrac{x}{2}$	2I
cb	d	10 to 20	$\left(20 + \dfrac{R_b}{2}\right)x - 20(x - 10)$	$\dfrac{x}{2}$	2I
			or $\dfrac{R_b x}{2} + 200$		

$$\int_0^{20} \frac{\{[20 + (R_b/2)]x - (x^2/2)\}x/2\,dx}{EI} + \int_0^{10} \frac{\{[20 + (R_b/2)]x\}x/2\,dx}{2EI}$$

$$+ \int_{10}^{20} \frac{\{(R_b x/2) + 200\}x/2\,dx}{2EI} = 0$$

$$R_b = -25.84 \text{ kips}$$

The negative sign indicates an upward reaction at support b.

To determine, by the method of least work, the *fixed-end moments* of beam AB due to general loading, as shown in Fig. 10-3(a), we select the left-end reaction components M_A and V_A as redundants, as shown in Fig. 10-3(b). The primary structure is a cantilever subjected to the original loads on the span together with the redundant forces M_A and V_A at the left end. Applying the

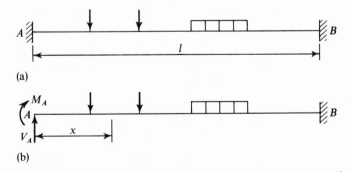

(a)

(b)

Fig. 10-3

method of least work, we obtain

$$\frac{\partial W}{\partial M_A} = \int_0^l \frac{M(\partial M/\partial M_A)\,dx}{EI} = 0 \qquad (10\text{-}3)$$

$$\frac{\partial W}{\partial V_A} = \int_0^l \frac{M(\partial M/\partial V_A)\,dx}{EI} = 0 \qquad (10\text{-}4)$$

Since the bending moment at any section of the primary structure is given by

$$M = M' + M_A + V_A x$$

where M' indicates the bending moment at the same section of the primary structure resulting from the original loads on the span, we have

$$\frac{\partial M}{\partial M_A} = 1 \quad \text{and} \quad \frac{\partial M}{\partial V_A} = x$$

Substituting these in Eqs. 10-3 and 10-4 results in the following two equations:

$$\int_0^l \frac{M\,dx}{EI} = 0 \qquad (10\text{-}5)$$

$$\int_0^l \frac{Mx\,dx}{EI} = 0 \qquad (10\text{-}6)$$

to solve for redundants M_A and V_A.

For a beam of uniform section with constant EI, the above two equations reduce to

$$\int_0^l M\,dx = 0 \qquad (10\text{-}7)$$

$$\int_0^l Mx\,dx = 0 \qquad (10\text{-}8)$$

Example 10-3. The fixed-end moments of the beam shown in Fig. 10-4 were solved by the method of consistent deformations in Example 9-4 and will be re-solved by using Eqs. 10-7 and 10-8.

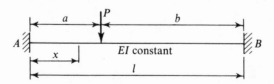

Fig. 10-4

Taking the origin at A, we note that

$$M = M_A + V_A x \qquad\qquad 0 \le x \le a$$

$$M = M_A + V_A x - P(x - a) \qquad a \le x \le l$$

Applying Eqs. 10-7 and 10-8 gives

$$\int_0^l M\,dx = \int_0^a (M_A + V_A x)\,dx + \int_a^l [M_A + V_A x - P(x-a)]\,dx$$

$$= \int_0^l (M_A + V_A x)\,dx + \int_a^l [-P(x-a)]\,dx = 0$$

or

$$M_A l + \frac{V_A l^2}{2} - \frac{Pb^2}{2} = 0 \qquad (10\text{-}9)$$

and

$$\int_0^l Mx\,dx = \int_0^a (M_A + V_A x)x\,dx + \int_a^l [M_A + V_A x - P(x-a)]x\,dx$$

$$= \int_0^l (M_A + V_A x)x\,dx + \int_a^l [-P(x-a)]x\,dx = 0$$

or

$$\frac{M_A l^2}{2} + \frac{V_A l^3}{3} - \frac{Pb^2(a+2l)}{6} = 0 \qquad (10\text{-}10)$$

Solving Eqs. 10-9 and 10-10 simultaneously, we obtain

$$M_A = -\frac{Pab^2}{l^2} \qquad V_A = \frac{Pb^2(l+2a)}{l^3}$$

Similarly,

$$M_B = -\frac{Pa^2 b}{l^2} \qquad V_B = \frac{Pa^2(l+2b)}{l^3}$$

10-3. ANALYSIS OF STATICALLY INDETERMINATE
RIGID FRAMES BY THE METHOD OF LEAST WORK

The procedure in analyzing a statically indeterminate rigid frame by the method of least work is similar to that described in the preceding section for statically indeterminate beams. The following examples serve to illustrate the application of this method.

Example 10-4. Consider the loaded frame shown in Fig. 10-5(a), which was analyzed by the method of consistent deformations (Example 9-7). Let us now analyze it by least work.

Let us take the reaction components M_a, H_a, and V_a at end a as redundants, as shown in Fig. 10-5(b). Because of symmetry, we may replace V_a with 6 kips so that there will be only two redundants M_a and H_a left, as indicated. The solutions for the equations of least work,

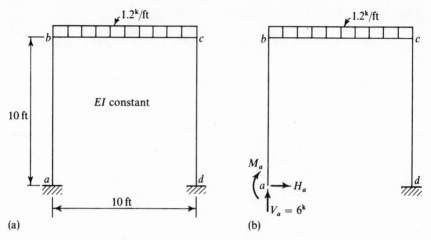

Fig. 10-5

$$\frac{\partial W}{\partial M_a} = 0 \quad \text{or} \quad \int_F \frac{M(\partial M/\partial M_a)\,dx}{EI} = 0$$

$$\frac{\partial W}{\partial H_a} = 0 \quad \text{or} \quad \int_F \frac{M(\partial M/\partial H_a)\,dx}{EI} = 0$$

are contained in Table 10-2.

TABLE 10-2

Member	Origin	Limit (ft)	M (ft-kips)	$\frac{\partial M}{\partial M_a}$	$\frac{\partial M}{\partial H_a}$ (ft)
ab	a	0 to 10	$M_a - H_a x$	1	$-x$
bc	b	0 to 10	$M_a - 10H_a + 6x - \dfrac{(1.2)x^2}{2}$	1	-10
cd	c	0 to 10	$M_a + H_a(x - 10)$	1	$x - 10$

$$\int_F M(\partial M/\partial M_a)\,dx$$

$$= \int_0^{10} (M_a - H_a x)(1)\,dx + \int_0^{10}\left[M_a - 10H_a + 6x - \frac{(1.2)x^2}{2}\right](1)\,dx$$

$$+ \int_0^{10} [M_a + H_a(x - 10)](1)\,dx$$

$$= (10M_a - 50H_a) + (10M_a - 100H_a + 100) + (10M_a - 50H_a)$$

$$= 0$$

or

$$3M_a - 20H_a + 10 = 0 \qquad (10\text{-}11)$$

$$\int_F M(\partial M/\partial H_a)\,dx$$

$$= \int_0^{10} (M_a - H_a x)(-x)\,dx + \int_0^{10}\left[M_a - 10H_a + 6x - \frac{(1.2)x^2}{2}\right](-10)\,dx$$

$$+ \int_0^{10}[M_a + H_a(x - 10)](x - 10)\,dx$$

$$= \left(-50M_a + \frac{1{,}000}{3}H_a\right) + (-100M_a + 1{,}000H_a - 1{,}000)$$

$$+ \left(-50M_a + \frac{1{,}000}{3}H_a\right) = 0$$

or

$$-3M_a + 25H_a - 15 = 0 \qquad (10\text{-}12)$$

Solving Eqs. 10-11 and 10-12 simultaneously, we obtain

$$H_a = 1.0 \text{ kip} \qquad M_a = 3.33 \text{ ft-kips}$$

Example 10-5. Analyze the frame in Fig. 10-5(a) by taking the internal shear, thrust, and moment in the midspan section of the beam as redundants.

Because of symmetry, the shear must be zero in the midspan section e of the beam, and only thrust and bending moment are left as redundants, as shown in Fig. 10-6(a). The solution can be simplified by considering only half of the frame, as shown in Fig. 10-6(b) and Table 10-3.

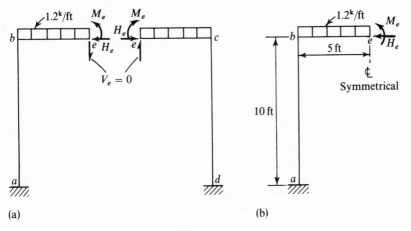

(a)   (b)

Fig. 10-6

TABLE 10-3

Member	Origin	Limit (ft)	M (ft-kips)	$\dfrac{\partial M}{\partial M_e}$	$\dfrac{\partial M}{\partial H_e}$ (ft)
eb	e	0 to 5	$M_e - \dfrac{(1.2)x^2}{2}$	1	0
ba	b	0 to 10	$M_e + H_e x - 15$	1	x

Applying

$$\frac{\partial W}{\partial M_e} = 0 \quad \text{or} \quad \int_F \frac{M(\partial M/\partial M_e)\, dx}{EI} = 0$$

we have

$$\frac{2}{EI}\left[\int_0^5 \left(M_e - \frac{(1.2)x^2}{2}\right)(1)\, dx + \int_0^{10}(M_e + H_e x - 15)(1)\, dx \right] = 0$$

or

$$3M_e + 10H_e - 35 = 0 \tag{10-13}$$

Applying

$$\frac{\partial W}{\partial H_e} = 0 \quad \text{or} \quad \int_F \frac{M(\partial M/\partial H_e)\, dx}{EI} = 0$$

we have

$$\frac{2}{EI}\left[\int_0^{10}(M_e + H_e x - 15)x\, dx \right] = 0$$

or

$$3M_e + 20H_e - 45 = 0 \tag{10-14}$$

Solving Eqs. 10-13 and 10-14 simultaneously gives

$$H_e = 1.0 \text{ kip} \qquad M_e = 8.33 \text{ ft-kips}$$

from which we obtain

$$H_a = 1.0 \text{ kip} \qquad M_a = 3.33 \text{ ft-kips}$$

as previously found.

In comparison with the preceding example, we see that, by proper selection of the redundants through taking advantage of the symmetry of the structure, we minimize considerably the numerical computations.

For a highly indeterminate rigid frame, such as the one shown in Fig. 10-7(a), the procedure of the analysis remains the same. The frame is statically indeterminate to the 24th degree. We may cut it back to three determinate structures and substitute the redundants $X_1, X_2, \ldots, X_{24}$ at the cut sections as shown in Fig. 10-7(b).

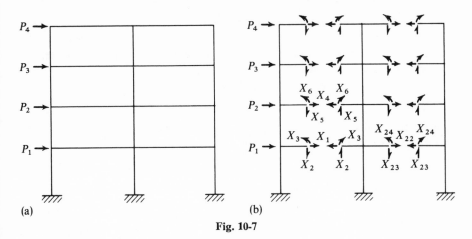

Fig. 10-7

From least work, we have 24 equations to solve for all the redundants simultaneously, namely,

$$\frac{\partial W}{\partial X_1} = 0$$

$$\frac{\partial W}{\partial X_2} = 0$$

$$\vdots$$

$$\frac{\partial W}{\partial X_{24}} = 0$$

where W is the total strain energy of the frame due to the external loads and redundant forces. The principle is neat and elegant, whereas the numerical calculations involved in the above equations are so cumbersome that it is almost impossible for a structural engineer to reach an exact solution for the system with only a slide rule or desk calculating machine. Formerly, to handle a practical problem like this, a grossly simplified model of the actual structure was often used. However, with the advent of the digital computer, this unsatisfactory situation underwent a drastic change, and the solving of simultaneous equations can now be performed in a matter of minutes.

10-4. ANALYSIS OF STATICALLY INDETERMINATE
TRUSSES BY THE METHOD OF LEAST WORK

Statically indeterminate trusses can be analyzed by the method of least work in a similar manner. Referring to Eq. 8-10, we find the total strain energy (internal work) of a truss is expressed by

$$W = \sum \frac{S^2 L}{2AE}$$

Setting the derivative of this expression with respect to any redundant X_i equal to zero gives the equation of least work as

$$\sum \frac{S(\partial S/\partial X_i)L}{AE} = 0$$

Thus, for a statically indeterminate truss with n redundant elements, we have a set of n simultaneous least-work equations available for their solution,

$$\sum \frac{S(\partial S/\partial X_1)L}{AE} = 0$$

$$\sum \frac{S(\partial S/\partial X_2)L}{AE} = 0 \qquad (10\text{-}15)$$

$$\vdots$$

$$\sum \frac{S(\partial S/\partial X_n)L}{AE} = 0$$

The application of this procedure is illustrated in the following example.

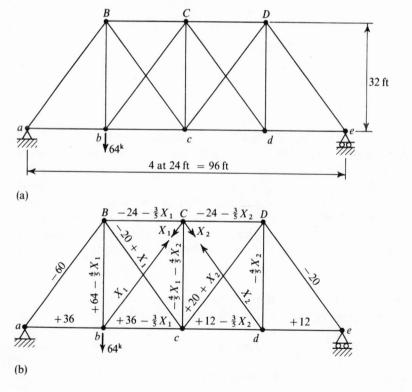

(a)

(b)

Fig. 10-8

Example 10-6. Analyze the truss in Fig. 10-8(a). Assume $E = 30,000$ kips per in.^{2} and $L(\text{ft})/A(\text{in.}^2) = 1$ for all members.

The truss is statically indeterminate to the second degree. We may take bars bC and Cd as redundant members. As shown in Fig. 10-8(b), these bars are cut and replaced by redundant axial forces X_1 and X_2 respectively. The internal force for each bar is then computed in terms of the external load and redundant forces as indicated. The unknowns X_1 and X_2 are then solved by the simultaneous equations:

$$\sum \frac{S(\partial S/\partial X_1)L}{AE} = 0 \qquad \sum \frac{S(\partial S/\partial X_2)L}{AE} = 0$$

as prepared in Table 10-4.

Setting

$$-78.4 + 4X_1 + 0.64X_2 = 0 \qquad (10\text{-}16)$$

$$27.2 + 0.64X_1 + 4X_2 = 0 \qquad (10\text{-}17)$$

and solving Eqs. 10-16 and 10-17 simultaneously, we obtain

$$X_1 = +21.2 \text{ kips} \qquad X_2 = -10.2 \text{ kips}$$

The answer for each of the bar forces is given in the last column of Table 10-4. Note that this procedure can be extended to trusses with many redundants.

10-5. ANALYSIS OF STATICALLY INDETERMINATE
COMPOSITE STRUCTURES BY THE METHOD OF
LEAST WORK

Structures made up of some members which are two-force members carrying only axial forces and others which are not are called composite structures. They are conveniently analyzed by the method of least work, as illustrated in the following examples.

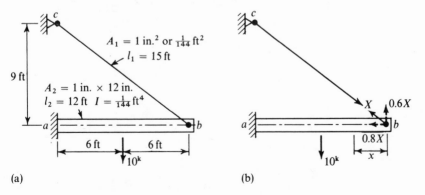

(a)   (b)

Fig. 10-9

TABLE 10-4

Member	$\frac{L}{A}$ (ft/in.²)	S (kips)	$\frac{\partial S}{\partial X_1}$	$\frac{\partial S}{\partial X_2}$	$\frac{S(\partial S/\partial X_1)L}{A}$ (ft-kips/in.²)	$\frac{S(\partial S/\partial X_2)L}{A}$ (ft-kips/in.²)	Answer (kips)
ab	1	$36 + 0 + 0$	0	0	0	0	$+36$
bc	1	$36 - (3/5)X_1 + 0$	$-3/5$	0	$-21.6 + (9/25)X_1 + 0$	0	$+23.3$
cd	1	$12 + 0 - (3/5)X_2$	0	$-3/5$	0	$-7.2 + 0 + (9/25)X_2$	$+18.2$
de	1	$12 + 0 + 0$	0	0	0	0	$+12$
BC	1	$-24 - (3/5)X_1 + 0$	$-3/5$	0	$14.4 + (9/25)X_1 + 0$	0	-36.7
CD	1	$-24 + 0 - (3/5)X_2$	0	$-3/5$	0	$14.4 + 0 + (9/25)X_2$	-17.8
aB	1	$-60 + 0 + 0$	0	0	0	0	-60
Bb	1	$64 - (4/5)X_1 + 0$	$-4/5$	0	$-51.2 + (16/25)X_1 + 0$	0	$+47$
Bc	1	$-20 + X_1 + 0$	$+1$	0	$-20 + X_1 + 0$	0	$+1.2$
bC	1	$0 + X_1 + 0$	$+1$	0	$0 + X_1 + 0$	0	$+21.2$
Cc	1	$0 - (4/5)X_1 - (4/5)X_2$	$-4/5$	$-4/5$	$0 + (16/25)X_1 + (16/25)X_2$	$0 + (16/25)X_1 + (16/25)X_2$	-8.8
Cd	1	$0 + 0 + X_2$	0	$+1$	0	$0 + 0 + X_2$	-10.2
cD	1	$20 + 0 + X_2$	0	$+1$	0	$20 + 0 + X_2$	$+9.8$
Dd	1	$0 + 0 - (4/5)X_2$	0	$-4/5$	0	$0 + 0 + (16/25)X_2$	$+8.2$
De	1	$-20 + 0 + 0$	0	0	0	0	-20
				Σ	$-78.4 + 4X_1 + 0.64X_2$	$27.2 + 0.64X_1 + 4X_2$	

Example 10-7. Fig. 10-9(a) shows a cantilever beam whose other end is supported by a rod. Find the force in the rod. $E = 30,000$ kips per in.^{2}

The structure is statically indeterminate to the first degree. Select the force in the tie rod as the redundant X as shown in Fig. 10-9(b). Then the internal work in the rod is

$$\frac{X^2 l_1}{2A_1 E}$$

and the internal work in the beam is equal to

$$\int_0^6 \frac{(0.6Xx)^2 \, dx}{2EI} + \int_6^{12} \frac{[0.6Xx - 10(x - 6)]^2 \, dx}{2EI} + \frac{(-0.8X)^2 l_2}{2A_2 E}$$

Applying $\partial W/\partial X = 0$ gives

$$\frac{X l_1}{A_1 E} + \int_0^6 \frac{(0.6Xx)(0.6x) \, dx}{EI} + \int_6^{12} \frac{[0.6Xx - 10(x - 6)][0.6x] \, dx}{EI}$$

$$+ \frac{(-0.8X)(-0.8)l_2}{A_2 E} = 0$$

or $\quad \dfrac{15X}{1/144} + \dfrac{207.4X - 3,024 + 1,944}{1/144} + \dfrac{(0.64X)(12)}{12/144} = 0$

After simplifying, we find

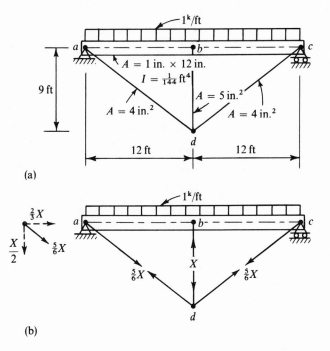

(a)

(b)

Fig. 10-10

$$15X + 207.4X - 1{,}080 + 0.64X = 0$$

which yields $\qquad X = 4.84 \text{ kips} \qquad$ (tension)

The effect of the axial force of beam on the strain energy is small and can be neglected.

Example 10-8. Analyze the trussed beam in Fig. 10-10(a). $E = 30{,}000$ kips per in.^{2}

The trussed beam shown is statically indeterminate to the first degree. This problem is most conveniently solved by choosing the internal force in rod *bd* as the redundant (see Fig. 10-10(b)) and by expressing the internal strain energy in terms of this unknown.

Referring to Fig. 10-10(b) and applying $\partial W / \partial X = 0$, we obtain

$$\frac{(X)(1)(9)}{5/144} + 2\left[\frac{(\frac{5}{6}X)(\frac{5}{6})(15)}{4/144}\right] + \frac{(\frac{2}{3}X)(\frac{2}{3})(24)}{12/144}$$

$$+ 2\int_0^{12} \frac{[(12 - X/2)x - (1)(x)^2/2][-x/2]\,dx}{1/144} = 0$$

or $\qquad 1.8X + 5.2X + 0.89X - 4{,}320 + 288X = 0$

from which $\qquad X = 14.6 \text{ kips} \qquad$ (compression)

After the compressive force in strut *bd* is determined, the rest of the analysis can be carried out without difficulty.

PROBLEMS

10-1. Solve Prob. 9-1 by the method of least work.

10-2. Solve Prob. 9-2 by the method of least work.

10-3. Solve Prob. 9-3 by the method of least work.

10-4. Solve Prob. 9-4 by the method of least work.

10-5. Solve Prob. 9-5 by the method of least work.

10-6. Solve Prob. 9-6 by the method of least work.

10-7. Find the internal force for the tie rod *ac* of the composite structure shown in Fig. 10-11, and sketch the moment diagram for member *ab*. $E = 30{,}000$ kips per in.^{2}

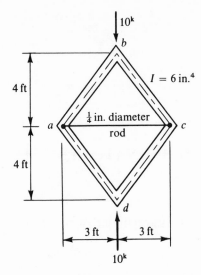

Fig. 10-11

10-8. Find the internal forces for all rods of the composite structure shown in Fig. 10-12, and sketch the moment diagram for beam *ab*. $E = 30,000$ kips per in.²

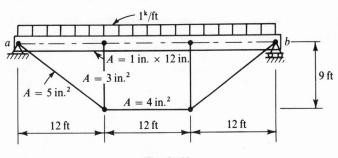

Fig. 10-12

11

INFLUENCE LINES FOR STATICALLY INDETERMINATE STRUCTURES

11-1. MAXWELL'S LAW OF RECIPROCAL DEFLECTIONS

Referring to Figs. 11-1(a) and (b), we note that the law simply states that

$$\Delta_{21} = \Delta_{12} \tag{11-1}$$

where

$\Delta_{21} =$ the deflection at point 2 due to the load P applied at point 1

$\Delta_{12} =$ the deflection at point 1 along the original line of action of P due to the same load applied at point 2 along the original deflection Δ_{21}

To prove this statement, both deflections are evaluated by the method of virtual work. Thus from Fig. 11-1(a) we have

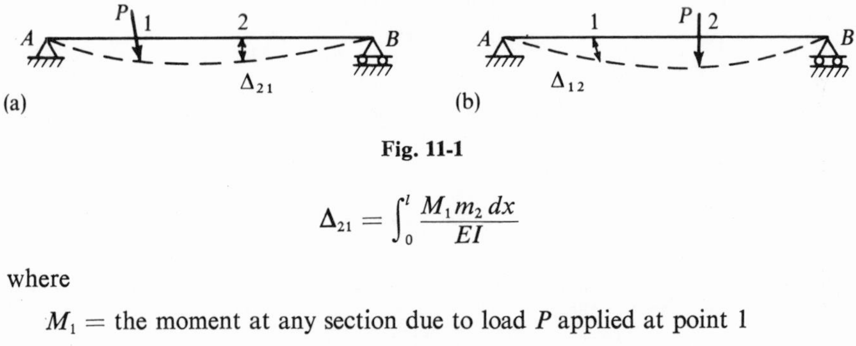

(a)   (b)

Fig. 11-1

$$\Delta_{21} = \int_0^l \frac{M_1 m_2 \, dx}{EI}$$

where

$M_1 =$ the moment at any section due to load P applied at point 1

$m_2 =$ the moment at the same section due to a unit load applied at point 2 along the desired deflection

Similarly, from Fig. 11-1(b) we obtain that

$$\Delta_{12} = \int_0^l \frac{M_2 m_1 \, dx}{EI}$$

but
$$M_1 = Pm_1 \quad \text{and} \quad M_2 = Pm_2$$

It is readily seen that

$$\Delta_{21} = \int_0^l \frac{(Pm_1)m_2 \, dx}{EI} = \int_0^l \frac{(Pm_2)m_1 \, dx}{EI} = \int_0^l \frac{M_2 m_1 \, dx}{EI} = \Delta_{12}$$

The special case is that $P = 1$ for which we can write

$$\delta_{21} = \delta_{12} \tag{11-2}$$

where

$\delta_{21} = $ the deflection at point 2 resulting from a unit load applied at point 1

$\delta_{12} = $ the deflection at point 1 along the original line of action due to a unit load applied at point 2 along the original deflection δ_{21}

We have hitherto demonstrated the law in regard to applied forces and their corresponding linear deflections. However, the reciprocity extends also to rotational displacement. For the case of two unit couples applied separately to any two points of a structure, the law is: *The rotational deflection at point 2 on a structure caused by a unit couple at point 1 is equal to the rotational deflection at point 1 due to a unit couple at point 2.*

By the reason of virtual work we also observe that the *rotational deflection at point 2 due to a unit force at point 1 is equal in magnitude to the linear deflection at point 1 along the original force due to a unit couple at point 2.*

Maxwell's law is perfectly general and is applicable to any type of structure so long as the material of the structure is elastic and follows Hooke's law.

11-2. BETTI'S LAW

Maxwell's theorem of reciprocity is a special case of *Betti's law* which may be expressed briefly by

$$W_{12} = W_{21} \tag{11-3}$$

where

$W_{12} = $ the external work done by a force system P_1 already on the structure during a distortion caused by a force system P_2 applied later

$W_{21} = $ the external work done by P_2 during a distortion caused by P_1 when P_1 is the second in order of application

To prove this theorem, let us consider the beam in Fig. 11-2(a) where P_1 (representing the force system 1) is applied first. The elastic curve caused by P_1 is indicated by the dotted line. During this operation the work done is

$$\tfrac{1}{2} P_1 (P_1 \delta_{11})$$

P_2 (representing the force system 2) is applied next, which causes further displacement of the beam (see Fig. 11-2(a)). The additional work done is

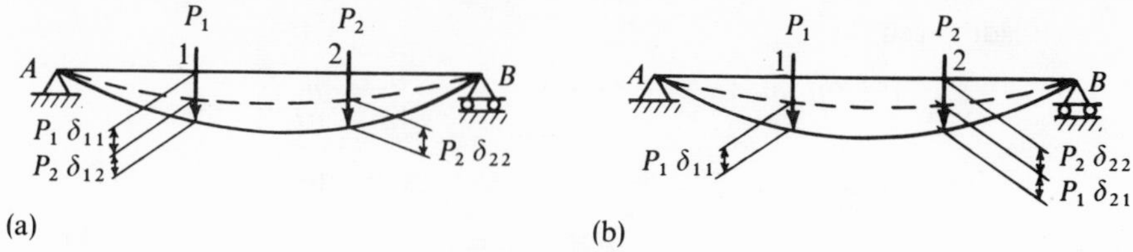

Fig. 11-2

$$P_1(P_2\,\delta_{12}) + \tfrac{1}{2}\,P_2(P_2\,\delta_{22})$$

The total work done is

$$\tfrac{1}{2}\,P_1(P_1\,\delta_{11}) + P_1(P_2\,\delta_{12}) + \tfrac{1}{2}\,P_2(P_2\,\delta_{22}) \qquad (11\text{-}4)$$

If we reverse the order of application of two systems of forces (see Fig. 11-2(b)) the total work, obtained similarly, can be expressed by

$$\tfrac{1}{2}\,P_2(P_2\,\delta_{22}) + P_2(P_1\,\delta_{21}) + \tfrac{1}{2}\,P_1(P_1\,\delta_{11}) \qquad (11\text{-}5)$$

Equating Eq. 11-4 to Eq. 11-5 gives

$$P_1(P_2\,\delta_{12}) = P_2(P_1\,\delta_{21}) \qquad (11\text{-}6)$$

as asserted by Eq. 11-3.

Equation 11-6 also indicates that

$$\delta_{12} = \delta_{21}$$

which is the *law of reciprocal deflections*.

Note that Betti's law depends on the principle of independence of effect of loads, which is true as long as the displacements are small and also independent. The law is applicable to any type of elastic structure with unyielding supports and at constant temperature. It holds for generalized forces and deflections, i.e., for couples and rotations as well as for forces and linear displacements.

11-3. INFLUENCE LINES AS DEFLECTED STRUCTURES: THE MÜLLER-BRESLAU PRINCIPLE

Suppose that we want the influence line for the reaction at support b of the indeterminate beam abc shown in Fig. 11-3(a). The influence ordinate at any point i, distance x from the left end, is obtained by placing a unit load at that point and computing the reaction at support b. The procedure for finding this reaction is contained in the following steps.

1. Remove the support at b and introduce in its place a redundant reaction called R_b.
2. Consider beam ac as the primary structure subjected to the combined effects of the unit force at i and R_b (see Fig. 11-3(b)).

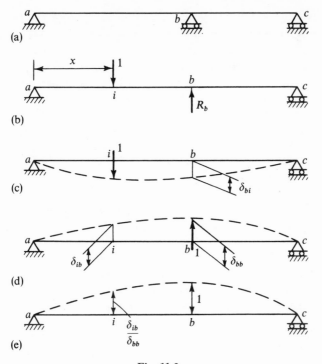

Fig. 11-3

3. Use the condition of compatibility that the total deflection at point b must be zero,

$$\Delta_b = R_b\,\delta_{bb} - \delta_{bi} = 0$$

See Fig. 11-3(c) for δ_{bi} and Fig. 11-3(d) for δ_{bb}.

$$R_b = \frac{\delta_{bi}}{\delta_{bb}}$$

4. Use the reciprocity,

$$\delta_{bi} = \delta_{ib}$$

to obtain

$$R_b = \frac{\delta_{ib}}{\delta_{bb}} \tag{11-7}$$

Note that in Eq. 11-7 the numerator δ_{ib} represents the ordinate of the deflection curve of the primary beam ac caused by applying a unit force at b. The denominator δ_{bb} is only a special case of δ_{ib}, i.e., $\delta_{bb} = \delta_{ib}$ if $i = b$, as shown in Fig. 11-3(d). Each ordinate of the curve of Fig. 11-3(d) divided by δ_{bb} will give the corresponding influence ordinate for R_b (see Eq. 11-7) as shown in Fig. 11-3(e).

Referring to Fig. 11-3(e), we note that at b, $\delta_{bb}/\delta_{bb} = 1$. Hence the influence line for R_b is nothing more than the deflected structure resulting from removal of the support at b and introduction in its place of a unit deflection along the line of reaction.

We have hitherto used the reaction of a support as illustration. This, then, is the Müller-Breslau principle and may be stated as:

The ordinates of the influence line for any element (reaction, axial force, shear, or moment) *of a structure equal those of the deflection curve obtained by removing the restraint corresponding to that element from the structure and introducing in its place a unit load divided by the deflection at the point of application of the unit load.*

This may be rephrased as:

The deflected structure resulting from a unit displacement corresponding to the action for which the influence line is desired gives the influence line for that action.

It may be noted that when applying the Müller-Breslau principle, if the actual structure is statically indeterminate to more than the first degree, the primary structure which remains after the removal of a redundant will still be statically indeterminate. This does not change our general procedure. However, the statically indeterminate primary structure must first be analyzed by one of the methods in Chs. 9 and 10 before we can compute the necessary deflections.

The Müller-Breslau principle is applicable to any type of structure—beam, truss, or rigid frame; statically determinate or indeterminate. The reader will find the virtual work method (Ch. 6) for obtaining influence lines for statically determinate beams and the Müller-Breslau principle are applied identically. In case of indeterminate structures this principle is limited to structures of elastic material obeying Hooke's law.

11-4. SKETCH OF INFLUENCE LINES

The Müller-Breslau principle provides a very convenient method for sketching qualitative influence lines for indeterminate structures and is the basis for certain experimental model analyses.

1. In the simplest case the influence line for a reaction component can be sketched by removing the restraint and allowing the reaction to move through a *unit displacement*. The deflected structure will then be the influence line for the reaction.

Thus the dotted line in Fig. 11-4(a) shows the influence line for the vertical reaction at support a of the three-span continuous beam. In Fig. 11-4(b) the dotted line indicates the influence line for the fixed-end moment at support a of the fixed-end beam ab. The curve is obtained by replacing the fixed support

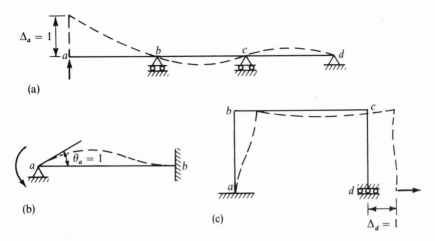

(a)

(b)

(c)

Fig. 11-4

at a with a hinge support and by introducing a unit rotation. In Fig. 11-4(c) is shown the construction of the influence line for the horizontal reaction component at the fixed support d of a portal frame. Note that the fixed support at d is replaced by a roller and slide acted on by a horizontal force so as to produce a unit horizontal displacement.

Note the following:

a. The vertical deflections of the structure will be influence line ordinates for the vertical loads on the structure.

b. The horizontal deflections of the structure will be influence line ordinates for the horizontal loads on the structure.

c. The rotation of the tangents of the structure will be influence line ordinates for the moment load on the structure.

2. The moment influence line for a section of a beam or rigid frame can be drawn by cutting the section and allowing a pair of equal and opposite moments to produce a *unit relative rotation* (but no relative translation) for the two sides of the section considered. The deflected structure will then be the influence line for the moment. Thus the influence line for the moment at the midspan section of a three-span continuous beam is the dotted line in Fig. 11-5.

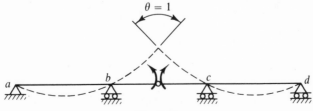

Fig. 11-5

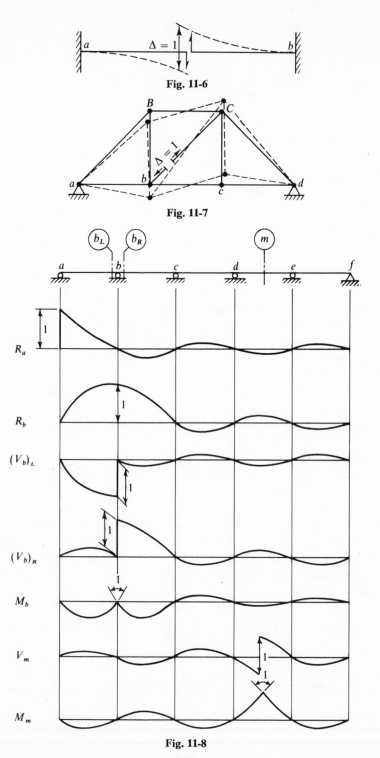

Fig. 11-6

Fig. 11-7

Fig. 11-8

228

3. The shear influence line for a section of a beam or rigid frame can be drawn by cutting the section and applying a pair of equal and opposite shearing forces to produce a *unit relative transverse displacement* (but no relative rotation) for the two sides of the section considered. The deflected structure will then be the influence line for the shear. The influence line for the shear at the midspan section of the fixed-end beam *ab* is shown in Fig. 11-6 by the dotted lines.

4. To obtain the influence line for the axial force in a bar, we cut the bar and apply a pair of equal and opposite axial forces so as to cause a *unit relative axial displacement* for the two cut ends. The deflected structure will give the desired influence line.

Figure 11-7 serves as a simple illustration of obtaining the influence line for the bar force in *bC* of the indeterminate truss. The vertical components of the panel point deflections are thus the influence ordinates for the vertical panel loads.

For highly indeterminate continuous beams or rigid frames the technique of sketching qualitative influence lines, based on the Müller-Breslau principle, is extremely useful in determining the loading patterns for design. Fig. 11-8 shows typical influence lines for a five-span continuous beam.

The sketches in Fig. 11-8 indicate that if a maximum R_a is desired, then spans *ab*, *cd*, and *ef* should be loaded; if the maximum values of reaction, of shear, and of bending moment at *b* are desired, then spans *ab*, *bc*, and *de* should be loaded.

Figure 11-9(a) shows the influence line for the positive moment at the midspan section of *A3-B3* of the frame shown. The uniform loading pattern for

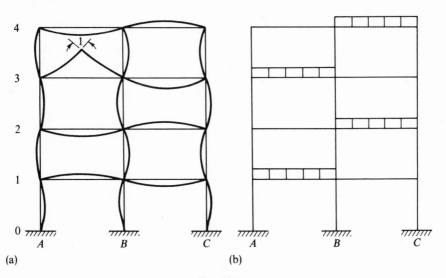

(a)   (b)

Fig. 11-9

obtaining the maximum positive moment of this section is shown in Fig. 11-9(b).

11-5. NUMERICAL EXAMPLES

Example 11-1. Construct by the Müller-Breslau principle the influence line for the reaction at support b of Fig. 11-10(a).

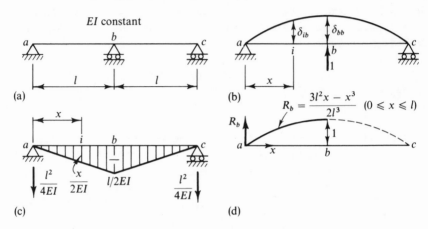

Fig. 11-10

We begin by removing the support at b and placing a unit load along the line of reaction, as shown in Fig. 11-10(b). The ordinates of the resulting deflection curve, in Fig. 11-10(b), divided by δ_{bb} give the corresponding influence ordinates for the reaction at b, called R_b.

Probably the easiest method for computing the ordinate of the curve in Fig. 11-10(b) at any point i, distance x from the left end, is the conjugate beam method, shown in Fig. 11-10(c).

Thus,

$$\delta_{ib} = \left(\frac{l^2}{4EI}\right)(x) - \left(\frac{x}{2EI}\right)\left(\frac{x}{2}\right)\left(\frac{x}{3}\right) = \frac{3l^2 x - x^3}{12EI} \qquad (0 \le x \le l)$$

Substituting $x = l$ in the above expression gives

$$\delta_{bb} = \frac{l^3}{6EI}$$

The influence ordinate for any point i $(0 \le x \le l)$ is governed by the equation

$$R_b = \frac{\delta_{ib}}{\delta_{bb}} = \frac{3l^2 x - x^3}{2l^3}$$

as shown in Fig. 11-10(d). Because of symmetry about the middle support, we can accomplish the other half, as shown by the dotted line in Fig. 11-10(d).

Example 11-2. The influence lines for other elements may easily be deduced from the influence line of R_b, previously obtained. For instance, let us find the influence line for the moment at b, called M_b, for the beam in Fig. 11-10(a) by using the results of the preceding example.

We observe that the influence line for M_b, which will be symmetrical about the middle support, may be obtained by solving the reaction at a and then taking the moment about b. For a unit load placed in the left span, as shown in Fig. 11-11(a), by taking

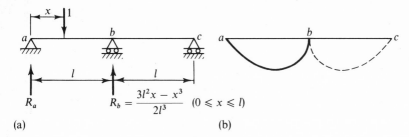

$$R_b = \frac{3l^2 x - x^3}{2l^3} \quad (0 \leqslant x \leqslant l)$$

(a)   (b)

Fig. 11-11

$$\sum M_c = 0 \qquad 2R_a l - (1)(2l - x) + \left(\frac{3l^2 x - x^3}{2l^3}\right)(l) = 0$$

we obtain $\qquad\qquad R_a = \dfrac{4l^3 - 5l^2 x + x^3}{4l^3} \qquad (0 \leq x \leq l)$

from which

$$M_b = R_a l - (1)(l - x) = -\frac{x(l^2 - x^2)}{4l^2} \qquad (0 \leq x \leq l)$$

This is plotted in Fig. 11-11(b) the other half is indicated by the dotted line.

Example 11-3. Without reference to Example 11-1 or 11-2 compute the ordinates at 2-ft intervals of the influence line for the moment at the midspan section d of ab for the beam shown in Fig. 11-12(a).

The ability to resist moment at section d is first destroyed by inserting a pin. Unit couples are applied on each side of the pin to produce certain relative rotation, denoted by θ, between the two sides, as shown in Fig. 11-12(b). The conjugate beam and elastic load are then obtained, as shown in Fig. 11-12 (c). Note that we assume $EI = 1$ (EI will be cancelled out in the final stage of calculation); hence, the elastic load of Fig. 11-12(c) is the moment diagram based on Fig. 11-12(b).

Referring to Fig. 11-12(c), we may solve reactions R_a, R_c, and R_d by the equilibrium equations and the condition equation $M_b = 0$. Thus

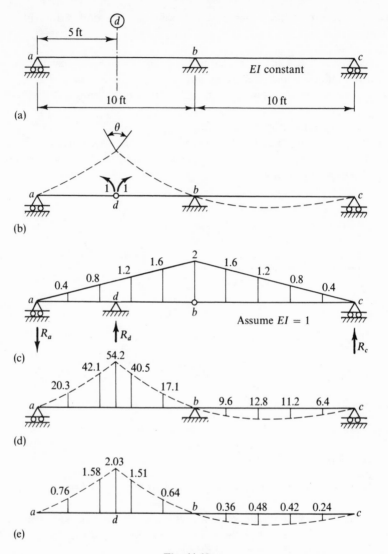

Fig. 11-12

$$\sum M_a = 0 \qquad 20R_c + 5R_d - \frac{(2)(20)}{2}(10) = 0 \qquad (11\text{-}8)$$

$$\sum M_c = 0 \qquad 15R_d - 20R_a - \frac{(2)(20)}{2}(10) = 0 \qquad (11\text{-}9)$$

$$M_b = 0 \qquad 10R_c - \frac{(2)(10)}{2}\left(\frac{10}{3}\right) = 0 \qquad (11\text{-}10)$$

Solving Eqs. 11-8, 11-9, and 11-10 simultaneously, we obtain

$$R_a = 10 \qquad R_c = 3.33 \qquad R_d = 26.67$$

Note that R_d is the shear difference between the left and right sides at section d of the conjugate beam and thus equals the relative rotation θ between corresponding portions of the distorted beam in Fig. 11-12(b). The various moments of the conjugate beam at 2-ft intervals and at point d which correspond to the deflections of the distorted beam in Fig. 11-12(b) are computed as shown in Fig. 11-12(d).

These values divided by 26.67 (so as to make $\theta = 1$) will give the ordinates of the influence line for the moment at section d, as shown in Fig, 11-12(e).

Example 11-4. Compute the influence ordinates at 2-ft intervals for the shear at the midspan section d of ab for the beam shown in Fig. 11-13(a) without reference to Example 11-1 or 11-2.

We start by removing the shearing resistance at section d without imparing the capacity for resistance to moment. This can be accomplished by cutting the beam and inserting a slide device. Next, we apply a pair of equal and opposite unit forces to produce certain relative vertical displacement, denoted by s, between the two cut ends without causing relative rotation, as shown in Fig. 11-13(b). Also indicated in Fig. 11-13(b) are the induced reactions and moments at the cut ends required by equilibrium.

The conjugate beam together with the elastic loads is shown in Fig. 11-13 (c). Attention should be paid to the elastic load M_d acting at d. This is necessary since the relative vertical displacement (without relative rotation) between the two cut ends at section d of the distorted beam in Fig. 11-13(b) requires a moment difference (without a shear difference) for the two sides at d of the conjugate beam shown in Fig. 11-13(c). This can be fulfilled only by applying a moment at d for the conjugate beam. Referring to Fig. 11-13(c), we solve reactions R_a, R_c, and M_d by the equilibrium equations and the condition equation $M_b = 0$. Thus

$$\sum M_a = 0 \qquad 20R_c + M_d - \frac{(10)(20)}{2}(10) = 0 \qquad (11\text{-}11)$$

$$\sum M_c = 0 \qquad 20R_a - M_d - \frac{(10)(20)}{2}(10) = 0 \qquad (11\text{-}12)$$

$$M_b = 0 \qquad 10R_c - \frac{(10)(10)}{2}\left(\frac{10}{3}\right) = 0 \qquad (11\text{-}13)$$

Solving Eqs. 11-11, 11-12, and 11-13 simultaneously, we obtain

$$R_a = 83.33 \qquad R_c = 16.67 \qquad M_d = 666.67$$

Note that M_d is the moment difference between the left and right sides at section d of the conjugate beam and, therefore, equals the relative deflection s between the corresponding portions of the distorted beam shown in Fig. 11-13 (b). The various moments of the conjugate beam at 2-ft intervals and at d are then computed, as shown in Fig. 11-13(d).

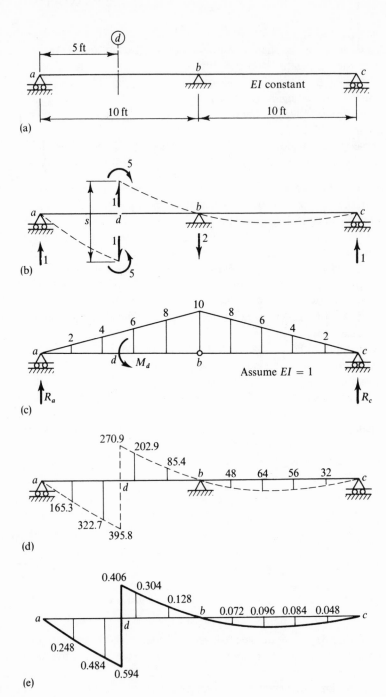

(a)

(b)

(c)

(d)

(e)

Fig. 11-13

234

These values divided by $M_d = 666.67$ (so as to make $s = 1$) will give the ordinates of the shear influence line for section d, as shown in Fig. 11-13(e).

Example 11-5. Construct the influence line for the reaction at support c of the truss shown in Fig. 11-14(a). Assume $E = $ a constant and $L(\text{ft})/A(\text{in.}^2) = 1$ for all members.

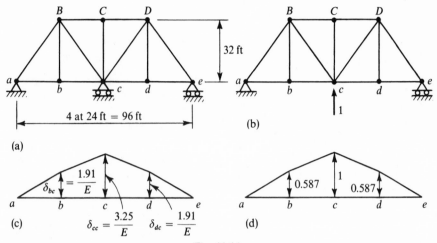

(a)

(b)

(c)

(d)

Fig. 11-14

We remove the support at c and introduce in its place a unit force, as shown in Fig. 11-14(b). The vertical deflections of all of the lower-chord panel points of the distorted truss are then computed as given in Table 11-1, in which u_b is

TABLE 11-1

Member	$\dfrac{L}{A}$	u_b	u_c	$\dfrac{u_b u_c L}{A}$	$\dfrac{u_c^2 L}{A}$
ab	1	$-9/16$	$-3/8$	$+27/128$	$+9/64$
bc	1	$-9/16$	$-3/8$	$+27/128$	$+9/64$
cd	1	$-3/16$	$-3/8$	$+9/128$	$+9/64$
be	1	$-3/16$	$-3/8$	$+9/128$	$+9/64$
BC	1	$+3/8$	$+3/4$	$+36/128$	$+36/64$
CD	1	$+3/8$	$+3/4$	$+36/128$	$+36/64$
aB	1	$+15/16$	$+5/8$	$+75/128$	$+25/64$
Bb	1	-1	0	0	0
Bc	1	$+5/16$	$-5/8$	$-25/128$	$+25/64$
Cc	1	0	0	0	0
cD	1	$-5/16$	$-5/8$	$+25/128$	$+25/64$
Dd	1	0	0	0	0
De	1	$+5/16$	$+5/8$	$+25/128$	$+25/64$
			Σ	$+1.91$	$+3.25$

the bar force in any member due to a unit load acting upward at joint b, and u_c is the bar force in the same member due to a unit load acting upward at joint c. The results are shown in Fig. 11-14(c). Note that in this particular case $\delta_{bc} = \delta_{dc}$.

$$\delta_{bc} = \sum \frac{u_b u_c L}{AE} = \frac{1.91}{E}$$

$$\delta_{cc} = \sum \frac{u_c^2 L}{AE} = \frac{3.25}{E}$$

The ordinates of the deflection curve in Fig. 11-14(c) divided by the value of δ_{cc} will give the influence ordinates for the reaction at support c, as shown in Fig. 11-14(d).

Example 11-6. Find the influence line for the bar force in member BC of the truss in Fig. 11-14(a).

We can accomplish this easily if we use the results of the preceding example. As a unit load is placed at panel point b (see Fig. 11-15(a)), we have

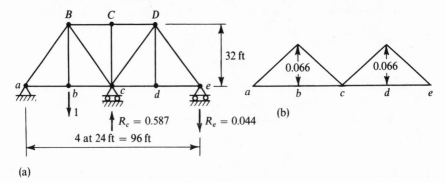

(a)

(b)

Fig. 11-15

$$R_c = 0.587 \quad \text{(up)}$$

From $\sum M_a = 0$, $R_e = 0.044 \quad \text{(down)}$

By taking the section and applying $\sum M_c = 0$, we obtain the bar force in BC as

$$S_{BC} = \frac{(0.044)(48)}{32} = +0.066 \quad \text{(tension)}$$

Because of symmetry, $S_{BC} = +0.066$ when the unit load is placed at panel point d. Note also that $S_{BC} = 0$ when the unit load is at points a, c, and e. We thus complete the influence line for S_{BC}, as shown in Fig. 11-15(b).

11-6. SOLVING INFLUENCE ORDINATES FOR
        STATICALLY INDETERMINATE STRUCTURES BY
        CONSISTENT DEFORMATIONS

For highly indeterminate structures the influence ordinates for various functions may be found by using a computer for a number of equations based on consistent deformations. Let us consider the four-span continuous beam shown in Fig. 11-16(a). Find the influence line for the reaction at support b, called R_b.

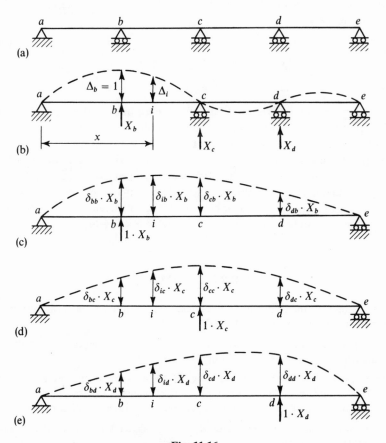

Fig. 11-16

We start by removing support b and applying to it a force X_b along the line of reaction so as to produce a unit displacement at b. Then, by the Müller-Breslau principle the elastic line of the distorted beam, shown by the dotted line in Fig. 11-16(b), will be the influence line for R_b. Also indicated in Fig. 11-16(b) are the induced reactions at supports c and d, called X_c and X_d, respectively, and the ordinate of the curve at any point i, called Δ_i, distance x from the left end.

To obtain the value of Δ_i by the method of consistent deformations, we regard the indeterminate beam in Fig. 11-16(b) as a simple beam ae (primary structure) subjected to forces X_b, X_c, and X_d the effects of which can be separated by the principle of superposition into three basic cases, as shown in Figs. 11-16(c), (d), and (e) respectively. Thus

$$\Delta_i = \delta_{ib} X_b + \delta_{ic} X_c + \delta_{id} X_d$$

in which the unknowns X_b, X_c, and X_d can be solved by three compatibility equations, namely:

$$\Delta_b = \delta_{bb} X_b + \delta_{bc} X_c + \delta_{bd} X_d = 1 \tag{11-14}$$

$$\Delta_c = \delta_{cb} X_b + \delta_{cc} X_c + \delta_{cd} X_d = 0 \tag{11-15}$$

$$\Delta_d = \delta_{db} X_b + \delta_{dc} X_c + \delta_{dd} X_d = 0 \tag{11-16}$$

In a similar manner, we can find the influence line for the reaction at each of the other supports. This done, the influence lines for the moment and shear at various points can be deduced from them.

Without the aid of a calculating machine, the Müller-Breslau procedures for quantitative influence lines are often tedious, especially if an indeterminate structure remains after establishing the initial displacement conditions along the given redundant. A much faster way by hand calculation is the moment-distribution procedure, which we shall discuss in Sec. 14-6.

PROBLEMS

11-1. Compute the ordinates at 2-ft intervals of the influence line for the reaction at a of the beam shown in Fig. 11-17(a).

11-2. Compute the ordinates at 3-ft intervals of the influence line for the end moment at a of the beam shown in Fig. 11-17(b).

11-3. By reference to the beam in Fig. 11-17(c), compute the ordinates at 2-ft intervals of the influence lines for the following elements: (a) the reaction at support a; (b) the moment at b; and (c) the shear at the midspan section of ab.

14-4. Sketch the influence lines for R_a, R_c, M_c, V_c (left), V_c (right), and M and V for the midspan section of bc of the six-span continuous beam shown in Fig. 11-17 (d).

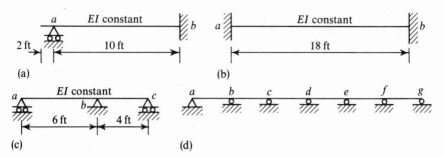

Fig. 11-17

11-5. (a) Assume L/A to be constant for all bars. Prepare an influence line for the bar force in cd of the truss shown in Fig. 11-18. (b) Use the results of (a) to prepare the influence line for the bar force in BC.

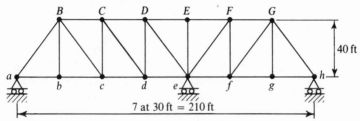

Fig. 11-18

11-6. Sketch the influence lines for the shear and moment in the midspan section of member ab of the rigid frame shown in Fig. 11-19.

11-7. Sketch the influence line for moment at end b of member ab in Fig. 11-19.

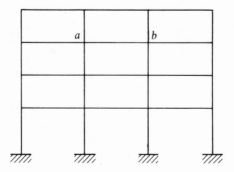

Fig. 11-19

12

ANALYSIS OF STATICALLY
INDETERMINATE BEAMS AND RIGID
FRAMES BY THE SLOPE-DEFLECTION METHOD

12-1. GENERAL

Throughout the previous chapters the methods of analyzing statically indeterminate structures have used forces (reactions or internal stresses) as the basic unknowns. These are often referred to as the force or action methods. Displacements, however, may be used equally well as unknowns. Methods using displacements as the basic unknowns are called displacement methods. One of the important displacement methods is the *method of slope deflection* based on determining the rotations and deflections of various joints from which the end moments for each member are found.

The slope-deflection method may be used in analyzing all types of statically indeterminate beams and rigid frames composed of prismatic or nonprismatic members. However, in this chapter we will discuss exclusively beams and rigid frames made of prismatic members.

12-2. BASIC SLOPE-DEFLECTION EQUATIONS

The basis of the slope-deflection method lies in the *slope-deflection equations*, which express the end moments of each member in terms of the end distortions of that member.

Consider member *ab* shown in Fig. 12-1, which is isolated from a loaded

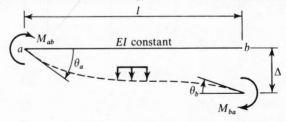

Fig. 12-1

240

statically indeterminate beam or rigid frame (not shown). The member is deformed (see the dashed line) with end rotations θ_a and θ_b and relative deflection Δ between the ends. Obviously the induced end moments at a and b, called M_{ab} and M_{ba}, are related to the elastic distortions at both ends as well as to the load on span ab, if any. Thus

$$M_{ab} = f(\theta_a, \theta_b, \Delta, \text{load on span}) \qquad (12\text{-}1)$$

$$M_{ba} = g(\theta_a, \theta_b, \Delta, \text{load on span}) \qquad (12\text{-}2)$$

where f and g are symbols for functions.

To find the expressions of Eqs. 12-1 and 12-2, let us first establish the following sign convention for slope deflection:

1. The moment acting on the end of a member (not joint) is positive when clockwise.
2. The rotation at the end of a member is positive when the tangent to the deformed curve at the end rotates clockwise from its original position.
3. The relative deflection between ends of a member is positive when it corresponds to a clockwise rotation of the member (the straight line joining the ends of the elastic curve).

All signs of end distortions and moments shown in Fig. 12-1 are positive. The sign conventions established here are purely arbitrary and could be replaced by any other convenient system; but once these conventions have been adopted, we will restrict ourselves to this system.

Next, let us refer to Fig. 12-1 and observe that the end moments M_{ab} and M_{ba} may be considered as the algebraic sum of four separate effects:

1. The moment due to end rotation θ_a while the other end b is fixed;
2. The moment due to end rotation θ_b while end a is fixed;
3. The moment due to a relative deflection Δ between the ends of the member without altering the existing slopes of tangents at the ends; and
4. The moment caused by placing the actual loads on the span without altering the existing end distortions.

In each of the above cases the corresponding end moments can by evaluated as follows:

1. Consider member ab supported as shown in Fig. 12-2(a). The dashed lines indicate the deformed shape. Note that end a is rotated through an angle θ_a, whereas end b is fixed ($\theta_b = 0$); there is no relative end displacement between a and $b(\Delta = 0)$. The corresponding end moments at a and b, denoted respectively by M'_{ab} and M'_{ba}, can best be found by the method of conjugate beam, which is shown in Fig. 12-2(b) with the moment diagram divided by EI as its elastic loads and θ_a as its reaction so that the positive shear in the con-

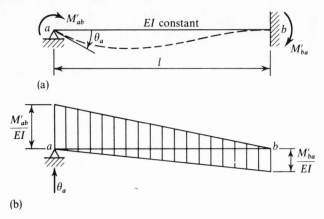

(a)

(b)

Fig. 12-2

jugate beam gives the desired positive slope in actual beam. From equilibrium conditions:

$$\sum M_a = 0 \qquad \left(\frac{M'_{ab}l}{2EI}\right)\left(\frac{l}{3}\right) - \left(\frac{M'_{ba}l}{2EI}\right)\left(\frac{2l}{3}\right) = 0 \tag{12-3}$$

$$\sum M_b = 0 \qquad (\theta_a l) - \left(\frac{M'_{ab}l}{2EI}\right)\left(\frac{2l}{3}\right) + \left(\frac{M'_{ba}l}{2EI}\right)\left(\frac{l}{3}\right) = 0 \tag{12-4}$$

From Eq. 12-3

$$M'_{ba} = \tfrac{1}{2} M'_{ab} \tag{12-5}$$

Substituting Eq. 12-5 in Eq. 12-4 gives

$$M'_{ab} = \frac{4EI\theta_a}{l} \tag{12-6}$$

Thus

$$M'_{ba} = \frac{2EI\theta_a}{l} \tag{12-7}$$

2. Consider member ab supported and deformed as shown in Fig. 12-3 where end b is rotated through an angle θ_b and end a is fixed. The corresponding end moment at b, called M''_{ba}, and the moment at a, called M''_{ab}, are obtained similarly:

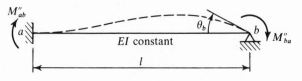

Fig. 12-3

$$M''_{ab} = \tfrac{1}{2} M''_{ba} \tag{12-8}$$

$$M''_{ab} = \frac{2EI\theta_b}{l} \tag{12-9}$$

$$M''_{ba} = \frac{4EI\theta_b}{l} \tag{12-10}$$

3. To find the moments developed at the ends as the result of a pure relative deflection Δ between the two ends without causing end rotations, let us consider the fixed-end beam in Fig. 12-4(a). Because of the symmetry of the deformation with reference to the midpoint of the member (see the dashed lines in Fig. 12-4(a)), the two end moments must be equal. Thus if we let M'''_{ab} and M'''_{ba} be the end moments at a and b respectively, we have

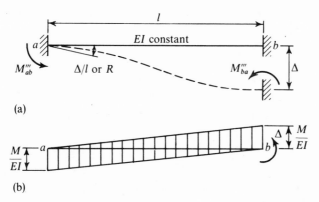

(a)

(b)

Fig. 12-4

$$M'''_{ab} = M'''_{ba} = -M$$

The negative sign indicates that M'''_{ab} and M'''_{ba} are counter-clockwise. The value of M may be found by the method of conjugate beam, as shown in Fig. 12-4(b). Note that, besides the distributed elastic loads of the M/EI diagram, a couple acts at end b equal to Δ corresponding to the deflection at b of the base structure. From $\sum M = 0$ we have

$$\left(\frac{Ml}{2EI}\right)\left(\frac{l}{3}\right) - \Delta = 0$$

or $$M = \frac{6EI\Delta}{l^2}$$

Thus the moments developed at the ends of a member due to a pure relative end displacement are given by

$$M_{ab}''' = M_{ba}''' = -\frac{6EI\Delta}{l^2} \tag{12-11}$$

4. Finally, the moments induced at the ends of a member without causing end distortions when the external loads are placed on the span are nothing more than the fixed-end moments, usually denoted by M_{ab}^F and M_{ba}^F.

Summing up the above four elements, we have:

$$M_{ab} = M_{ab}' + M_{ab}'' + M_{ab}''' \pm M_{ab}^F$$

$$M_{ba} = M_{ba}' + M_{ba}'' + M_{ba}''' \pm M_{ba}^F$$

Using Eqs. 12-6, 12-7, 12-9, 12-10, and 12-11, we find

$$M_{ab} = \frac{4EI\theta_a}{l} + \frac{2EI\theta_b}{l} - \frac{6EI\Delta}{l^2} \pm M_{ab}^F$$

$$M_{ba} = \frac{2EI\theta_a}{l} + \frac{4EI\theta_b}{l} - \frac{6EI\Delta}{l^2} \pm M_{ba}^F$$

Rearranging the above gives:

$$M_{ab} = 2E\frac{I}{l}\left(2\theta_a + \theta_b - 3\frac{\Delta}{l}\right) \pm M_{ab}^F \tag{12-12}$$

$$M_{ba} = 2E\frac{I}{l}\left(2\theta_b + \theta_a - 3\frac{\Delta}{l}\right) \pm M_{ba}^F \tag{12-13}$$

which are the basic equations of slope deflection for a general deformed member of uniform cross section. The equations express end moment M_{ab} and M_{ba} in terms of the end slopes (θ_a, θ_b), the relative deflection between the two ends (Δ), and the loading on the span ab.

If we let

$$\frac{I}{l} = K \qquad \frac{\Delta}{l} = R$$

K being the *stiffness factor of the member* and R the *rotation of the member* (see Fig. 12-4(a)), the equations become:

$$M_{ab} = 2EK(2\theta_a + \theta_b - 3R) \pm M_{ab}^F \tag{12-14}$$

$$M_{ba} = 2EK(2\theta_b + \theta_a - 3R) \pm M_{ba}^F \tag{12-15}$$

The signs and values of M_{ab}^F and M_{ba}^F depend on the loading condition on span ab. If the member ab carries no load itself, then $M_{ab}^F = M_{ba}^F = 0$. See Secs. 9-2 and 10-2 for the evaluation of the fixed-end moments. For reference, the fixed-end moments for a straight member of constant EI due to the common types of loading are given in Table 12-1.

TABLE 12-1

M_{ab}^F	Loading Case	M_{ba}^F
$-\dfrac{Pl}{8}$		$+\dfrac{Pl}{8}$
$-\dfrac{Pcd^2}{l^2}$		$+\dfrac{Pc^2d}{l^2}$
$-\alpha(1-\alpha)Pl$		$+\alpha(1-\alpha)Pl$
$-\dfrac{wl^2}{12}$		$+\dfrac{wl^2}{12}$
$-\dfrac{wl^2}{20}$		$+\dfrac{wl^2}{30}$
$-\dfrac{11wl^2}{192}$		$+\dfrac{5wl^2}{192}$
$-\dfrac{wl^2}{15}$		$+\dfrac{wl^2}{15}$
$+\dfrac{Md(2c-d)}{l^2}$		$+\dfrac{Mc(2d-c)}{l^2}$

12-3. PROCEDURE OF ANALYSIS BY THE SLOPE-DEFLECTION METHOD

The slope-deflection method consists of writing a series of slope-deflection equations expressing the end moments for all members in terms of the slope (rotation) and the deflection (relative translation) of various joints, or of quantities proportional to them, and of solving these unknown displacements by a number of equilibrium equations, which these end moments must satisfy. Once the displacements have been determined, the end moments for each member may be figured back. The solution thus obtained is unique, since it satisfies the equilibrium equations and end conditions (compatibility conditions) embedded in the slope-deflection equations.

To illustrate, let us consider the frame in Fig. 12-5(a). First we draw the free-body diagrams for all members as shown in Fig. 12-5(b), where the unknown end moments for each member are assumed positive, i.e., acting clockwise according to our sign convention.

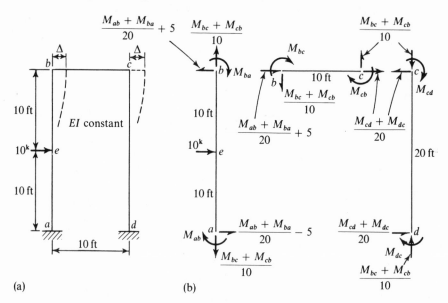

(a)   (b)

Fig. 12-5

Next, we observe that ends a and d of the frame are fixed and will undergo no rotation ($\theta_a = \theta_d = 0$) nor linear displacement. Joint b, owing to the restriction of length ab (we neglect the small change in length in ab due to the axial forces) and the support a, cannot move otherwise but rotates about a. However, since the deformations of the frame are extremely small as compared to the length, we may replace arc length with tangent length without appreciable error. With joint b, and therefore c, moving a horizontal distance Δ to the right (see Fig. 12-5(a)), there is a relative deflection Δ between joints a and b and also joints c and d. There is no relative deflection between joints b and c if the small lengthening or shortening in ab or cd, caused primarily by axial forces, are neglected. There is some joint rotations at b and c. Attention should be paid to the fact that, when a rigid frame is deformed, each rigid joint is considered to rotate as a whole. For instance, members ba and bc rotate the same angle θ_b at joint b. Similarly, members cb and cd rotate the same angle θ_c at joint c.

To determine the end moments for each member shown in Fig. 12-5(b), we write a series of slope-deflection equations as follows:

$$M_{ab} = 2EK_{ab}(\theta_b - 3R) - M_{ab}^F = 2E\left(\frac{I}{20}\right)(\theta_b - 3R) - 25$$

$$M_{ba} = 2EK_{ab}(2\theta_b - 3R) + M_{ba}^F = 2E\left(\frac{I}{20}\right)(2\theta_b - 3R) + 25$$

$$M_{bc} = 2EK_{bc}(2\theta_b + \theta_c) = 2E\left(\frac{I}{10}\right)(2\theta_b + \theta_c)$$

$$M_{cb} = 2EK_{bc}(2\theta_c + \theta_b) = 2E\left(\frac{I}{10}\right)(2\theta_c + \theta_b) \qquad (12\text{-}16)$$

$$M_{cd} = 2EK_{cd}(2\theta_c - 3R) = 2E\left(\frac{I}{20}\right)(2\theta_c - 3R)$$

$$M_{ac} = 2EK_{cd}(\theta_c - 3R) = 2E\left(\frac{I}{20}\right)(\theta_c - 3R)$$

Involved in the above expressions are three unknowns: $\theta_b, \theta_c,$ and R (or $\Delta/20$). These can be solved by the three equations of statics the end moments must satisfy.

By taking joints b and c as free bodies, we obtain immediately two equilibrium equations:

$$\sum M_{\text{joint } b} = 0 \quad \text{or} \quad M_{ba} + M_{bc} = 0 \qquad (12\text{-}17)$$

$$\sum M_{\text{joint } c} = 0 \quad \text{or} \quad M_{cb} + M_{cd} = 0 \qquad (12\text{-}18)$$

Usually we have as many joint equilibrium equations as the number of joint rotations θs involved. Now, with the member rotation R unknown, a third equation must be secured from the equilibrium of the structure. Referring to Fig. 12-5, by taking the whole frame as free body, we see that the horizontal shear in ends a and d must balance the horizontal external force acting on the frame. Thus,

$$10 + \left(\frac{M_{ab} + M_{ba}}{20} - 5\right) + \left(\frac{M_{cd} + M_{dc}}{20}\right) = 0$$

or

$$M_{ab} + M_{ba} + M_{cd} + M_{dc} + 100 = 0 \qquad (12\text{-}19)$$

Before we try to substitute the expressions of Eq. 12-16 in Eqs. 12-17, 12-18, and 12-19, we should note that if our purpose is to determine the end moments, but not to obtain the exact values of the slope and deflection of each joint, then we may substitute the relative values for the coefficients $2EI/l$, usually some simple integers, in the slople-deflection equations in order to facilitate the calculation. This can be done because such a substitution will only magnify the values of θ and Δ but will not affect the final result of the end moments. Thus if we set

$$2E\frac{I}{20} = 1$$

accordingly, $$2E\left(\frac{I}{10}\right) = 2$$

then the moment expressions of Eq. 12-16 become

$$M_{ab} = \theta_b - 3R - 25$$
$$M_{ba} = 2\theta_b - 3R + 25$$
$$M_{bc} = 2(2\theta_b + \theta_c)$$
$$M_{cb} = 2(2\theta_c + \theta_b) \qquad (12\text{-}20)$$
$$M_{cd} = 2\theta_c - 3R$$
$$M_{dc} = \theta_c - 3R$$

Substituting these in Eqs. 12-17, 12-18, and 12-19 yields

$$6\theta_b + 2\theta_c - 3R + 25 = 0 \qquad (12\text{-}21)$$
$$2\theta_b + 6\theta_c - 3R = 0 \qquad (12\text{-}22)$$
$$3\theta_b + 3\theta_c - 12R + 100 = 0 \qquad (12\text{-}23)$$

Solving Eqs. 12-21, 12-22, and 12-23 simultaneously, we obtain

$$\theta_b = -1.20 \qquad \theta_c = 5.05 \qquad R = 9.30$$

Note that the values thus obtained are only the relative values of the slope and deflection for the various joints. They must be divided by the factor $2EI/20$ to give the absolute values of the slope and deflection.

To determine the end moments for each member of the frame, we put $\theta_b = -1.20, \theta_c = 5.05, R = 9.30$ in Eq. 12-20 and obtain

$$M_{ab} = -54.10 \text{ ft-kips} \qquad \text{(counter-clockwise)}$$
$$M_{ba} = -5.30 \text{ ft-kips} \qquad \text{(counter-clockwise)}$$
$$M_{bc} = +5.30 \text{ ft-kips} \qquad \text{(clockwise)}$$
$$M_{cb} = +17.80 \text{ ft-kips} \qquad \text{(clockwise)}$$
$$M_{cd} = -17.80 \text{ ft-kips} \qquad \text{(counter-clockwise)}$$
$$M_{dc} = -22.80 \text{ ft-kips} \qquad \text{(counter-clockwise)}$$

The answer diagram for the end actions for each member of the frame is shown in Fig. 12-6(a), which is based on Fig. 12-5(b).

The moment diagram for the frame is drawn as in Fig. 12-6(b). The moment is plotted on the compressive side of each member. In this particular case

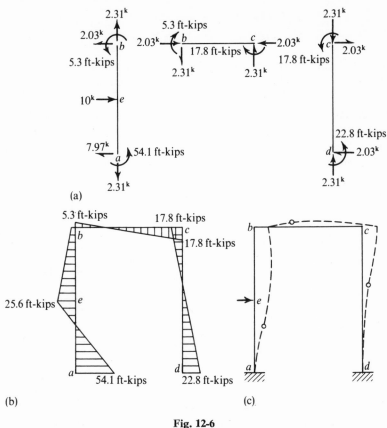

Fig. 12-6

each member has one point of inflection corresponding to the point of zero moment.

Finally, we can sketch the elastic curve of the deformed structure, as shown in Fig. 12-6(c), by using the values (or relative values) of the joint rotations and deflections together with the bending-moment diagram. Note particularly the following:

1. The elastic curve of the deformed frame bends according to the bending-moment diagram.

2. Both joints b and c deflect to the right the same horizontal distance.

3. Joint b rotates counter-clockwise while joint c rotates clockwise.

4. Since joints b and c are rigid, the tangents to the elastic curves ba and bc at b and the tangents to the elastic curves cb and cd at c should be perpendicular to each other so as to maintain the original formation at the joints of the unloaded frame.

12-4. ANALYSIS OF STATICALLY INDETERMINATE
BEAMS BY THE SLOPE-DEFLECTION METHOD

The application of the slope-deflection method in solving statically in-determinate beams will be illustrated in the following examples.

Example 12-1. Figure 12-7(a) shows a continuous, two-section beam, all the supports of which are immovable. We wish to draw the shear and bending-moment diagrams for the beam. The solution for the end moments and shears is contained in the following steps:

1. Since $2EK_{ab} = 2E(2I)/16 = EI/4$ and $2EK_{bc} = 2EI/12 = EI/6$, if we

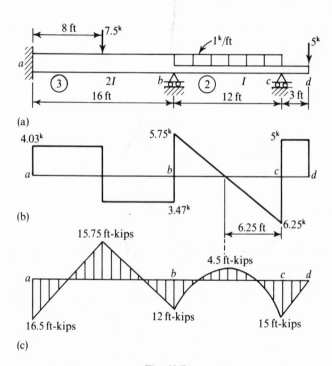

(a)

(b)

(c)

Fig. 12-7

let $2EK_{ab} = 3$, then $2EK_{bc} = 2$ relatively. The relative values of $2EK$ are shown encircled in Fig. 12-7(a).

2. By inspection $\theta_a = 0$ (fixed end at a) and $R_{ab} = R_{bc} = 0$ (immovable supports at a, b, c).

3. Calculate the fixed-end moments as follows:

$$M^F_{ab} = -\frac{(7.5)(16)}{8} = -15 \text{ ft-kips}$$

$$M^F_{ba} = +\frac{(7.5)(16)}{8} = +15 \text{ ft-kips}$$

$$M^F_{bc} = -\frac{(1)(12)^2}{12} = -12 \text{ ft-kips}$$

$$M^F_{cb} = +\frac{(1)(12)^2}{12} = +12 \text{ ft-kips}$$

4. Write the slope-deflection equations using the relative $2EK$ values.

$$M_{ab} = (3)(\theta_b) - 15 \tag{12-24}$$

$$M_{ba} = (3)(2\theta_b) + 15 = 6\theta_b + 15 \tag{12-25}$$

$$M_{bc} = (2)(2\theta_b + \theta_c) - 12 = 4\theta_b + 2\theta_c - 12 \tag{12-26}$$

$$M_{cb} = (2)(2\theta_c + \theta_b) + 12 = 2\theta_b + 4\theta_c + 12 \tag{12-27}$$

5. Involved in the above equations are two unknowns θ_b and θ_c that can be solved by two joint conditions; i.e.,

$$\sum M_{\text{joint } b} = M_{ba} + M_{bc} = 0 \tag{12-28}$$

$$M_{cb} = (5)(3) = 15 \tag{12-29}$$

Substituting Eqs. 12-25 and 12-26 in Eq. 12-28, and Eq. 12-27 in Eq. 12-29 yields

$$10\theta_b + 2\theta_c + 3 = 0 \tag{12-30}$$

and

$$2\theta_b + 4\theta_c - 3 = 0 \tag{12-31}$$

Solving for
$$\theta_b = -\tfrac{1}{2} \qquad \theta_c = 1$$

6. Substituting the above values in Step 4, we obtain

$$M_{ab} = -16.5 \text{ ft-kips}$$

$$M_{ba} = +12.0 \text{ ft-kips}$$

$$M_{bc} = -12.0 \text{ ft-kips}$$

$$M_{cb} = +15.0 \text{ ft-kips}$$

7. Having determined the end moments for each member, we can find the end shears and, therefore, the reactions.

$$R_a = V_{ab} = \frac{7.5}{2} + \frac{16.5 - 12}{16} = 4.03 \text{ kips} \qquad \text{(up)}$$

$$V_{ba} = \frac{-7.5}{2} + \frac{16.5 - 12}{16} = -3.47 \text{ kips} \qquad \text{(up)}$$

$$V_{bc} = \frac{12}{2} - \frac{15 - 12}{12} = 5.75 \text{ kips} \qquad \text{(up)}$$

$$R_b = 3.47 + 5.75 = 9.22 \text{ kips} \qquad \text{(up)}$$

$$V_{cb} = -\frac{12}{2} - \frac{15 - 12}{12} = -6.25 \text{ kips} \qquad \text{(up)}$$

$$V_{cd} = 5 \text{ kips} \qquad \text{(up)}$$

$$R_c = 6.25 + 5 = 11.25 \text{ kips} \qquad \text{(up)}$$

8. The shear and moment diagrams are now drawn, as shown in Figs. 12-7(b) and (c) respectively.

Example 12-2. Find the end moments for the four-span continuous beam shown in Fig. 12-8 resulting from the effect of 0.4 in. settlement at support c. Assume that $E = 30,000$ kips per in.², $I = 1,000$ in.⁴

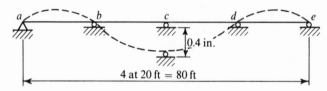

Fig. 12-8

Because of symmetry,

$$\theta_c = 0 \qquad \theta_e = -\theta_a \qquad \theta_d = -\theta_b \qquad R_{cd} = -R_{bc} \text{ (known)}$$

This problem involves only two unknowns, θ_a and θ_b, which can be determined by the two joint conditions

$$M_{ab} = 0 \qquad\qquad\qquad\qquad 2EK(2\theta_a + \theta_b) = 0 \quad (12\text{-}32)$$

$$M_{ba} + M_{bc} = 0 \qquad 2EK(2\theta_b + \theta_a) + 2EK\left[2\theta_b - (3)\frac{0.4}{(20)(12)}\right] = 0 \quad (12\text{-}33)$$

Solving Eqs. 12-32 and 12-33 gives

$$\theta_a = -0.0007 \qquad \theta_b = 0.0014$$

The moments at b and c can now be evaluated:

$$M_{ba} = (2)(30,000)\left(\frac{1,000}{240}\right)[(2)(0.0014) - 0.0007] \div 12$$

$$= 43.75 \text{ ft-kips}$$

$$M_{cb} = (2)(30,000)\left(\frac{1,000}{240}\right)\left[0.0014 - (3)\left(\frac{0.4}{240}\right)\right] \div 12$$

$$= -75 \text{ ft-kips}$$

12-5. ANALYSIS OF STATICALLY INDETERMINATE
RIGID FRAMES WITHOUT JOINT TRANSLATION
BY THE SLOPE-DEFLECTION METHOD

Some rigid frames, such as those shown in Figs. 12-9(a), (b), (c), and (d), are so constructed that translations of joints are prevented. Others, although capable of joint translation in construction, will undergo no joint translation because of the symmetry of the structure and the loading about a certain axis, such as those shown in Figs. 12-9(e), (f), and (g).

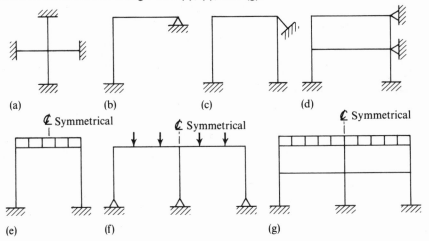

Fig. 12-9

In both cases

$$R = 0$$

in the equations of slope deflection so that the analysis is considerably simplified, as we shall see in the following examples.

Example 12-3. The end moments for the frame shown in Fig. 12-10 were solved by the methods of consistent deformation (Example 9-7) and least work (Examples 10-4 and 10-5) and will be re-solved by slope deflection.

The analysis is contained in the following steps:

1. The relative $2EK$ values for all members are shown encircled.
2. $\theta_a = \theta_d = R = 0$, and because of symmetry, $\theta_c = -\theta_b$.
3. $M_{bc}^F = -M_{cb}^F = -(1.2)(10)^2/12 = -10$ ft-kips.
4. Equations of slope-deflection are then given by

$$M_{ab} = -M_{dc} = (1)(\theta_b) \tag{12-34}$$

$$M_{ba} = -M_{cd} = (1)(2\theta_b) \tag{12-35}$$

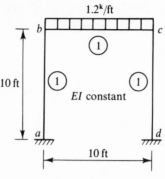

Fig. 12-10

$$M_{bc} = -M_{cb} = (1)(2\theta_b - \theta_b) - 10 \qquad (12\text{-}36)$$

5. There is only one unknown, θ_b, involved in this analysis, and it can be solved by

$$\sum M_{\text{joint } b} = M_{ba} + M_{bc} = 0 \qquad (12\text{-}37)$$

Substituting Eqs. 12-35 and 12-36 in Eq. 12-37 gives

$$3\theta_b - 10 = 0$$

from which $$\theta_b = 3.33$$

6. Going back to Step 4, we obtain

$$M_{ab} = -M_{dc} = 3.33 \text{ ft-kips}$$

$$M_{ba} = -M_{cd} = 6.67 \text{ ft-kips}$$

$$M_{bc} = -M_{cb} = -6.67 \text{ ft-kips}$$

Example 12-4. Analyze the frame in Fig. 12-11 if the support at a yields 0.0016 radian clockwise. Assume that $EI = 10,000 \text{ kips-ft.}^2$

$$M_{ab} = \frac{2EI}{10}(2\theta_a + \theta_b) = \frac{EI}{5}[(2)(0.0016) + \theta_b]$$

$$M_{ba} = \frac{2EI}{10}(2\theta_b + \theta_a) = \frac{EI}{5}(2\theta_b + 0.0016)$$

$$M_{bc} = \frac{2E(2I)}{20}(2\theta_b + \theta_c) = \frac{EI}{5}(2\theta_b) \qquad (\theta_c = 0)$$

$$M_{cb} = \frac{2E(2I)}{20}(2\theta_c + \theta_b) = \frac{EI}{5}(\theta_b)$$

The unknown θ_b is solved by

$$\sum M_{\text{joint } b} = M_{ba} + M_{bc} = 0$$

or $$4\theta_b + 0.0016 = 0$$

from which $$\theta_b = -0.0004$$

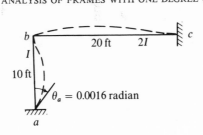

Fig. 12-11

With θ_b determined, all end moments can be figured as:

$$M_{ab} = \frac{10,000}{5}(0.0032 - 0.0004) = 5.6 \text{ ft-kips}$$

$$M_{ba} = \frac{10,000}{5}(-0.0008 + 0.0016) = 1.6 \text{ ft-kips}$$

$$M_{bc} = \frac{10,000}{5}(-0.0008) = -1.6 \text{ ft-kips}$$

$$M_{cb} = \frac{10,000}{5}(-0.0004) = -0.8 \text{ ft-kips}$$

The deformed structure is indicated by the dashed lines in Fig. 12-11.

12-6. ANALYSIS OF STATICALLY INDETERMINATE RIGID FRAMES WITH ONE DEGREE OF FREEDOM OF JOINT TRANSLATION BY THE SLOPE-DEFLECTION METHOD

Figure 12-12 shows several examples of rigid frames with one degree of freedom of joint translation. In each of these examples, if the translation of one joint is given or assumed, the translation of all other joints can be deduced from it.

For instance, suppose that, in the frame in Fig. 12-12(a), joint a moves to a', a distance Δ. Since we neglect any slight change in the length of a member due to axial forces, and since the rotations of members are small, joint a moves essentially perpendicular to member Aa, and in this case, horizontally. Similarly, joint b at the top of column Bb must move horizontally. Furthermore, as b is the end of member ab and the change of axial length of ab is neglected, the horizontal movement of b (see bb' in Fig. 12-12(a)) must also be Δ; i.e., $aa' = bb' = \Delta$.

In the same manner, we reason that if joint a in Fig. 12-12(b) moves a horizontal distance Δ to a', the tops of all other columns must have the same horizontal displacement.

Let us now consider the case of Fig. 12-12(c) in which the frame is acted on by a lateral force at the top of column Aa. Joint a cannot move otherwise

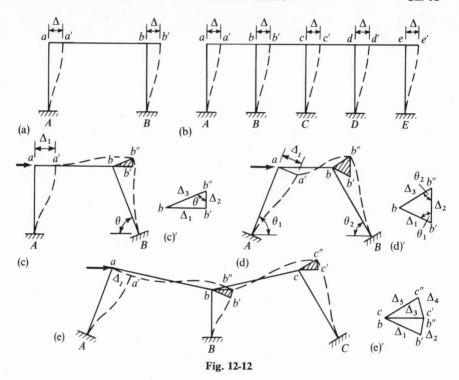

Fig. 12-12

but horizontally, say, a distance Δ_1 to a'. To find the final location of joint b, we imagine that the frame is temporarily disconnected at b. Point b, being the end of member ab, will move to b' the displacement Δ_1, if free from other effects except that due to the movement of a. However, b is also the end of member Bb; the final position of b, called b'', must be determined by two arcs that restrict the motion of b'—one from a' with a radius equal to ab (or $a'b'$) and the other from B with a radius equal to Bb. Since the deformations of the frame are very small in proportion to the length, it is permissible to substitute the tangents for the arcs, as shown in Fig. 12-12(c). The displacement diagram is shown separately in Fig. 12-12(c)′ for which we note that Δ_1 is the relative displacement between ends of member Aa; Δ_2 that between the ends of ab; and Δ_3 that between the ends of Bb. The relationship between Δ_1, Δ_2, and Δ_3 can be expressed by sine law; i.e.,

$$\frac{\Delta_1}{\sin \theta} = \frac{\Delta_2}{\sin (90° - \theta)} = \frac{\Delta_3}{\sin 90°}$$

from which $\Delta_1 = \Delta_2 \tan \theta = \Delta_3 \sin \theta$

The joint displacements of Fig. 12-12(d) are similar to that described for Fig. 12-12(c) noting that joint a should move perpendicularly to member Aa. From the displacement diagram shown in Fig. 12-12(d)′ we see that

$$\frac{\Delta_1}{\sin \theta_2} = \frac{\Delta_2}{\sin (\theta_1 + \theta_2)} = \frac{\Delta_3}{\sin \theta_1}$$

The procedure described above is now extended to a two-span frame such as the one shown in Fig. 12-12(e) together with the joint-displacement diagram in Fig. 12-12(e)'. It may be extended to any number of spans. With the relative deflection between the ends for each member clarified, it becomes a simple matter to apply the slope-deflection equations.

Example 12-5. Find the end moments for each member of the portal frame shown in Fig. 12-13(a) resulting from a lateral force P acting on the top of the

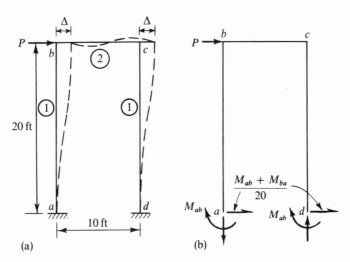

(a)   (b)

Fig. 12-13

column. Assume constant EI throughout the entire frame.

Because of the lateral force P acting at b, the frame deflects to the right. Both joints b and c move the same horizontal distance Δ, as indicated. There are also rotations θ_b at joint b and θ_c at joint c. Now, since an equal and opposite force P acting at c would completely balance the original force at b, and would thus return the structure to the original position (except some small change of length in bc), θ_b must be equal to θ_c. Thus,

$$M_{ab} = M_{dc} = (1)(\theta_b - 3R) \qquad (\theta_a = \theta_d = 0)$$
$$M_{ba} = M_{cd} = (1)(2\theta_b - 3R)$$
$$M_{bc} = M_{cb} = (2)(2\theta_b + \theta_b) = 6\theta_b$$

This special case in which the end moments and joint rotations of one side of the center line axis of the structure are the same as those of other side is termed *antisymmetry* in contrast to the case of *symmetry* in which the values of the end moments and joint rotations of one side of the center line axis of the structure are equal but opposite to those of the other side, according to the sign convention of slope deflection.

The unknowns, θ_b and R, are then solved by two equilibrium equations, one for joint b and the other for the entire frame.

$$\sum M_{\text{joint } b} = M_{ba} + M_{bc} = 0$$

or

$$8\theta_b - 3R = 0 \qquad (12\text{-}38)$$

By isolating the frame from the supports (Fig. 12-13(b)),

$$\sum F_x = (2)\frac{M_{ab} + M_{ba}}{20} + P = 0$$

or

$$3\theta_b - 6R + 10P = 0 \qquad (12\text{-}39)$$

Solving Eqs. 12-38 and 12-39 simultaneously, we obtain

$$\theta_b = \frac{10}{13}P \qquad 3R = \frac{80}{13}P$$

Substituting these in equations for end moments, we obtain

$$M_{ab} = M_{dc} = -\frac{70}{13}P$$

$$M_{ba} = M_{cd} = -\frac{60}{13}P$$

$$M_{bc} = M_{cb} = +\frac{60}{13}P$$

Example 12-6. Draw the bending-moment diagram for the frame shown in Fig. 12-14(a). The relative values of $2EK$ are encircled.

To find the end moments, we begin by sketching the relative displacement diagram, as shown in Fig. 12-14(b). Thus

$$R_{ab} = \frac{\Delta}{l_{ab}} = \frac{\Delta}{15}$$

$$R_{bc} = -\frac{(3/4)\Delta}{l_{bc}} = -\frac{(3/4)\Delta}{15} = -\frac{\Delta}{20}$$

$$R_{cd} = \frac{(5/4)\Delta}{l_{cd}} = \frac{(5/4)\Delta}{25} = \frac{\Delta}{20}$$

If we let $R_{ab} = R$, then $R_{bc} = -3R/4$ and $R_{cd} = 3R/4$. The expressions for the various end moments are now written

$$M_{ab} = (1)(\theta_b - 3R) \qquad (\theta_a = 0)$$

$$M_{ba} = (1)(2\theta_b - 3R)$$

$$M_{bc} = (2)[2\theta_b + \theta_c + (3)(\tfrac{3}{4}R)]$$

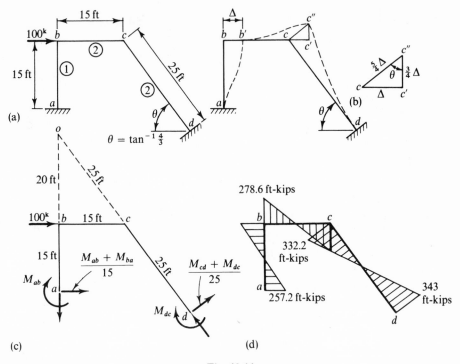

Fig. 12.14

$$M_{cb} = (2)[2\theta_c + \theta_b + (3)(\tfrac{3}{4}R)]$$
$$M_{cd} = (2)[2\theta_c - (3)(\tfrac{3}{4}R)] \qquad (\theta_d = 0)$$
$$M_{dc} = (2)[\theta_c - (3)(\tfrac{3}{4}R)]$$

Two of the three condition equations required to evaluate the three independent unknowns θ_b, θ_c, and R are from $\sum M = 0$ for joints b and c. Thus

$$M_{ba} + M_{bc} = 0$$
$$6\theta_b + 2\theta_c + 1.5R = 0 \qquad (12\text{-}40)$$
$$M_{cb} + M_{cd} = 0$$
$$\theta_b + 4\theta_c = 0 \qquad (12\text{-}41)$$

The third condition equation can best be found by expressing $\sum M_o = 0$ for the entire frame, o being the center of moment chosen at the intersection of the two legs (see Fig. 12-14(c)), since this eliminates the axial forces from the equation. Thus

$$M_{ab} + M_{dc} - (100)(20) - \left(\frac{M_{ab} + M_{ba}}{15}\right)(35) - \left(\frac{M_{cd} + M_{dc}}{25}\right)(50) = 0$$
$$-6\theta_b - 10\theta_c + 24.5R - 2{,}000 = 0$$
$$(12\text{-}42)$$

Solving Eqs. 12-40, 12-41, and 12-42 simultaneously, we obtain

$$\theta_b = -2.14 \qquad \theta_c = 0.536 \qquad R = 7.86$$

Substituting these values in the moment expressions yields

$$M_{ab} = -257.2 \text{ ft-kips} \qquad M_{ba} = -278.6 \text{ ft-kips}$$

$$M_{bc} = 278.6 \text{ ft-kips} \qquad M_{cb} = 332.2 \text{ ft-kips}$$

$$M_{cd} = -332.2 \text{ ft-kips} \qquad M_{dc} = -343.0 \text{ ft-kips}$$

as shown in Fig. 12-14(d).

12-7. ANALYSIS OF STATICALLY INDETERMINATE RIGID FRAMES WITH TWO DEGREES OF FREEDOM OF JOINT TRANSLATION BY THE SLOPE-DEFLECTION METHOD

The number of degrees of freedom for joint translation in a rigid frame equals the number of independent joint translations which can be given to the frame. Figure 12-15 shows several examples of rigid frames with two degrees of freedom of joint translation.

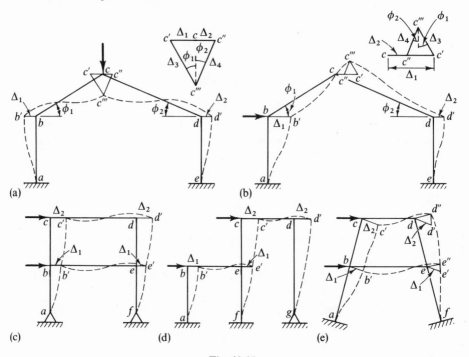

Fig. 12-15

In Fig. 12-15(a) is shown an unsymmetrical gable bent subjected to a vertical load at the top. Under this pressure joint b will move a distance Δ_1 to b' and joint d a distance Δ_2 to d', as indicated. To locate the position of c, let us imagine that joint c is temporarily disconnected. Joint c, being the end of member bc, will move to c' ($cc' = bb' = \Delta_1$). Joint c, being the end of member cd, will move to c'' ($cc'' = dd' = \Delta_2$). The final position of c is c''', which is the intersection of the line perpendicular to bc drawn at c' and the line perpendicular to cd drawn at c'', as indicated in Fig. 12-15(a). From the displacement diagram Δ_3 (the relative displacement between ends b and c) and Δ_4 (that between c and d) is related to Δ_1 (that between a and b) and Δ_2 (that between d and e) by the sine law,

$$\frac{\Delta_1 + \Delta_2}{\sin(\phi_1 + \phi_2)} = \frac{\Delta_3}{\sin(90° - \phi_2)} = \frac{\Delta_4}{\sin(90° - \phi_1)}$$

or

$$\frac{\Delta_1 + \Delta_2}{\sin(\phi_1 + \phi_2)} = \frac{\Delta_3}{\cos\phi_2} = \frac{\Delta_4}{\cos\phi_1}$$

so that Δ_3 and Δ_4 can be deduced from Δ_1 and Δ_2.

The case shown in Fig. 12-15(b) is similar to that of Fig. 12-15(a) except that we assume the tops of the two legs move in the same direction as the result of a lateral force applied at b. The relationships of joint displacements are expressed by

$$\frac{\Delta_1 - \Delta_2}{\sin(\phi_1 + \phi_2)} = \frac{\Delta_3}{\sin(90° - \phi_2)} = \frac{\Delta_4}{\sin(90° - \phi_1)}$$

or

$$\frac{\Delta_1 - \Delta_2}{\sin(\phi_1 + \phi_2)} = \frac{\Delta_3}{\cos\phi_2} = \frac{\Delta_4}{\cos\phi_1}$$

Figures 12-15(c) and (e) show the joint displacements in two-story frames, and Fig. 12-15(d) shows the joint displacements in a two-stage frame. In each of these cases, the arrangement of the structure is such that the joint translations of the first floor are not required to be in any fixed relationship to those of the second floor; i.e., the translations of joints b and e do not have fixed ratio to the translations of joints c and d. The procedure for finding joint displacements on each of the floors is the same as that discussed in Sec. 12-6 for one-story frames. With the relative end displacement for each member consistently determined, it becomes rather easy to analyze the frame by the method of slope deflection.

Example 12-7. Find all the end moments for the gable bent shown in Fig. 12-16. Assume constant EI throughout the entire frame so that the relative $2EK$ values are the encircled numbers.

We start by sketching the joint displacements of the frame as shown in Fig. 12-16(b) for which we note that

$$\Delta_3 = \frac{(\Delta_1 + \Delta_2)\sin(90° - \phi_2)}{\sin(\phi_1 + \phi_2)} = \frac{(\Delta_1 + \Delta_2)\cos\phi_2}{\sin[180° - (\phi_1 + \phi_2)]} = \frac{4}{5}(\Delta_1 + \Delta_2)$$

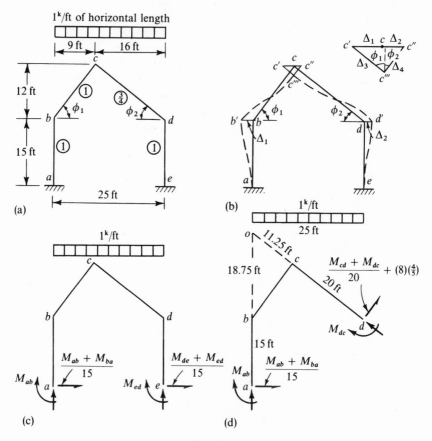

Fig. 12-16

since, if we refer to Fig. 12-16(a), $\cos \phi_2 = 4/5$ and $180° - (\phi_1 + \phi_2) = \sphericalangle c$ $= 90°$ in this case. Similarly,

$$\Delta_4 = \frac{(\Delta_1 + \Delta_2) \cos \phi_1}{\sin [180° - (\phi_1 + \phi_2)]} = \frac{3}{5}(\Delta_1 + \Delta_2)$$

Thus the rotation of each member can be expressed in terms of Δ_1 and Δ_2.

$$R_{ab} = -\frac{\Delta_1}{15} \qquad\qquad R_{bc} = \left(\frac{1}{15}\right)\left(\frac{4}{5}\right)(\Delta_1 + \Delta_2)$$

$$R_{cd} = -\left(\frac{1}{20}\right)\left(\frac{3}{5}\right)(\Delta_1 + \Delta_2) \qquad R_{de} = \frac{\Delta_2}{15}$$

If we let $\Delta_1/15 = R_1$ and $\Delta_2/15 = R_2$, we have

$$R_{ab} = -R_1 \qquad\qquad R_{bc} = 0.8(R_1 + R_2)$$

$$R_{cd} = -0.45(R_1 + R_2) \qquad R_{de} = R_2$$

The fixed-end moments are found to be

$$M_{bc}^F = -M_{cb}^F = -\frac{(1)(9)^2}{12} = -6.750 \text{ ft-kips}$$

$$M_{cd}^F = -M_{dc}^F = -\frac{(1)(16)^2}{12} = -21.333 \text{ ft-kips}$$

The expressions for various end moments are then written

$$M_{ab} = (1)(\theta_b + 3R_1) \qquad (\theta_a = 0)$$
$$M_{ba} = (1)(2\theta_b + 3R_1)$$
$$M_{bc} = (1)[2\theta_b + \theta_c - 2.4(R_1 + R_2)] - 6.750$$
$$M_{cb} = (1)[2\theta_c + \theta_b - 2.4(R_1 + R_2)] + 6.750$$
$$M_{cd} = (\tfrac{3}{4})[2\theta_c + \theta_d + 1.35(R_1 + R_2)] - 21.333$$
$$M_{dc} = (\tfrac{3}{4})[2\theta_d + \theta_c + 1.35(R_1 + R_2)] + 21.333$$
$$M_{de} = (1)(2\theta_d - 3R_2) \qquad (\theta_e = 0)$$
$$M_{ed} = (1)(\theta_d - 3R_2)$$

Five statical equations are needed to solve the independent unknowns $\theta_b, \theta_c, \theta_d, R_1,$ and R_2—three from $\sum M = 0$ for joints $b, c,$ and d; one from shear balance ($\sum F_x = 0$) for the entire frame (see Fig. 12-16(c)); and one from $\sum M_o = 0$ for the portion of the frame shown in Fig. 12-16(d). Thus,

$$\sum M_{\text{joint } b} = M_{ba} + M_{bc} = 0$$
$$4\theta_b + \theta_c + 0.6R_1 - 2.4R_2 - 6.750 = 0 \qquad (12\text{-}43)$$
$$\sum M_{\text{joint } c} = M_{cb} + M_{cd} = 0$$
$$\theta_b + 3.5\theta_c + 0.75\theta_d - 1.387(R_1 + R_2) - 14.583 = 0 \qquad (12\text{-}44)$$
$$\sum M_{\text{joint } d} = M_{dc} + M_{de} = 0$$
$$0.75\theta_c + 3.5\theta_d + 1.013R_1 - 1.987R_2 + 21.333 = 0 \qquad (12\text{-}45)$$

$\sum F_x = 0$ for the frame of Fig. 12-16(c); i.e.,

$$\frac{M_{ab} + M_{ba}}{15} + \frac{M_{de} + M_{ed}}{15} = 0$$
$$\theta_b + \theta_d + 2R_1 - 2R_2 = 0 \qquad (12\text{-}46)$$

$\sum M_o = 0$ for the portion of the frame (Fig. 12-16(d)); i.e.,

$$M_{ab} - \left(\frac{M_{ab} + M_{ba}}{15}\right)(33.75) - \left(\frac{M_{cd} + M_{dc}}{20}\right)(31.25) - (8)\left(\frac{4}{5}\right)(31.25)$$
$$+ M_{dc} + \frac{(1)(25)^2}{2} = 0$$

$(\theta_b + 3R_1) - 2.25(3\theta_b + 6R_1) - 1.5625[2.25\theta_c + 2.25\theta_d + 2.025(R_1 + R_2)]$
$- 200 + [0.75\theta_c + 1.5\theta_d + 1.013(R_1 + R_2) + 21.333] + 312.5 = 0$

or

$$5.75\theta_b + 2.766\theta_c + 2.016\theta_d + 12.651R_1 + 2.151R_2 - 133.833 = 0$$
$$(12\text{-}47)$$

Solving Eqs. 12-43, 12-44, 12-45, 12-46, and 12-47 simultaneously, we obtain

$$\theta_b = 0.341 \qquad \theta_c = 11.039 \qquad \theta_d = -8.342$$
$$R_1 = 8.543 \qquad R_2 = 4.561$$

Substituting these in the expressions for the various end moments, we obtain

$$M_{ab} = 26 \text{ ft-kips}$$
$$M_{ba} = -M_{bc} = 26.3 \text{ ft-kips}$$
$$M_{cb} = -M_{cd} = -2.3 \text{ ft-kips}$$
$$M_{dc} = -M_{de} = 30.3 \text{ ft-kips}$$
$$M_{ed} = -22 \text{ ft-kips}$$

Example 12-8. Analyze the frame in Fig. 12-17(a). Assume that all members are of uniform cross section so that the relative values of $2EK$ are the

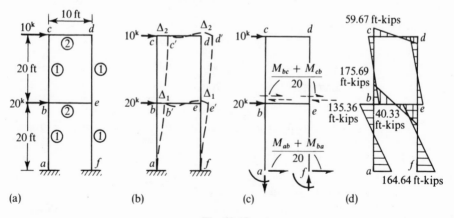

(a)   (b)   (c)   (d)

Fig. 12-17

encircled numbers. Because of the action of the lateral forces, the frame will deflect to the right. Assume that joint b and joint e, on the first floor level, move a horizontal distance Δ_1; joint c and joint d, on the second floor level, move a horizontal distance Δ_2, as shown in Fig. 12-17(b). Antisymmetry exists in this case. Thus,

$$\theta_b = \theta_e \qquad \theta_c = \theta_d \qquad R_{ab} = R_{ef} = \frac{\Delta_1}{20} \qquad R_{bc} = R_{de} = \frac{\Delta_2 - \Delta_1}{20}$$

Now since $\qquad\qquad \theta_a = \theta_f = R_{be} = R_{cd} = 0$

the problem involves a total of four unknowns, $\theta_b, \theta_c, R_{ab}$, and R_{bc}, which are to be solved by four equations of statics, two from $\sum M = 0$ for joints b and c and the other two from shear balance for the frame.

The equations expressing end moments are then written

$$M_{ab} = M_{fe} = (1)(\theta_b - 3R_{ab}) \qquad (\theta_a = 0)$$
$$M_{ba} = M_{ef} = (1)(2\theta_b - 3R_{ab})$$
$$M_{bc} = M_{ed} = (1)(2\theta_b + \theta_c - 3R_{bc})$$
$$M_{be} = M_{eb} = (2)(2\theta_b + \theta_e) = 6\theta_b \qquad (\theta_b = \theta_e)$$
$$M_{cb} = M_{de} = (1)(2\theta_c + \theta_b - 3R_{bc})$$
$$M_{cd} = M_{dc} = (2)(2\theta_c + \theta_d) = 6\theta_c \qquad (\theta_c = \theta_d)$$

The equations from joint equilibrium in moment are then established;

$$M_{ba} + M_{bc} + M_{be} = 0$$
$$10\theta_b + \theta_c - 3R_{ab} - 3R_{bc} = 0 \tag{12-48}$$
$$M_{cb} + M_{cd} = 0$$
$$\theta_b + 8\theta_c - 3R_{bc} = 0 \tag{12-49}$$

The third equation is from shear balance for the entire frame isolated from the supports (see Fig. 12-17(c));

$$(2)\left(\frac{M_{ab} + M_{ba}}{20}\right) + 20 + 10 = 0$$
$$\theta_b - 2R_{ab} + 100 = 0 \tag{12-50}$$

The fourth equation results from considering the free body cut out by a horizontal section just above be; the shear in the two legs (see the dash lines in Fig. 12-17(c)) must balance the lateral force of 10 kips. Thus,

$$(2)\left(\frac{M_{bc} + M_{cb}}{20}\right) + 10 = 0$$
$$3\theta_b + 3\theta_c - 6R_{bc} + 100 = 0 \tag{12-51}$$

Solving Eqs. 12-48, 12-49, 12-50, and 12-51 simultaneously, we obtain

$$\theta_b = 29.282 \qquad \theta_c = 9.945$$
$$R_{ab} = 64.641 \qquad R_{cd} = 36.282$$

Substituting these in the moment equations, we arrive at

$$M_{ab} = M_{fe} = -164.64 \text{ ft-kips}$$
$$M_{ba} = M_{ef} = -135.36 \text{ ft-kips}$$
$$M_{bc} = M_{ed} = -40.33 \text{ ft-kips}$$
$$M_{be} = M_{eb} = 175.69 \text{ ft-kips}$$
$$M_{cb} = M_{de} = -59.67 \text{ ft-kips}$$
$$M_{cd} = M_{dc} = +59.67 \text{ ft-kips}$$

as plotted in Fig. 12-17(d).

12-8. ANALYSIS OF STATICALLY INDETERMINATE RIGID FRAMES WITH SEVERAL DEGREES OF FREEDOM OF JOINT TRANSLATION BY THE SLOPE-DEFLECTION METHOD

The slope-deflection procedure, described in the preceding sections, may be extended to the analysis of frames with more than two degrees of freedom of joint translation. Solving simultaneous slope-deflection equations in terms of joint rotations and translations, although time consuming, is frequently simpler and more easily applied than any method previously discussed. Consider the four-story, two-bay building frame shown in Fig. 12-18. It would require the solu-

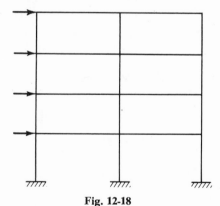

Fig. 12-18

tion of twenty-four simultaneous equations in terms of the total internal work of the frame by the method of least work (see Sec. 10-3) as compared with sixteen equations needed by the method of slope deflection; i.e., twelve equations expressing $\sum M = 0$ for each of the twelve joints having rotations and four equations expressing the shear balance for each floor level of the four stories.

On the other hand, in the analysis of certain types of frames, such as the one shown in Fig. 12-19, the slope-deflection method becomes difficult to apply,

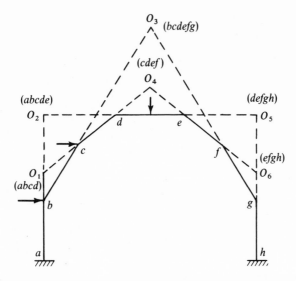

Fig. 12-19

since it would require the solution of thirteen simultaneous equations—six equations expressing $\sum M = 0$ for joints b, c, d, e, f, and g; one equation expressing $\sum F_x = 0$ for the entire frame; and six equations expressing $\sum M = 0$ about the six moment centers O_1, O_2, O_3, O_4, O_5, and O_6 for the portions of frame indicated for each moment center in Fig. 12-19. However, if we use the method of least work, we must solve only three simultaneous equations in terms of the total internal work of the structure. Many engineering short cuts, such as the *elastic-center method* and *column-analogy method*, used in solving this type of structure are the direct outcome of least work or consistent deformation. Because of their limited use, they are not treated in this book.

PROBLEMS

12-1. Analyze the beam shown in Fig. 12-20 by slope deflection. Draw the shear and moment diagrams.

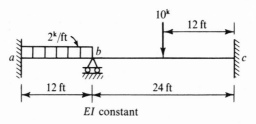

Fig. 12-20

12-2. Figure 12-21 shows a frame of uniform cross section. Find all the end moments by slope deflection, and sketch the deformed structure.

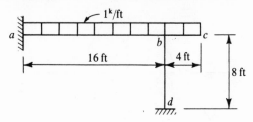

Fig. 12-21

12-3. Analyze the beam shown in Fig. 12-22 by slope deflection. Draw the shear and moment diagrams.

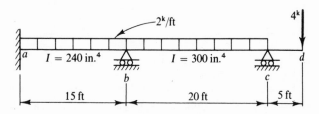

Fig. 12-22

12-4. Find all the end moments by slope deflection for the rigid frame shown in Fig. 12-23. Draw the moment diagram, and sketch the deformed structure.

12-5. In Fig. 12-22 remove all the loads, and assume that the support b settles vertically 0.5 in. Find all the end moments by slope deflection. $E = 30,000$ kips per in.^{2}

12-6. In Fig. 12-23 remove the load, and assume that a rotational yield of 0.002 radian clockwise and a linear yield downwards of 0.1 in. occur at support a. Find the moment diagram. $EI = 10,000$ kips-ft.^{2}

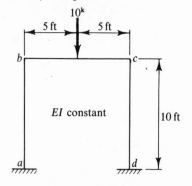

Fig. 12-23

12-7. Analyze each of the frames shown in Fig. 12-24 by slope deflection, and draw the moment diagram.

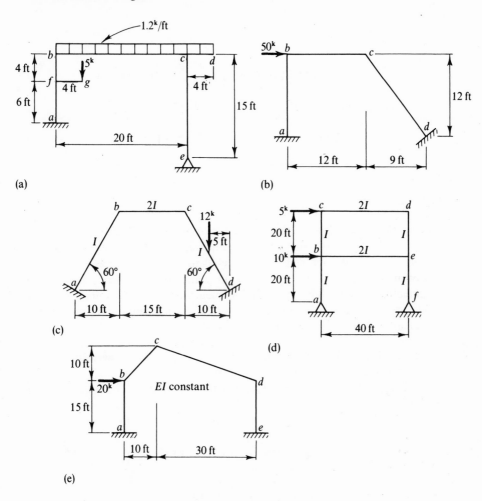

(a)

(b)

(c)

(d)

(e)

Fig. 12-24

13

AN INTRODUCTION TO MOMENT
DISTRIBUTION: MOMENT DISTRIBUTION
WITHOUT JOINT TRANSLATION

13-1. GENERAL

The method of *moment distribution* was originated by Hardy Cross in 1932 in a paper entitled "Analysis of Continuous Frames by Distributing Fixed-End Moments." It is the method normally used to analyze all types of statically indeterminate beams and rigid frames in which the members are primarily subjected to bending. The process of moment distribution is initiated by the basic slope-deflection equation (see Sec. 12-2); the moment acting on the end of a member is the algebraic sum of four effects:

1. The moment due to the loads on the member if the member is considered as a fixed-end beam (the fixed-end moment).
2. The moment due to the rotation of the near end (this end) while the far end (the other end) is fixed.
3. The moment due to the rotation of the far end, the near end being fixed.
4. The moment due to the relative translation between the two ends of the member.

This suggests that one line of attack might be to allow these effects to take place separately through a series of steps, first *locking* the joints and then *unlocking* them. For instance, in a rigid-joint structure without joint translation, once the joints are locked (held against rotation), each member is in the state of a fixed-end beam. By unlocking (releasing) a joint, we find that resisting moments will be developed or distributed at the near ends of the members meeting at the joint in proportion to their stiffnesses or according to their distribution factors. At the same time moments will be induced or carried over to the far ends of these members according to their carry-over factors. Joints may be successively released and reheld, one by one, as many times as necessary until each joint will have rotated into its actual, or nearly actual, position. Thus the process is essentially one of successive approximations which can be carried to any degree of accuracy desired. We shall define the terms *stiffness*, *distribution factor*, and *carry-over factor* and explain them step by step in the following sections.

It must be noted that the method of moment distribution, although it depends on solving slope-deflection equations, is nevertheless a new approach in structural analysis. For instance, determining the end moments of a highly indeterminate rigid frame with joint translations prevented, does not require solving any simultaneous equations in contrast to the previously discussed methods. Even in analyzing rigid frames having joint translations, the method of moment distribution usually does not involve as many simultaneous equations as are required by any of the methods already discussed. Furthermore, this method is adaptable to computer programming, since it is cyclic and the same operations are repeated several times over. Owing to these tremendous advantages, it is regarded as the most ingenious and convenient method contributed to structural engineering.

The method of moment distribution can be applied to structures composed of prismatic members or nonprismatic members—with or without joint translation. To simplify the discussion, the present chapter will be confined to beams and frames of prismatic members without joint translation.

The sign convention adopted is the same as that previously suggested for the slope-deflection method; clockwise end moment and rotation of a member are considered positive.

13-2. FIXED-END MOMENT

The application of the method of moment distribution requires knowledge of the moments developed at the ends of loaded beams with both ends built-in. These moments are called the fixed-end moments often denoted by the symbol M^F or F.E.M. in the table and illustrations.

The determination of fixed-end moments was discussed in Sec. 9-2 (consistent deformations) and in Sec. 10-2 (least work). For a straight prismatic member the fixed-end moments due to common types of loading were given in Table 12-1.

13-3. STIFFNESS, DISTRIBUTION FACTOR, AND
DISTRIBUTION OF EXTERNAL MOMENT APPLIED
TO A JOINT

For a member of uniform section (constant EI), the *stiffness* (or more specifically the *rotational stiffness*) is defined as the end moment required to produce a unit rotation at one end of the member while the other end is fixed.

Consider member ab in Fig. 13-1 with a constant section. End b is fixed, and end a is allowed to rotate. The end moment required at end a to rotate

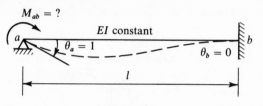

Fig. 13-1

$\theta_a = 1$ while $\theta_b = 0$ is given by

$$M_{ab} = 2E\frac{I}{l}(2\theta_a + \theta_b) = 2E\frac{I}{l}(2 + 0) = 4E\frac{I}{l}$$

$$= 4EK$$

This moment is defined as *stiffness* and will be denoted by S. Thus

$$S = 4E\frac{I}{l}$$

$$= 4EK \tag{13-1}$$

I/l or K being the *stiffness factor*.

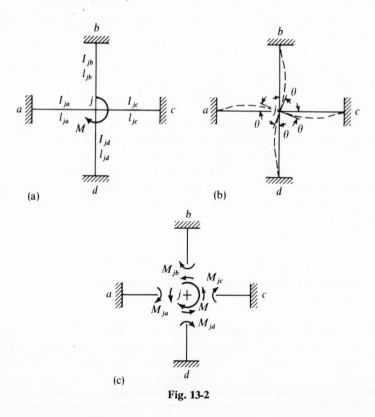

(a)   (b)

(c)

Fig. 13-2

Let us turn to Fig. 13-2(a), which shows a frame composed of four members, each with one end fixed and the other end rigidly connected at joint j whose translation is prevented. If a clockwise moment M is applied to the joint, it will cause the joint to rotate clockwise through an angular deformation θ, as shown in Fig. 13-2(b). Since j is a rigid joint, each tangent to the elastic curve of the connected end rotates the same angle θ. The applied moment M is resisted by the four members meeting at the joint. The resisting moments M_{ja}, M_{jb}, M_{jc}, and M_{jd} will be induced at the ends of the four members to balance the effect of the external moment M, as shown in Fig. 13-2(c).

Equilibrium of the joint requires that

$$M_{ja} + M_{jb} + M_{jc} + M_{jd} = M \tag{13-2}$$

and the slope-deflection equations for the four members give

$$M_{ja} = 4E\frac{I_{ja}}{l_{ja}}\theta = 4EK_{ja}\theta = S_{ja}\theta$$

$$M_{jb} = 4E\frac{I_{jb}}{l_{jb}}\theta = 4EK_{jb}\theta = S_{jb}\theta$$

$$M_{jc} = 4E\frac{I_{jc}}{l_{jc}}\theta = 4EK_{jc}\theta = S_{jc}\theta \tag{13-3}$$

$$M_{jd} = 4E\frac{I_{jd}}{l_{jd}}\theta = 4EK_{jd}\theta = S_{jd}\theta$$

Equation 13-3 shows that, when an external moment is applied to a joint, the resisting moments developed at the near ends of the members meeting at the joint, while the other ends are all fixed, are in direct proportion to the rotational stiffnesses.

Substituting Eq. 13-3 in Eq. 13-2, we obtain

$$(S_{ja} + S_{jb} + S_{jc} + S_{jd})\theta = M$$

or

$$4E(K_{ja} + K_{jb} + K_{jc} + K_{jd})\theta = M$$

Thus,

$$\theta = \frac{M}{4E\sum K} \tag{13-4}$$

where $\sum K = K_{ja} + K_{jb} + K_{jc} + K_{jd}$.

From Eqs. 13-3 and 13-4, we see that

$$M_{ja} = \frac{K_{ja}}{\sum K}M = D_{ja}M$$

$$M_{jb} = \frac{K_{jb}}{\sum K}M = D_{jb}M$$

$$M_{jc} = \frac{K_{jc}}{\sum K}M = D_{jc}M \tag{13-5}$$

$$M_{jd} = \frac{K_{jd}}{\sum K}M = D_{jd}M$$

in which the ratio $K_{ji}/\sum K$ or D_{ji} ($i = a, b, c, d$) is defined as the *distribution factor*. Thus, a moment resisted by a joint will be distributed among the connecting members in proportion to their distribution factors. In determining the distribution factors, only the relative K values for connected members are needed. Thus, in most cases we are concerned with the *relative stiffness* rather than the absolute stiffness (Eq. 13-1).

13-4. CARRY-OVER FACTOR AND CARRY-OVER MOMENT

Referring to Fig. 13-2, we evaluate the moments at the far ends (fixed ends) of the four members by the slope-deflection method as

$$M_{aj} = 2E\frac{I_{ja}}{l_{ja}} = \left(\frac{1}{2}\right)M_{ja}$$

$$M_{bj} = 2E\frac{I_{jb}}{l_{jb}} = \left(\frac{1}{2}\right)M_{jb}$$

$$M_{cj} = 2E\frac{I_{jc}}{l_{jc}} = \left(\frac{1}{2}\right)M_{jc} \qquad (13\text{-}6)$$

$$M_{dj} = 2E\frac{I_{jd}}{l_{jd}} = \left(\frac{1}{2}\right)M_{jd}$$

Equation 13-6 indicates that the moment induced at the far end (fixed) of a prismatic member equals *one-half* the distributed moment at the near end. The ratio $(1/2)$ is called the *carry-over factor*, and the induced moments M_{aj}, M_{bj}, M_{cj} and M_{dj} are called the *carry-over moments*.

In general, the *carry-over factor* may be defined as the ratio of the induced moment at the far end, which is fixed, to the applied moment at the near end, which is prevented from translation but allowed to rotate. Consider Fig. 13-3.

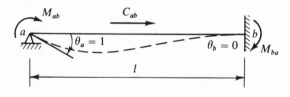

Fig. 13-3

If an end moment M_{ab} is applied at the near end a, then the moment induced at the far end b, called M_{ba}, is given by

$$M_{ba} = C_{ab}M_{ab} \qquad (13\text{-}7)$$

where C_{ab} is the carry-over factor from a to b. For a member of uniform EI,

$$M_{ab} = 2E\frac{I}{l}(2\theta_a)$$

$$M_{ba} = 2E\frac{I}{l}(\theta_a) = \frac{1}{2}M_{ab}$$

Therefore,　　　　　　　　$C_{ab} = \frac{1}{2}$

If we consider end a as the far end (fixed), and end b as the near end (allowed to rotate), we can, in a like manner, prove that

$$C_{ba} = \frac{1}{2}$$

Thus for a member of uniform section,

$$C_{ab} = C_{ba} = \frac{1}{2} \tag{13-8}$$

We recapitulate some of the main points of Secs. 13-3 and 13-4 as follows:

When an external moment is applied to a joint for which translation is prevented, the joint rotates; but the rotation is checked by members meeting at the joint. The resisting moment is then distributed to the near ends of the connected members according to their distribution factors, provided that all the far ends of these members are fixed. The distributed moment to the near end for each member based on the free body of the member (not the joint) equals the applied moment times the distribution factor bearing the same sign as that of the applied moment. Meanwhile, moment is carried over to the far end of each member, which equals one-half the distributed moment to the near end and bears the same sign.

In the subsequent illustrations and tables we often use the symbols D.M. to denote the distributed moment; D.F., the distribution factor; C.O.M., the carry-over moment; and C.O.F., the carry-over factor.

Example 13-1. For the loaded frame shown in Fig. 13-4(a), find the end moments at a and c.

We begin by putting the portion abc of Fig. 13-4(a) into its equivalent (Fig.

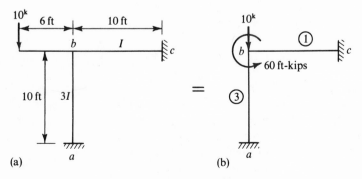

Fig. 13-4

13-4(b)) and obtaining the relative K values for members ab and bc (encircled). It is then readily seen that the end moments at a and c are the carry-over moments due to the external moment 60 ft-kips applied to joint b. Thus,

$$M_{ab} = \frac{1}{2} M_{ba} = \left(\frac{1}{2}\right)(-60)\left(\frac{3}{3+1}\right) = -22.5 \text{ ft-kips}$$

$$M_{cb} = \frac{1}{2} M_{bc} = \left(\frac{1}{2}\right)(-60)\left(\frac{1}{3+1}\right) = -7.5 \text{ ft-kips}$$

The negative signs indicate counter-clockwise moments.

13-5. THE PROCESS OF LOCKING AND UNLOCKING: ONE JOINT

The essence of moment distribution lies in locking and unlocking the joints based on the principle of superposition; i.e., the effect of an artificial moment applied to a rigid joint of the frame and then eliminated is the same as no effect on the actual structure, since the two actions are neutralized. In this section our attention will be confined to only those frames that have all joints (including supported ends) fixed except one, which is allowed to rotate. Figure 13-5 (a) shows the case which may then be considered as the superposition of effects of Figs. 13-5(b) and (c).

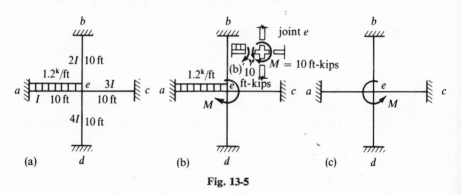

Fig. 13-5

In connection with the set-up in Fig. 13-5, we note the following:

1. Suppose that the artificial moment M imposed on joint e (see Fig. 13-5 (b)) is so chosen as just to lock the joint against rotation (keeping $\theta_e = 0$) under the original loading (in present case, the uniform load over span ae). Each member of the frame is then in the state of a fixed-end beam, and consequently, fixed-end moments will be developed at the ends of member ae;

$$M_{ea}^F = -M_{ae}^F = \frac{(1.2)(10)^2}{12} = 10 \text{ ft-kips}$$

all other member ends being subjected to no moment. The results are shown in Row 1 of Table 13-1.

TABLE 13-1

		\multicolumn{8}{c}{End Moment (ft-kips)}
		\multicolumn{2}{c}{ae}	\multicolumn{2}{c}{be}	\multicolumn{2}{c}{ce}	\multicolumn{2}{c}{de}
Row	Step	M_{ae}	M_{ea}	M_{be}	M_{eb}	M_{ce}	M_{ec}	M_{de}	M_{ed}
1	F.E.M.	−10	+10	0	0	0	0	0	0
2	D.M.	0	−1	0	−2	0	−3	0	−4
3	C.O.M.	−0.5	0	−1	0	−1.5	0	−2	0
4	Σ	−10.5	+9	−1	−2	−1.5	−3	−2	−4

Referring to Fig. 13-5(b)′, we find the equilibrium condition

$$\sum M_{\text{joint } e} = 0$$

requires that the locking moment

$$M = 10 \text{ ft-kips}$$

acting clockwise on joint e.

2. Next, let us release the joint e from the artificial restraint, i.e., apply to it an unlocking moment equal and opposite to the locking moment. Referring to Fig. 13-5(c), we see a counter-clockwise M equal to the value of 10 ft-kips, i.e.,

$$M = -10 \text{ ft-kips}$$

is thus applied to joint e.

As a result of this unlocking moment, resisting moment will be distributed at near ends and carried over to the far ends, as described in preceding sections. Thus,

$$M_{ea} = (-10)\left(\frac{1}{10}\right) = -1.0 \text{ ft-kips} \qquad M_{ae} = (-1)\left(\frac{1}{2}\right) = -0.5 \text{ ft-kips}$$

$$M_{eb} = (-10)\left(\frac{2}{10}\right) = -2.0 \text{ ft-kips} \qquad M_{be} = (-2)\left(\frac{1}{2}\right) = -1.0 \text{ ft-kips}$$

$$M_{ec} = (-10)\left(\frac{3}{10}\right) = -3.0 \text{ ft-kips} \qquad M_{ce} = (-3)\left(\frac{1}{2}\right) = -1.5 \text{ ft-kips}$$

$$M_{ed} = (-10)\left(\frac{4}{10}\right) = -4.0 \text{ ft-kips} \qquad M_{de} = (-4)\left(\frac{1}{2}\right) = -2.0 \text{ ft-kips}$$

They are shown in Rows 2 and 3 of Table 13-1, respectively.

3. The sum of the results from the above steps gives the solution, as shown in Row 4 of Table 13-1.

In analyzing problems like this, with all ends fixed except one rigid joint allowed to rotate, the method of moment distribution provides a very rapid tool, since it involves only one round of locking and unlocking. More examples to illustrate this process are given as follows:

Example 13-2. The end moments at a, b, and c for the beam and loading shown in Fig. 13-6(a) may be obtained by locking joint b and then unlocking it, as indicated in Figs. 13-6(b) and (c). The complete analysis is shown in Fig. 13-6(d), which contains the following steps:

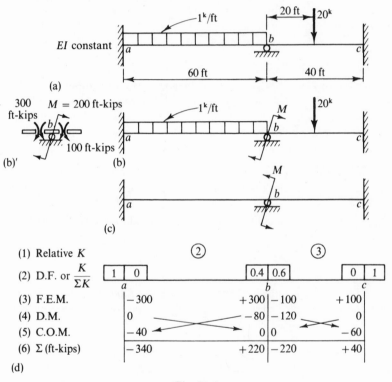

Fig. 13-6

1. The values of stiffness for members ab and bc are found to be

$$K_{ab} = \frac{4EI}{60} \qquad K_{bc} = \frac{4EI}{40}$$

Multiplying each by $30/EI$ gives the relative K values (encircled).

2. The distribution factors are computed according to $K/\sum K$. Thus,

$$D_{ba} = \frac{2}{2 + 3} = 0.4 \qquad D_{bc} = \frac{3}{2 + 3} = 0.6$$

We consider the immovable supports at a and c with infinite stiffness so that

$$D_{ab} = D_{cb} = 0$$

The distribution factors are indicated in the attached box at each of the joints.

3. The locking joint b artificially puts members ab and bc in the state of fixed-end beams. We write the fixed-end moments as

$$M_{ab}^F = -M_{ba}^F = -\frac{(1)(60)^2}{12} = -300 \text{ ft-kips}$$

$$M_{bc}^F = -M_{cb}^F = -\frac{(20)(40)}{8} = -100 \text{ ft-kips}$$

Note that locking joint b means applying an external clockwise moment equal to 200 ft-kips, required by the equilibrium of moments for that joint (see Fig. 13-6(b)').

4. Unlocking joint b (i.e., eliminating the artificial restraint acting on joint b) means applying an external counter-clockwise moment equal to 200 ft-kips. We write the distributed moment for each of near ends according to the distribution factor.

5. Write down the carry-over moment for each of the far ends equal to one-half the distributed moment of the near end.

6. The sum of the results from Steps 3, 4, and 5 gives the solution.

Example 13-3. Find the end moments for the frame shown in Fig. 13-7(a) resulting from the rotational yield of support a 0.0016 radian clockwise. $EI = 10,000$ kips-ft^2. Note that this problem was solved by slope deflection in Example 12-4.

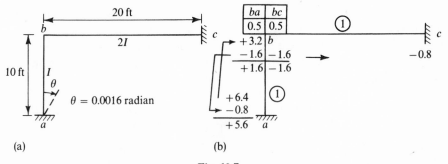

(a)   (b)

Fig. 13-7

The moment required to produce a rotation of 0.0016 radian at a is given by

$$M = \frac{4EI_{ab}\theta_a}{l_{ab}} = \frac{(4)(10,000)(0.0016)}{10} = 6.4 \text{ ft-kips}$$

if joint b is temporarily fixed. Half the amount of this moment will be carried over to end b of member ab. By releasing joint b, a process of distribution and carry-over takes place, as recorded in Fig. 13-7(b). This gives

$$M_{ab} = 5.6 \text{ ft-kips} \qquad M_{ba} = -M_{bc} = 1.6 \text{ ft-kips} \qquad M_{cb} = -0.8 \text{ ft-kips}$$

13-6. THE PROCESS OF LOCKING AND UNLOCKING: TWO OR MORE JOINTS

If a rigid frame or continuous beam involves more than one joint permitted to rotate, then the process of moment distribution consists of the repeated application of the principle of superposition, as briefly stated in the following steps:

1. The joints are first locked; all members, accordingly, are fixed-end. Write the fixed-end moments for all members.

2. The joints are then unlocked. Only one joint at a time is selected to be unlocked. While one joint is unlocked, the rest of joints are assumed to be held against rotation.

Calculate the unlocking moment at this joint, and write distributed moments for the near ends of the members meeting at this joint.

3. Also write down the carry-over moments at the far ends of these members. Note that the carry-over moments constitute a new set of fixed-end moments for the far ends.

4. Relock the joint, and select the next joint to be unlocked. Repeat Steps 2 and 3.

Note that after a joint is unlocked and the moments at a joint distributed, the joint is in balance, or in equilibrium, since the artificial restraint is removed. However, there are other joints still locked by external means; hence, the next step is to relock the joint and then proceed to unlock the next joint. The process of locking and unlocking each joint only once constitutes *one cycle* of moment distribution.

5. Joints are unlocked and relocked one by one; therefore Steps 2 and 3 are repeated several times. The process can be halted as soon as the carry-over moments are so small that we are willing to neglect them.

6. Sum up the moments to obtain the final result.

We see that the analysis starts from an alteration of the original structure by locking all joints against rotation. This means that artificial restraints are actually applied to the original structure. The altered structure, consisting of a number of fixed-end members, is then modified by unlocking and relocking joints one by one until all artificial restraints are removed or diminished to a sufficiently small amount. Thus moment distribution is a method of successive approximations by which the exact results can be approached with the desired degree of precision.

The complete analysis of a loaded three-span continuous beam by moment distribution, shown in Fig. 13-8, will serve to illustrate the above procedure.

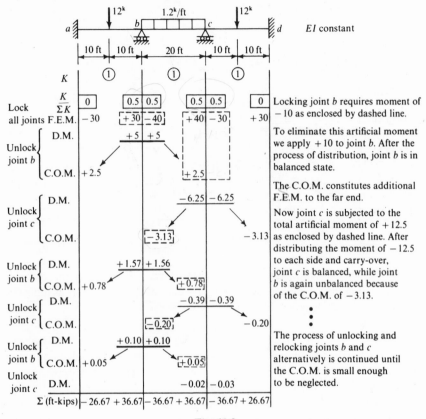

Fig. 13-8

The presentation of moment distribution for the preceding illustration may be rearranged as shown in Fig. 13-9. At first glance it seems as if joints b and c were locked and then unlocked simultaneously. However, the performance can still be considered under the restriction of unlocking one joint at a time. For the loads given, the fixed-end moments are recorded in Step 1 (see Fig. 13-9). Next, we may consider joint c as being held against rotation and joint b as being unlocked first. The unlocking moment $+10$ at b is then equally distributed to the near ends of members ba and bc as indicated in Step 2, and one-half of the amount is carried over to the far ends of these members as indicated in Step 3. Next, we consider joint b as locked and joint c as released, but only partially, by applying an unlocking moment of $-(40-30)$ or -10 to it, since the complete releasing of joint c would require an unlocking moment of $-(40-30+2.5)$ or -12.5 for the time being. This unlocking moment

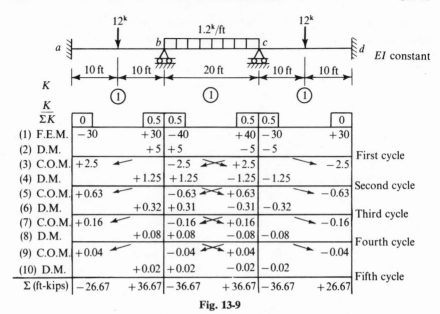

K	0		0.5	0.5		0.5	0.5		0	
$\dfrac{K}{\Sigma K}$										
(1) F.E.M.	−30		+30	−40		+40	−30		+30	
(2) D.M.			+5	+5		−5	−5			First cycle
(3) C.O.M.	+2.5			−2.5		+2.5			−2.5	
(4) D.M.			+1.25	+1.25		−1.25	−1.25			Second cycle
(5) C.O.M.	+0.63			−0.63		+0.63			−0.63	
(6) D.M.			+0.32	+0.31		−0.31	−0.32			Third cycle
(7) C.O.M.	+0.16			−0.16		+0.16			−0.16	
(8) D.M.			+0.08	+0.08		−0.08	−0.08			Fourth cycle
(9) C.O.M.	+0.04			−0.04		+0.04			−0.04	
(10) D.M.			+0.02	+0.02		−0.02	−0.02			Fifth cycle
Σ (ft-kips)	−26.67		+36.67	−36.67		+36.67	−36.67		+26.67	

Fig. 13-9

— 10 is equally distributed to the near ends of members cb and cd, and one-half of the amount is carried over to the far ends of these members as indicated in Steps 2 and 3, leaving the just received carry-over moment $+2.5$ to be handled later.

Referring to Step 3 of Fig. 13-9, we find the carry-over moments form a new set of fixed-end moments for the beam, and a process of unlocking joints can be carried out in a similar manner. The process will thus be repeated in a cyclic fashion until the carry-over moments are neglected.

In fact, if the carry-over moments are neglected, the joints, after being unlocked, are in balance; i.e., no external constraint exists. This gives the approximate solution for the analysis. For instance, the first approximation may be obtained from the sum of Steps 1 and 2, the result of the first cycle; the second approximation may be obtained from the sum of Steps 1 to 4, the result up to the second cycle; and so on. It may be interesting to point out that, in this particular problem, even the first cycle yields a good approximation of the exact solution. After two or three cycles, the carry-over values become negligible.

Note that the beam and the loading shown in Fig. 13-9 are symmetrical, the data presented on each side of the line of symmetry are equal in magnitude but opposite in sign. This special display suggests that some modification could be made in order to facilitate the process of moment distribution by working with only half the structure. See Sec. 13-7 for *modified stiffness*. This problem will be re-solved in Example 13-6 by using modified stiffness.

More examples are given to illustrate the cyclic process.

Example 13-4. Analyze the two-span continuous beam shown in Fig. 13-10 by moment distribution.

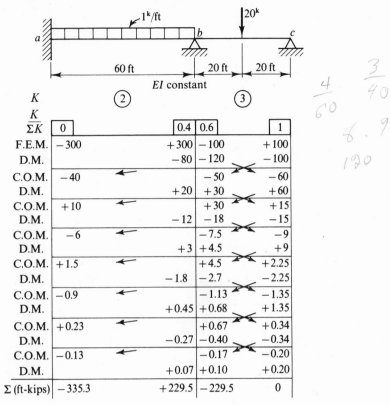

Fig. 13-10

The convergence in the solution shown in Fig. 13-10 is rather slow because the hinged end c continuously sends back sizable carry-over moments to joint b. However, since the final result $M_c = 0$ is known beforehand, the convergence may be improved by using modified stiffness as suggested in Sec. 13-7. This problem will be re-solved in Example 13-8.

Example 13-5. Analyze the frame in Fig. 13-11 by moment distribution. The relative stiffnesses for the frame members are computed first as

$$K_{\substack{ab \\ (cd)}} = \frac{2I}{30} \qquad \text{say} \quad 4$$

$$K_{bc} = \frac{2I}{40} \qquad \text{say} \quad 3$$

$$K_{\substack{be \\ (cf)}} = \frac{I}{20} \qquad \text{say} \quad 3$$

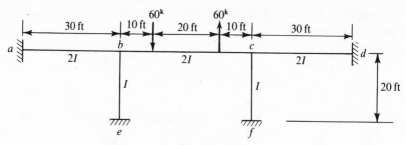

Fig. 13-11

Next, the fixed-end moments for the loaded member bc are found to be

$$M_{bc}^F = -\frac{(60)(10)(30)^2}{(40)^2} + \frac{(60)(30)(10)^2}{(40)^2} = -225 \text{ ft-kips}$$

$$M_{cb}^F = +\frac{(60)(30)(10)^2}{(40)^2} - \frac{(60)(10)(30)^2}{(40)^2} = -225 \text{ ft-kips}$$

The complete analysis is shown in Fig. 13-12. Note that the intermediate

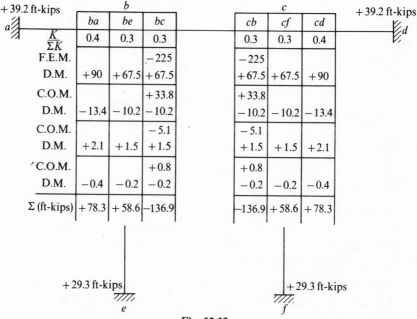

Fig. 13-12

values of the moments at ends a, d, e, and f are not shown. Since members ab, cd, be, and cf are not loaded, it is evident that

$$M_{ab} = \tfrac{1}{2}M_{ba} = 39.2 \text{ ft-kips} \qquad M_{dc} = \tfrac{1}{2}M_{cd} = 39.2 \text{ ft-kips}$$

$$M_{eb} = \tfrac{1}{2}M_{be} = 29.3 \text{ ft-kips} \qquad M_{fc} = \tfrac{1}{2}M_{cf} = 29.3 \text{ ft-kips}$$

From Fig. 13-12 the values of the moments obtained on the left side of the center line of the structure are exactly the same as those on the right. Such a special display is referred to as *antisymmetry*, which yields

$$\theta_b = \theta_c$$

An adjustment can be made to the stiffness of the center beam bc that will permit the analyst to work with only half the structure. See Sec. 13-7 for modified stiffness. This problem will be re-solved in Example 13-9.

13-7. MODIFIED STIFFNESSES

The examples given in Sec. 13-6 have illustrated three special cases that suggest that some modifications for simplifying the moment-distribution process might be found by recognizing certain known conditions.

Consider the frame subjected to a clockwise external moment M applied to the connecting joint as shown in Fig. 13-13(a) for which we note the following:

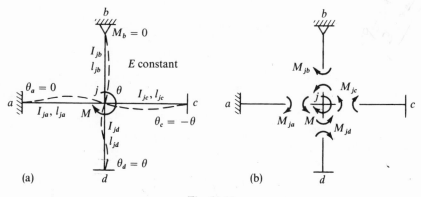

Fig. 13-13

1. $\theta_a = 0$; i.e., the member is fixed at end a.
2. $M_b = 0$; i.e., the member is simply supported at end b.
3. $\theta_c = -\theta$; i.e., the member rotates through an equal but opposite angle at the other end c as in the case of symmetry.
4. $\theta_d = \theta$; i.e., the member rotates through the same angle at the other end d as in the case of antisymmetry.

Referring to Fig. 13-13(b), we find the equilibrium of joint j requires

$$M_{ja} + M_{jb} + M_{jc} + M_{jd} = M \tag{13-9}$$

From the slope-deflection equations we obtain

$$M_{ja} = 4EK_{ja}\theta \tag{13-10}$$

$$M_{jb} = 2EK_{jb}(2\theta + \theta_b)$$

but $\qquad M_{bj} = 2EK_{jb}(2\theta_b + \theta) = 0$

or $\qquad\qquad \theta_b = -\dfrac{\theta}{2}$

Hence $\qquad M_{jb} = 2EK_{jb}\left(2\theta - \dfrac{\theta}{2}\right) = 4E\left(\dfrac{3}{4}K_{jb}\right)\theta$

$$= 4EK'_{jb}\theta \tag{13-11}$$

where we let

$$K'_{jb} = \tfrac{3}{4}K_{jb} \tag{13-12}$$

K'_{jb} being called the *modified stiffness factor* for member jb. Similarly,

$$M_{jc} = 2EK_{jc}(2\theta - \theta) = 4E(\tfrac{1}{2}K_{jc})\theta$$

$$= 4EK'_{jc}\theta \tag{13-13}$$

where

$$K'_{jc} = \tfrac{1}{2}K_{jc} \tag{13-14}$$

$$M_{jd} = 2EK_{jd}(2\theta + \theta) = 4E(\tfrac{3}{2}K_{jd})\theta$$

$$= 4EK'_{jd}\theta \tag{13-15}$$

where

$$K'_{jd} = \tfrac{3}{2}K_{jd} \tag{13-16}$$

Substituting Eqs. 13-10, 13-11, 13-13, and 13-15 in Eq. 13-9 yields:

$$4E(K_{ja} + K'_{jb} + K'_{jc} + K'_{jd})\theta = M$$

or

$$\theta = \dfrac{M}{4E\sum K'} \tag{13-17}$$

where $\qquad \sum K' = K_{ja} + K'_{jb} + K'_{jc} + K'_{jd}$

Substituting Eq. 13-17 in Eqs. 13-10, 13-11, 13-13, and 13-15 yields

$$M_{ja} = \dfrac{K_{ja}}{\sum K'}M$$

$$M_{jb} = \dfrac{K'_{jb}}{\sum K'}M$$

$$M_{jc} = \dfrac{K'_{jc}}{\sum K'}M \tag{13-18}$$

$$M_{jd} = \dfrac{K'_{jd}}{\sum K'}M$$

Thus it is seen that, when an external moment is applied to joint j, the distributed moments to the near ends of the members meeting at the joint are in direct proportion to their modified stiffness factors if the conditions of the far ends are known. By using the modified stiffness factor for one end, we actually eliminate the carry over to the other end except writing down the final result for that part.

Let us recapitulate the modified stiffness factors for various end conditions:

1. If the one end is simply supported, the modified stiffness factor for the other end is given by

$$K' = \tfrac{3}{4}K \tag{13-19}$$

2. If the one end is symmetrical to the other end, then

$$K' = \tfrac{1}{2}K \tag{13-20}$$

3. If the one end is antisymmetrical to the other end, then

$$K' = \tfrac{3}{2}K \tag{13-21}$$

K' denoting the modified stiffness factor. In parallel with Eq. 13-1, we may have

$$S' = 4EK' \tag{13-22}$$

S' being called the *modified stiffness*. A general definition for *modified stiffness* is the end moment required to produce a unit rotation at this end (simple end), while the other end remains in the actual condition. The modified stiffnesses, hence the modified stiffness factors, for various end conditions may be secured by direct application of this definition, as will be discussed in Sec. 15-6.

The following examples illustrate the process of moment distribution by using the modified K values.

Example 13-6. Analyze the symmetrical beam in Fig. 13-14 by using modified K value in the center span.

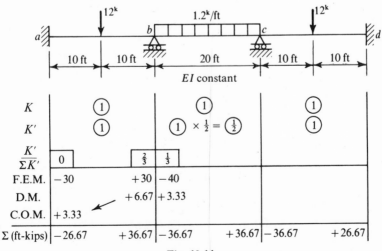

Fig. 13-14

Example 13-7. Analyze the symmetrical frame in Fig. 13-15, which was solved by the methods of consistent deformations, least work, and slope deflection. Once again the method of moment distribution demonstrates its superiority over any other method previously discussed.

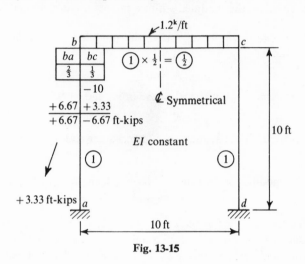

Fig. 13-15

Example 13-8. Analyze the beam in Fig. 13-16 by using modified K for end b of member bc.

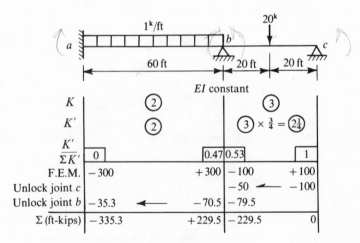

Fig. 13-16

Example 13-9. Analyze the frame in Fig. 13-11 by using the modified K value in the center span for the antisymmetrical system. The solution is given in Fig. 13-17.

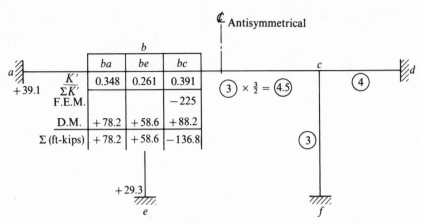

Fig. 13-17

PROBLEMS

13-1. Solve the end moments for Prob. 12-1 by moment distribution.

13-2. Solve the end moments for Prob. 12-2 by moment distribution.

13-3. Solve the end moments for Prob. 12-3 by moment distribution.

13-4. Solve the end moments for Prob. 12-4 by moment distribution: (1) not using modified stiffness; (2) using modified K value in the center span.

13-5. Analyze the box shown in Fig. 13-18 by moment distribution. Assume constant EI.

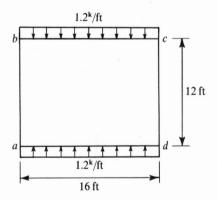

Fig. 13-18

13-6. Analyze the continuous beam in Fig. 13-19 by moment distribution. Assume constant *EI*.

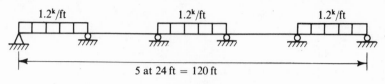

Fig. 13-19

13-7. Analyze the continuous beam in Fig. 13-20 by moment distribution. Take advantage of modified stiffnesses, replacing the unsymmetrical loading system with a symmetrical and an antisymmetrical system. Assume constant *EI*.

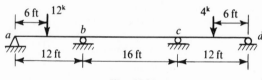

Fig. 13-20

14

MOMENT DISTRIBUTION WITH JOINT

TRANSLATIONS

14-1. GENERAL

The procedure of moment distribution discussed in Ch. 13 is based on the restriction that the joints of the structure do not move. However, many frames encountered in practice undergo joint translations, and therefore, the process of moment distribution, previously described, cannot be directly applied without certain adjustments. There are, in general, two methods available for frames with joint translations in an analysis by moment distribution. One is the direct method, which includes the balancing of end shears; the other is the indirect method, which requires the set up and elimination of restraints to joint translation and, thus, involves the solution of simultaneous equations for frames with several degrees of joint translation. However, it is the latter method which seems, in most cases, more acceptable to a beginner and, therefore, will be discussed in this chapter.

Let us consider two types of loaded frames mounted on nonyielding supports in which joint translations are involved. The first is a frame under the action of a lateral force applied at a joint, such as the one shown in Fig. 14-1 (a); the second is a frame that, together with loads acting on its members,

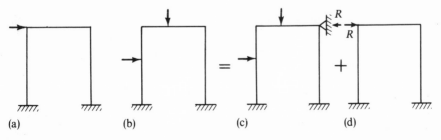

(a) (b) (c) (d)

Fig. 14-1

forms an unsymmetrical system, such as the one shown in Fig. 14-1(b). To handle the latter type of frame, we may resort to the principle of superposition.

291

Artificial holding force to prevent joint translation is imposed on the structure and subsequently eliminated. Thus the frame in Fig. 14-1(b) can be considered as the superposed effect of the two separate systems indicated in Figs. 14-1(c) and (d). In the first place (Fig. 14-1(c)), the translation of joints is prevented by providing an artificial support at the top of the column so that moment distribution can be carried out in the usual manner. The required holding force R is then obtained by statics. The next step (Fig. 14-1(d)) is to eliminate the artificial restraint by applying to the top of column a lateral force equal to R but opposite in direction. The resulting configuration is the same as that shown in Fig. 14-1(a). Therefore, the problem now reduces to dealing with frame under lateral forces applied at the joints.

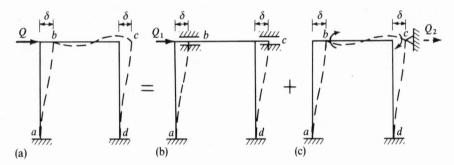

Fig. 14-2

To handle this type of frame, we consider the frame in Fig. 14-2(a) in which joints a and d are fixed, whereas joints b and c undergo both translation and rotation because of the lateral force Q applied at the top of column. It is obvious that if the joint translation of b is specified as δ, then the joint translation of c is determined and, in this case, is also δ. Now the distortion imposed on joints b and c may be regarded as the superposed effect of the two following, separate steps:

1. Translation without rotation (see Fig. 14-2(b)).
2. Rotation without translation (see Fig. 14-2(c)).

In Step 1 joints b and c are locked against rotation ($\theta_b = \theta_c = 0$), and the joint translation δ is produced by applying a lateral force Q_1. It is clear that some external restraints, i.e., *locking moments*, are required at joints b and c in order to hold both joints against rotation. Also, end moments will be induced in the columns from which the lateral force Q_1 can be figured.

In Step 2 further joint translations are checked by providing an artificial support at the top of the column. Joints b and c are then unlocked so that they finally rotate to their actual position. Recall that to unlock a joint means to apply a moment equal but opposite to the locking moment at the joint. From the resulting end moments, the holding force Q_2 at the artificial support can be figured.

From the point of view of moment distribution Step 1, which consists of locking all joints against rotation and performing joint translations by the force applied to joint, gives the fixed-end moment; and Step 2, unlocking all joints and allowing no further joint translation, gives distributed moments and carry-over moments. These two steps parallel the procedures for analyzing frames without joint translation discussed in the preceding chapter, which involve first locking all joints and applying loads to the members—fixed-end moment—and then unlocking all joints—distributed moments and carry-over moments.

Refer to Fig. 14-2. Since the analysis of the frame in Fig. 14-2(a) can be carried out by superposing the effects given in Figs. 14-2(b) and (c), the values of Q_1 and Q_2, determined from the resulting end moments in each of the separate cases, must satisfy $Q_1 + Q_2 = Q$. However, we usually do not know the value of δ from which the fixed-end moment is derived at the outset. To solve this difficulty, we actually start by applying an arbitrary lateral joint translation, say Δ, to the same structure. From this a set of fixed-end moments can be determined (Fig. 14-3(a)). After the joints are unlocked, the distributed

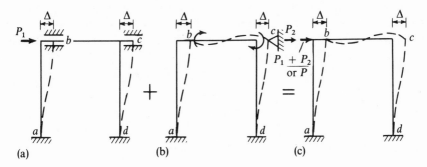

Fig. 14-3

moments and carry-over moments are obtained (Fig. 14-3(b)). In this manner, we reach an independent set of balanced force systems corresponding to the assumed joint translation Δ or the laterally applied force P (Fig. 14-3(c)) which is figured from the end moments obtained in Figs. 14-3(a) and (b). With this done, and since the structure is linear, the solution for the frame in Fig. 14-2(a) can be found by the direct proportion Q/P. We shall discuss the fixed-end moment due to joint translation and illustrate the procedures mentioned above in the sections to follow.

14-2. FIXED-END MOMENT DUE TO JOINT TRANSLATION

Figure 14-4(a) shows a frame with the length and moment of inertia for each member indicated. We begin by locking all joints against rotation so that

$$\theta_a = \theta_b = \theta_c = \theta_d = 0$$

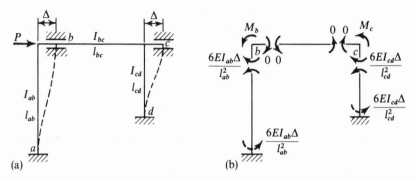

Fig. 14-4

Next, we apply a lateral force P to the top of column under which force the frame will deflect sidewise, and the column tops will move a horizontal distance Δ. End moments are then set up in the columns ab and cd because of the relative end deflection Δ. These end moments, often referred to as *fixed-end moments due to joint translation* (sidesway), can be obtained easily by applying the slope-deflection equation; i.e.,

$$M_{ab} = M_{ba} = 2E\frac{I_{ab}}{l_{ab}}\left(-\frac{3\Delta}{l_{ab}}\right) = -\frac{6EI_{ab}\Delta}{l_{ab}^2} \tag{14-1}$$

$$M_{cd} = M_{dc} = 2E\frac{I_{cd}}{l_{cd}}\left(-\frac{3\Delta}{l_{cd}}\right) = -\frac{6EI_{cd}\Delta}{l_{cd}^2} \tag{14-2}$$

Replacing I/l with K and Δ/l with R, we obtain

$$M_{ab} = M_{ba} = -6EK_{ab}R_{ab} \tag{14-1a}$$

$$M_{cd} = M_{dc} = -6EK_{cd}R_{cd} \tag{14-2a}$$

It is apparent that $M_{bc} = M_{cb} = 0$

since $\theta_b = \theta_c = 0$ and there is no member rotation for bc.

These end moments are shown in Fig. 14-4(b) for which we note that the equilibrium of the moments at joint b requires a counter-clockwise external moment at b, called M_b, of the value $6EI_{ab}\Delta/l_{ab}^2$ in order to lock joint b against rotation. Similarly, a counter-clockwise external moment M_c equal to $6EI_{cd}\Delta/l_{cd}^2$ is needed at c to keep $\theta_c = 0$ during sidesway of the bent.

In general, the fixed-end moments developed in a member as the result of a pure relative end displacement Δ are given by

$$M^F = -\frac{6EI\Delta}{l^2} \qquad (14\text{-}3)$$

or

$$M^F = -6EKR \qquad (14\text{-}3a)$$

if we assume that the member rotates in a positive direction (clockwise).

It should be pointed out that, when applying Eq. 14-3, in most cases the exact value of Δ is not necessary, and the relative value of I/l^2 or K/l is usually used in frame analysis.

14-3. ANALYSIS OF STATICALLY INDETERMINATE RIGID FRAMES WITH ONE DEGREE OF FREEDOM OF JOINT TRANSLATION BY MOMENT DISTRIBUTION

Figure 14-5(a) shows a frame composed of members of uniform cross section and subjected to a transverse load of 30 kips acting to the right. The analysis of it by the method of moment distribution is contained in the following steps:

1. Suppose that the structure of Fig. 14-5(a) is held against rotation at all joints. An arbitrary lateral force P_1 is then introduced so that the frame deflects to the right with a horizontal displacement Δ at the column tops as indicated in Fig. 14-5(b). The fixed-end moments resulting from sidesway are then induced in columns ab and cd according to Eq. 14-3. Hence

$$M^F_{ab} = M^F_{ba} = -\frac{6EI\Delta}{400}$$

$$M^F_{cd} = M^F_{dc} = -\frac{6EI\Delta}{100}$$

Note that Δ is an arbitrary horizontal displacement corresponding to some lateral force P_1; the value of Δ is not known at the outset. Let us obtain a set (any set) of balanced force systems. Then we are free to assign any convenient value to Δ or in this case any convenient value to $6EI\Delta$. In other words, we assume any set of fixed-end moments inversely proportional to the square length of the member. Thus, if we choose

$$M^F_{ab} = M^F_{ba} = -100 \text{ ft-kips}$$

then $\qquad M^F_{cd} = M^F_{dc} = -400 \text{ ft-kips}$

Having recorded these as in Fig. 14-5(c), we can calculate the end shears at a and d as indicated and figure

$$P_1 = 90 \text{ kips}$$

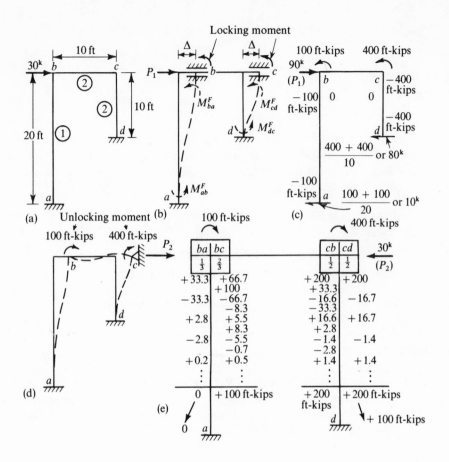

(a)

Unlocking moment (b)

(c)

(d)

(e)

(f)

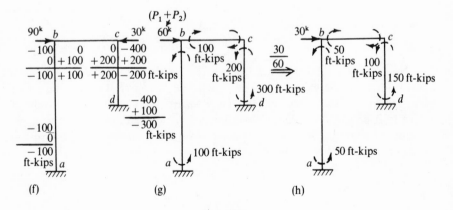

(g)

(h)

Fig. 14-5

from the equilibrium condition $\sum F_x = 0$ for the whole frame. Meanwhile, from the equilibrium of joints b and c, we observe that we must apply an artificial locking moment of -100 ft-kips at joint b and -400 ft-kips at joint c to keep $\theta_b = \theta_c = 0$. In this manner, we reach a system of balanced forces under the condition of joint translation without rotation, as shown in Fig. 14-5(c).

2. Next, let us introduce an imaginary lateral support at c so as to prevent the frame from further joint translation. We unlock joints b and c by applying a couple of $+100$ ft-kips at joint b and a couple of $+400$ ft-kips at joint c, as shown in Fig. 14-5(d). We then distribute moments at the near ends and carry over moments to the far ends—the result of unlocking moments following the usual moment-distribution procedures, as recorded in Fig. 14-5(e). The resulting moments are found to be consistent with a lateral force P_2 equal to -30 kips (acting to the left at c) in order to keep the frame from joint translation during the process of unlocking. In this manner, we obtain a set of balanced forces under the condition of joint rotation without translation.

3. The sum of the results of Steps 1 and 2 is given in Fig. 14-5(f) or (g).

4. Multiplying this set of balance forces (Fig. 14-5(g)) with a ratio of 30/60 will give the solution for the frame of Fig. 14-5(a), as shown in Fig. 14-5(h).

This procedure for solving the frame in Fig. 14-5(a) is rather lengthy; however, the data can be presented in a more compact form as in Fig. 14-6 which involves the following:

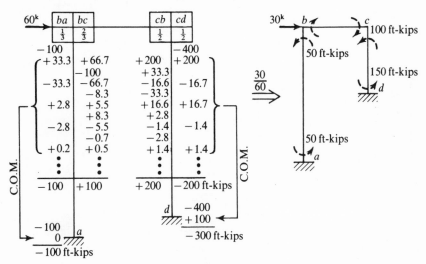

Fig. 14-6

1. Assume fixed-end moments for those members having relative end deflections in proportion to their values of $-6EI\Delta/l^2$. In this problem, we use

$$M_{ab}^F = M_{ba}^F = -100 \qquad M_{cd}^F = M_{dc}^F = -400$$

2. Distribute, carry over, and add together to obtain all the end moments.

3. Find the lateral force consistent with the end moments obtained.

4. Proportioning the result with a ratio of *given lateral force* to *obtained lateral force* will give the solution.

In case one of the ends, say end d, of the frame is hinged (see Fig. 14-7(a)),

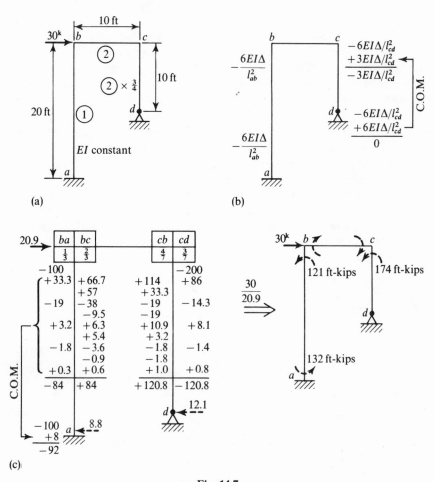

Fig. 14-7

then $M_d = 0$. The procedure of moment distribution can be simplified by using modified stiffness for member cd; i.e.,

$$K'_{cd} = \tfrac{3}{4} K_{cd}$$

as indicated.

Meanwhile, by releasing end d and rendering it to its actual condition $M_d = 0$, one-half the amount of the distributed moment will be carried over to end c. The modified fixed-end moment at c of member cd due to joint translation Δ is therefore equal to $-3EI\Delta/l^2_{cd}$ (see Fig. 14-7(b)).

Generally, for a member pin-connected at one end, the modified fixed-end moment at the other end due to lateral joint translation Δ is given by

$$M^F = -\frac{3EI\Delta}{l^2} = -3EKR \tag{14-4}$$

which is one-half the value previously obtained by assuming both ends fixed.

The analysis of the frame in Fig. 14-7(a) is shown in Fig. 14-7(c). We start by assuming $M^F_{ab} = M^F_{ba} = -100$ and $M^F_{cd} = -200$ consistent with the values given in Fig. 14-7(b). From the resulting end moments, we find that the end shear at a is 8.8 kips and that at d is 12.1 kips, both acting to the left. These correspond to a lateral force of 20.9 kips on the top of column acting to the right. Multiplying the obtained set of balanced force system by the ratio 30/20.9 gives the solution. More illustrative examples follow.

Example 14-1. In Fig. 14-8(a) is shown a loaded one-story bent with an inclined leg, the relative K value for each member is encircled. The end moments were solved by slope deflection in Example 12-6. Let us now re-solve them by moment distribution.

We begin by finding the relative end displacement for each member as shown in Fig. 14-8(b). Next, we assume fixed-end moments, due to consistent joint translations, in proportion to the value of $-6EK\Delta/l$,

$$M^F_{ab} = M^F_{ba} = -\frac{6E(1)(\Delta)}{15} \qquad \text{say} \quad -100$$

$$M^F_{bc} = M^F_{cb} = +\frac{6E(2)[(3/4)\Delta]}{15} \qquad \text{say} \quad +150$$

$$M^F_{cd} = M^F_{dc} = -\frac{6E(2)[(5/4)\Delta]}{25} \qquad \text{say} \quad -150$$

The process of moment distribution for this particular set up is shown in Fig. 14-8(c). Referring to Fig. 14-8(d), by applying $\sum M_o = 0$ for the entire frame, we find the resulting moments are consistent with a horizontal force $P = 42.5$ kips acting at b; i.e.,

$$(15.15)(35) + (11.5)(50) - 109.1 - 145.5 - 20P = 0$$

$$P = 42.5 \text{ kips}$$

The final result, which is obtained by multiplying all the moments in Fig. 14-8(d) by the ratio 100/42.5, is shown in Fig. 14-8(e).

Example 14-2. Determine all the end moments of the loaded frame in Fig. 14-9(a).

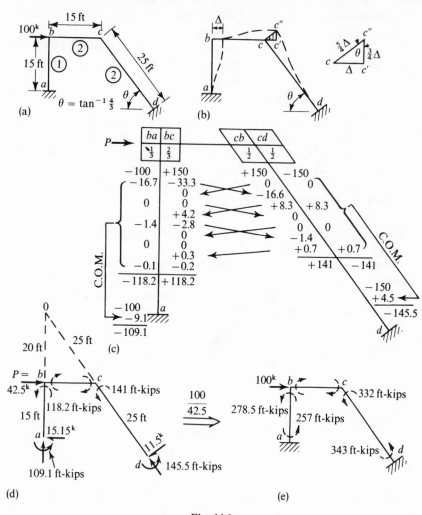

Fig. 14-8

The complete analysis is contained in the following steps:

1. Hold the loaded frame at the top of the column, say at joint c, against sidesway, and obtain the end moments by the usual moment-distribution procedures (Fig. 14-9(b)).

2. From $\sum F_x = 0$ for the entire frame, calculate the holding force needed to prevent sidesway. In this case, it is

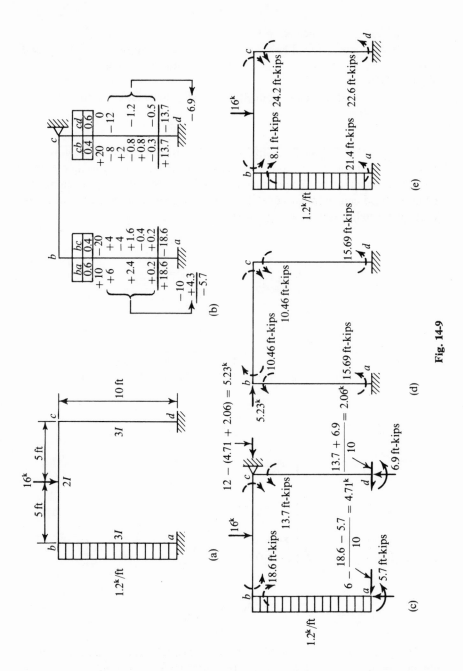

Fig. 14-9

301

$$12 - (4.71 + 2.06) = 5.23 \text{ kips}$$

acting to the left, as indicated in Fig. 14-9(c).

3. Remove the artificial holding force by the application of an equal and opposite force at the top of the column, and find the resulting end moments (see Fig. 14-9(d)).

4. Add the end moments from Steps 1 and 3 to obtain the final solution (see Fig. 14-9(e)).

Note that the analysis for Step 3 (Fig. 14-9(d)) can easily be accomplished by taking advantage of the antisymmetry of the system, i.e., by using modified stiffness for member bc. The complete analysis, which begins by assuming an arbitrary fixed-end moment equal to -100 ft-kips for the column because of joint translation, is shown in Fig. 14-10.

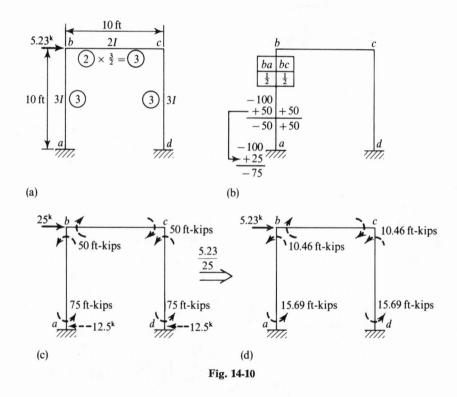

Fig. 14-10

The technique employed in the preceding examples can be used in analyzing any frame having one degree of freedom of joint translation.

14-4. ANALYSIS OF STATICALLY INDETERMINATE RIGID
FRAMES WITH TWO DEGREES OF FREEDOM OF
JOINT TRANSLATION BY MOMENT DISTRIBUTION

A rigid frame having two degrees of freedom of joint translation can be analyzed by breaking it down to two independent cases in each of which only one degree of freedom of joint translation is allowed to occur.

Consider the two-story frame in Fig. 14-11(a). To handle it, let us refer to

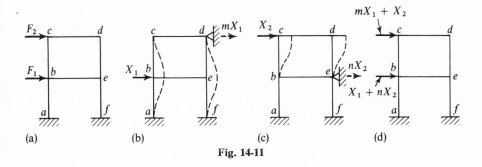

Fig. 14-11

two separate cases, as shown in Figs. 14-11(b) and (c), each involving only one degree of freedom of joint translation. Each of these cases can be analyzed by the method of moment distribution previously described. For example, in the case in Fig. 14-11(b) joints c and d are held from translation by providing a lateral support at d, whereas joints b and e are displaced horizontally because of a force X_1 applied laterally at b. A moment-distribution solution can then be obtained, and all the end moments, shearing forces, and reactions are in terms of X_1. Let mX_1 denote the corresponding reaction of the lateral support at d, m being a constant of proportionality. Note that the deflected shape (dotted line) indicates the initial position of the frame where the joint displacement of b and e has been introduced with all joints held against rotation. The final elastic curve after the joints have been released is not shown.

A similar solution can be carried out for the case in Fig. 14-11(c) where the pushing force is X_2 and the reaction of the lateral support at e is nX_2, n being a constant of proportionality.

The superposition of effects in the cases shown in Figs. 14-11(b) and (c) results in the case shown in Fig. 14-11(d) in which all the internal forces and reactions can be found as the linear combination of X_1 and X_2. Comparing Fig. 14-11(d) with Fig. 14-11(a), we see that (d) will be the solution of (a) by solving X_1 and X_2 from

$$X_1 + nX_2 = F_1 \tag{14-5}$$

$$mX_1 + X_2 = F_2 \qquad (14\text{-}6)$$

A similar procedure can be used to analyze the two-stage bent shown in Fig. 14-12(a) or the gable bent shown in Fig. 14-12(b).

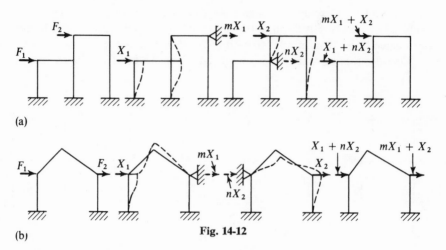

(a)

(b)   **Fig. 14-12**

Example 14-3. Find all the end moments for the frame in Fig. 14-13. Assume the same EI for all members. This is the same problem as Example 12-8, which was solved by the method of slope deflection.

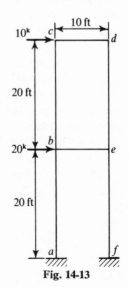

Fig. 14-13

To determine all the end moments by the method of moment distribution, we begin with the analysis of the frame in Fig. 14-14(a). The antisymmetry of

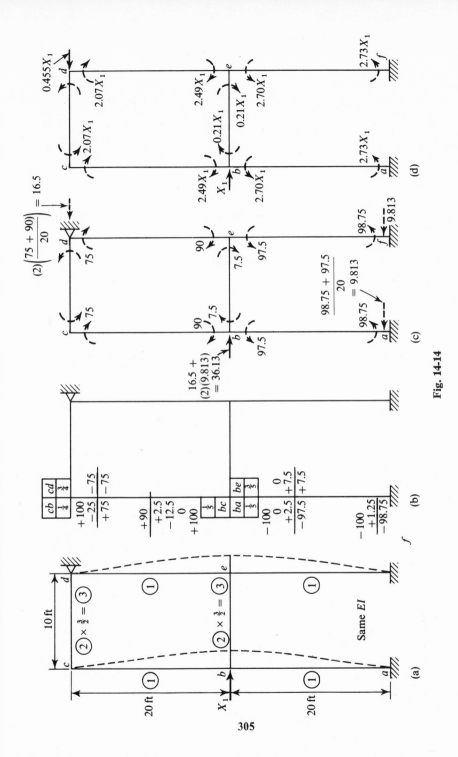

Fig. 14-14

305

the system permits us to use modified stiffnesses for members be and cd and to work with only half the structure while balancing moments, as shown in Fig. 14-14(b). In this case joints c and d are held from translation by the lateral support at d, and joints b and e are displayed horizontally an equal distance because of the lateral action applied at b. Consistent with the joint translation and the property of structure, we may assume the fixed-end moments for the columns as follows:

$$M_{ab}^F = M_{ba}^F = M_{ef}^F = M_{fe}^F = -100$$

$$M_{bc}^F = M_{cb}^F = M_{de}^F = M_{ed}^F = +100$$

After releasing the joints and balancing the moments, we obtain a set of end moments (see Fig. 14-14(b)) consistent with a pushing force of 36.13 applied at b and a lateral reaction of 16.5 acting to the left at d (see Fig. 14-14 (c)). Note that the reaction at d is first obtained from the balance of shear of the upper story; i.e., from $\sum F_x = 0$ for the portion of frame just above the level be. The pushing force at b is then determined by applying $\sum F_x = 0$ for the entire frame.

Multiplying the result of Fig. 14-14(c) by the ratio $X_1/36.13$ gives the solution of Fig. 14-14(a), as shown in Fig. 14-14(d).

Next, let us follow the same procedure and analyze the frame in Fig. 14-15 (a) in which joints b and e are held from translation owing to the lateral support at e, whereas joints c and d are displayed horizontally an equal distance due to a pushing force applied at c. Consistent with the sidesway and the property of structure, the fixed-end moments for the columns bc and de may be assumed to be

$$M_{bc}^F = M_{cb}^F = M_{de}^F = M_{ed}^F = -100$$

Taking advantage of antisymmetry by using modified stiffnesses in members be and cd and working with only half the structure, we complete the process of moment distribution as shown in Fig. 14-15(b). The resulting end moments are found to be consistent with a pushing force of 13.75 applied at c and a lateral reaction of 16.38 acting to the left at e (see Fig. 14-15(c)). In this case the force at c is first obtained from the balance of shear of the upper story, and the reaction at e is then determined from $\sum F_x = 0$ for the entire frame.

Multiplying the result of Fig. 14-15(c) by the ratio $X_2/13.75$ gives the solution for the frame in Fig. 14-15(a), as shown in Fig. 14-15(d).

Let us now sum up the findings in the above two steps (Fig. 14-14(d) and Fig. 14-15(d)) as shown in Fig. 14-16(a). Moments and forces are taken positive following the indicated directions.

For solving the frame in Fig. 14-13, we set

$$X_1 - 1.19X_2 = 20 \qquad (14-7)$$

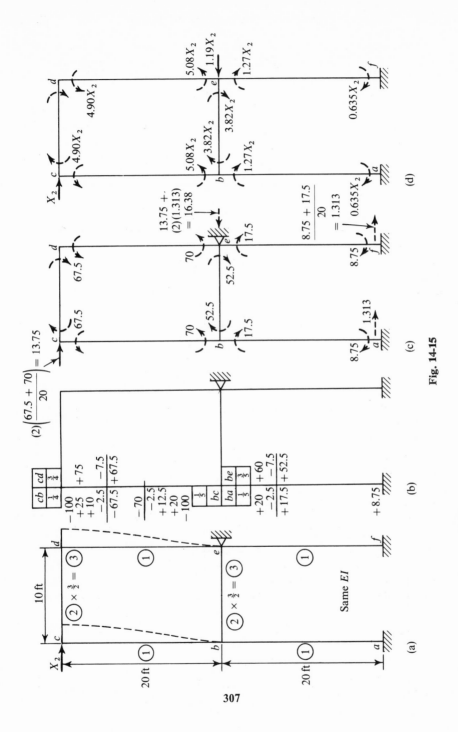

307

Fig. 14-15

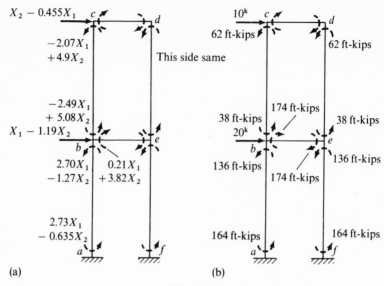

Fig. 14-16

$$X_2 - 0.455X_1 = 10 \qquad (14\text{-}8)$$

from which
$$X_1 = 69.55$$
$$X_2 = 41.64$$

Substituting these in Fig. 14-16(a) yields the desired solution, as shown in Fig. 14-16(b).

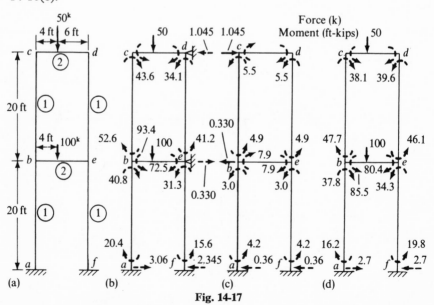

Fig. 14-17

The small discrepancy between the above result and the more exact solution obtained by the slope-deflection method (see Example 12-8) is due to the fact that we complete the moment distribution in each of above cases with only two cycles. However, two cycles have provided sufficient accuracy for practical purposes.

Example 14-4. Find all the end moments for the frame in Fig. 14-17(a). Assume the same EI for all members.

The analysis is contained in the following steps:

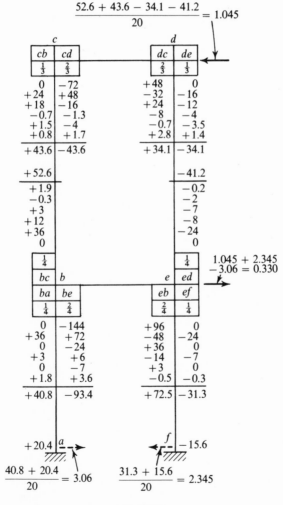

$$\frac{52.6 + 43.6 - 34.1 - 41.2}{20} = 1.045$$

c			d	
cb	cd		dc	de
$\frac{1}{3}$	$\frac{2}{3}$		$\frac{2}{3}$	$\frac{1}{3}$
0	−72		+48	0
+24	+48		−32	−16
+18	−16		+24	−12
−0.7	−1.3		−8	−4
+1.5	−4		−0.7	−3.5
+0.8	+1.7		+2.8	+1.4
+43.6	−43.6		+34.1	−34.1
+52.6				−41.2
+1.9				−0.2
−0.3				−2
+3				−7
+12				−8
+36				−24
0				0

$$\frac{1.045 + 2.345}{-3.06} = 0.330$$

bc	b		e	ed
ba	be		eb	ef
$\frac{1}{4}$	$\frac{2}{4}$		$\frac{2}{4}$	$\frac{1}{4}$
0	−144		+96	0
+36	+72		−48	−24
0	−24		+36	0
+3	+6		−14	−7
0	−7		+3	0
+1.8	+3.6		−0.5	−0.3
+40.8	−93.4		+72.5	−31.3

$+20.4$ *a* →   *f* ← -15.6

$$\frac{40.8 + 20.4}{20} = 3.06$$ $$\frac{31.3 + 15.6}{20} = 2.345$$

Fig. 14-18

1. The frame is completely prevented from sidesway by introducing lateral supports at joints d and e. Following the usual procedures for moment distribution, we determine all the end moments and find that they are consistent with a lateral holding force of 1.045 kips acting to the left at d and a lateral holding force of 0.330 kips acting to the right at e. The results are recorded, as in Fig. 14-17(b), by the dashed symbols. (See also Fig. 14-18 for the process of moment distribution in this step.)

2. Eliminate these artificial restraints by applying a set of equal and opposite forces to the corresponding joints, as shown in Fig. 14-17(c). To analyze it, we take advantage of the results of the previous example (see Fig. 14-16(a)) and set

$$X_1 - 1.190X_2 = -0.330 \tag{14-9}$$

$$-0.455X_1 + X_2 = 1.045 \tag{14-10}$$

and solve for $X_1 = 1.99$ $X_2 = 1.95$

Substituting these in Fig. 14-16(a) yields the result recorded in Fig. 14-17(c) by the dashed symbols.

3. The superposition of Steps 1 and 2 will give the final solution of the moments, as in Fig. 14-17(d).

14-5. ANALYSIS OF STATICALLY INDETERMINATE RIGID FRAMES WITH SEVERAL DEGREES OF FREEDOM OF JOINT TRANSLATION BY MOMENT DISTRIBUTION

The procedures of moment distribution used in the analysis of frames having two degrees of joint translation can be extended to the analysis having multiple degrees of freedom with respect to sidesway. For a frame with n degrees of freedom of joint translation, the moment-distribution solution may be broken down into $(n + 1)$ separate cases including:

1. a case in which the frame is held from joint translation by introducing n artificial supports;

2. n independent cases in each of which only one degree of freedom of joint translation is allowed to occur by maintaining $(n - 1)$ holding forces at other points where sidesway would take place.

The superposition of these $(n + 1)$ cases will give the final solution provided that the artificial holding forces in Step 1 are all eliminated by applying the results obtained in Step 2. This necessitates the solution of n simultaneous equations.

To illustrate this procedure, consider the four-story, two-bay building bent, which has four degrees of freedom of joint translation, shown in Fig. 14-19(a).

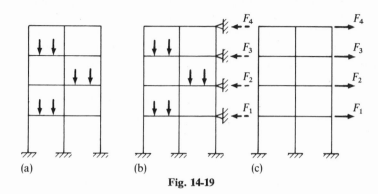

Fig. 14-19

The first step is to introduce artificial lateral supports at all the floors, as in Fig. 14-19(b), to prevent the frame from sidesway. All the loads are then applied, and a regular moment distribution without joint translation is performed to obtain all the end moments from which the reactions (the holding forces) at these supports can be found. We let F_1, F_2, F_3 and F_4 denote these forces.

The next step is to eliminate all of these artificial forces by applying a set of equal and opposite forces at the corresponding points, as shown in Fig. 14-19(c). The sum of the two steps of Figs. 14-19(b) and (c) will give the final solution.

However, to obtain the solution for the frame in Fig. 14-19(c) would require the complete analysis of four independent cases, as shown in Figs. 14-20 (a), (b), (c), and (d). In each of the four cases, only one degree of freedom of

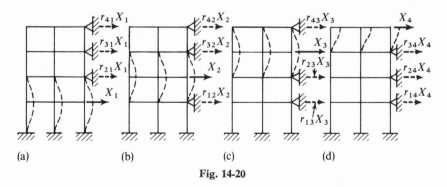

Fig. 14-20

sidesway is involved. We have, therefore, no difficulty in determining all the end moments and the reactions of the frame in terms of the lateral force applied to the floor level where the sidesway is not inhibited. For instance, in Fig. 14-

20(a) a lateral force X_1 is applied to the first floor level which deflects sidewise. Note that the deflected shape (dotted line) indicates only the initial position of the frame before the balancing moments. Following the procedure described in Sec. 14-4, we find the reactions of the lateral supports at the second, third, and fourth floor levels to be $r_{21}X_1$, $r_{31}X_1$, and $r_{41}X_1$ respectively; r_{21}, r_{31}, r_{41} being constants of proportionality. The other cases shown in Fig. 14-20 are similarly conducted. It now remains to find X_1, X_2, X_3, and X_4 by solving the four simultaneous equations,

$$X_1 + r_{12}X_2 + r_{13}X_3 + r_{14}X_4 = F_1 \tag{14-11}$$

$$r_{21}X_1 + X_2 + r_{23}X_3 + r_{24}X_4 = F_2 \tag{14-12}$$

$$r_{31}X_1 + r_{32}X_2 + X_3 + r_{34}X_4 = F_3 \tag{14-13}$$

$$r_{41}X_1 + r_{42}X_2 + r_{43}X_3 + X_4 = F_4 \tag{14-14}$$

With X_1, X_2, X_3, and X_4 determined and substituted in the cases of Figs. 14-20 (a), (b), (c), and (d) respectively, we superpose the results to reach the solution for the frame in Fig. 14-19(c).

14-6. INFLUENCE LINES BY MOMENT DISTRIBUTION

The Müller-Breslau procedures for constructing quantitative influence lines of indeterminate structures given in Secs. 11-5 and 11-6 are often laborious without the aid of a calculating machine. The method of moment distribution, which provides an easy way of determining the end moments of indeterminate beams or frames, can, therefore, be used to establish the influence lines for the end moments in all members and from them deduce all the other influence lines (reaction, shear, bending moment, etc.) by statics.

Consider the three-span uniform beam shown in Fig. 14-21(a). Find the

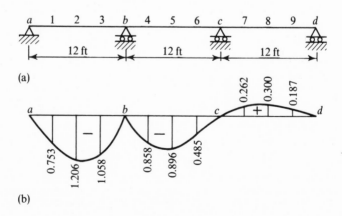

(a)

(b)

Fig. 14-21

ordinates, at intervals of 3 ft, of the influence line for M_{bc} (moment over support b).

The first step is to determine the fixed-end moment coefficients, the end moments in all members caused by applying a unit fixed-end moment at the end of each of the members. For example, when the fixed-end moment of unity occurs to end a, the resulting end moments in all members can be obtained by moment-distribution procedures, as shown in Fig. 14-22(a). Likewise,

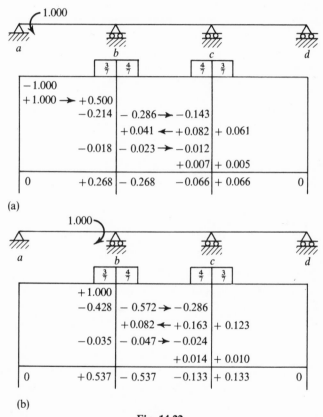

(a)

(b)

Fig. 14-22

when a unit fixed-end moment is applied at end b of member ab, all the end moments are computed as in Fig. 14-22(b).

Once these are done, the effects of a fixed-end moment of unity applied at each of the other ends may be obtained by inspection for the given beam. The complete fixed-end moment coefficients are shown in Fig. 14-23.

With the fixed-end moment coefficients determined, we may rather easily find any of the end moments of the beam due to any loading acting on the span. First compute the fixed-end moments caused by the loading; then

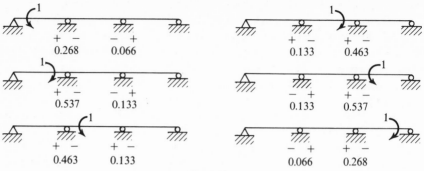

Fig. 14-23

multiply each of them by the corresponding fixed-end moment coefficient for the desired end moment; and finally sum up to obtain the result.

For instance, to find the influence ordinate for M_{bc} at point 1, i.e., 3 feet from the left end (Fig. 14-21(a)), we place a unit load there and compute the fixed-end moments for span ab as

$$M_{ab}^F = \frac{(1)(3)(9)^2}{(12)^2} = 1.69 \qquad \text{(counter-clockwise)}$$

$$M_{ba}^F = \frac{(1)(9)(3)^2}{(12)^2} = 0.56 \qquad \text{(clockwise)}$$

Multiplying each of these by the corresponding fixed-end moment coefficient for M_{bc} and summing up gives

$$M_{bc} = (1.69)(-0.268) + (0.56)(-0.537) = -0.753$$

which is the influence ordinate for M_{bc} at point 1.

Likewise, to obtain the influence ordinate for M_{bc} at point 8, i.e., 6 feet from the right end, we place a unit load at the midpoint of span cd and compute

$$M_{cd}^F = \frac{(1)(12)}{8} = 1.5 \qquad \text{(counter-clockwise)}$$

$$M_{dc}^F = 1.5 \qquad \text{(clockwise)}$$

Thus, $M_{bc} = (1.5)(+0.133) + (1.5)(+0.066) = +0.300$

In this manner, we complete the influence line for M_{bc} as shown in Fig. 14-21(b).

14-7. SECONDARY STRESSES IN TRUSSES SOLVED BY MOMENT DISTRIBUTION

Recall that the elementary stress analysis of a truss (Ch. 4) is based on the following assumptions:

1. The ends of all members of a truss are jointed by smooth pins.

2. The external loads and reactions lie in the same plane as the truss and act only at the pins.

3. The centroidal axis of each member is straight and coincides with the line connecting the joint centers.

4. The weight of each member is neglected.

From these assumptions, it follows that each member in a truss is under purely axial load—tension or compression. The stresses so obtained are called *primary stresses*. In computing primary stresses, we use the original dimensions of the truss since the changes in the bar length are small and negligible. Stresses caused by conditions not taken in the primary stress analysis are called *secondary stresses*. Of these the most significant is due to the fact that in the frameworks used in present-day construction, the joints are seldom pinned but riveted to gusset plates or welded. The angles between the members cannot change freely; hence, the members are actually bent. As a result of this, end moments are set up and thus give rise to the secondary stresses.

In most light frameworks, the effect of the rigidity of the joints is not sufficient to cause any appreciable modification in the axial forces, and it is common practice to analyze them as if they were pin connected. However, for the heavy trusses of bridges or industrial buildings, the stiffness of the joints often introduces secondary stresses too great to be neglected in the design. One way of obtaining approximate values of the secondary stresses is by the moment-distribution procedures, which may be outlined as follows:

1. Calculate the primary stresses by assuming the framework to be pin connected.

2. Find the relative end deflection Δ or rotation R for each member (see Example 8-9 for the rotation of a member).

3. Assume all joints are locked against rotation, and compute the fixed-end moment set up in each member based on $-6EI\Delta/l^2$ or $-6EKR$.

4. As in the moment-distribution method, the joints are then released until joint balance is obtained.

5. The sum of the results obtained for Steps 3 and 4 gives the first approximation of the secondary end moments from which the secondary stresses can be determined.

6. The new stresses so obtained will cause additional joint deflections which cause additional end moments, stresses, etc. However, in most practical cases, only the first approximation of secondary stresses is significant and is used in the design.

Other secondary effects, such as the weight of members and the eccentricity of bar forces, can be handled similarly by moment distribution, i.e., by first locking all joints (obtaining fixed-end moments) then releasing them (balancing moments).

PROBLEMS

14-1. Analyze the frame in Fig. 14-24 by moment distribution.

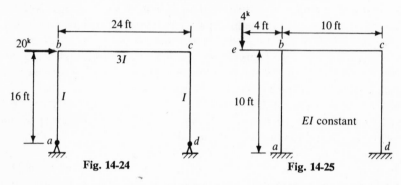

Fig. 14-24   Fig. 14-25

14-2. In the frame of the preceding problem, remove the lateral load of 20 kips, and place a downward vertical load of 12 kips acting on member bc 8 ft from b. Find the end moments for member bc by moment distribution.

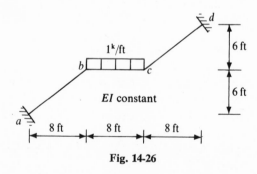

Fig. 14-26

14-3. Analyze the frame in Fig. 14-25 by moment distribution.

14-4. Solve the frame in Fig. 14-26 by moment distribution.

14-5. Solve each frame of Prob. 12-7 by moment distribution.

15-6. Analyze the frame in Fig. 14-27 by moment distribution.

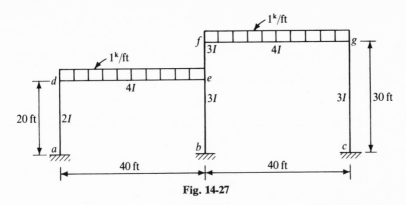

Fig. 14-27

14-7. Use the moment-distribution procedures to compute the ordinates, at 4-ft intervals, of the influence line for the moment over the center support of the four-span uniform beam. The length of each span equals 16 ft.

15

ANALYSIS OF STATICALLY INDETERMINATE BEAMS AND RIGID FRAMES COMPOSED OF NONPRISMATIC MEMBERS

15-1. GENERAL

In this chapter we shall demonstrate that the moment-distribution method can be used in analyzing statically indeterminate beams and frames having non-prismatic members, such as those shown in Fig. 15-1. The fundamental concepts of the moment-distribution procedures are exactly the same as if the structure were composed of prismatic members; however, the expressions for the fixed-end moments, stiffness, and carry-over factors derived specifically for prismatic members are not valid for nonprismatic members. Methods and formulas for obtaining these data for nonprismatic members will first be developed. This done, the analysis of structures is carried out in the usual manner of moment distribution.

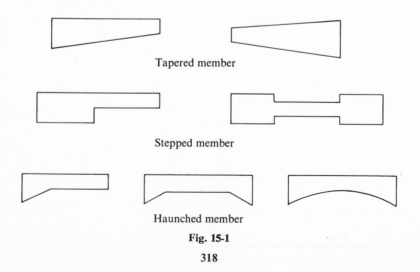

Tapered member

Stepped member

Haunched member

Fig. 15-1

Also the standard slope-deflection equation, derived for application in instances of uniform members, may be modified so as to suit the cases of non-uniform members, as will be discussed in Sec. 15-9.

15-2. INTEGRAL EXPRESSIONS FOR FIXED-END MOMENTS, STIFFNESS, AND CARRY-OVER FACTORS

For a beam of varying EI, with both ends fixed and subjected to bending action caused by loads on the span, the general expressions for the fixed-end moments may be found by the method of least work, using Eqs. 10-5 and 10-6:

$$\int_0^l \frac{M\,dx}{EI} = 0$$

$$\int_0^l \frac{Mx\,dx}{EI} = 0$$

If the beam is made of the same material, then we can assume that E is a constant. The above expressions thus become

$$\int_0^l \frac{M\,dx}{I} = 0 \qquad\qquad (15\text{-}1)$$

$$\int_0^l \frac{Mx\,dx}{I} = 0 \qquad\qquad (15\text{-}2)$$

where both M and I are functions of x.

As an illustration, let us find the fixed-end moment at a for the beam shown in Fig. 15-2 due to a uniform load over the entire span. The unknown reaction components M_{ab}^F and V_a at end a are shown by the dotted line.

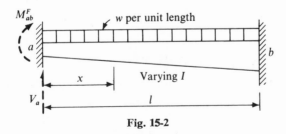

Fig. 15-2

The moment at any section distance x from the left end is

$$M = M_{ab}^F + V_a x - \frac{wx^2}{2}$$

Substituting M in Eqs. 15-1 and 15-2 gives

$$M_{ab}^F \int_0^l \frac{dx}{I} + V_a \int_0^l \frac{x\,dx}{I} - \frac{w}{2}\int_0^l \frac{x^2\,dx}{I} = 0 \qquad\qquad (15\text{-}3)$$

$$M_{ab}^F \int_0^l \frac{x\,dx}{I} + V_a \int_0^l \frac{x^2\,dx}{I} - \frac{w}{2} \int_0^l \frac{x^3\,dx}{I} = 0 \tag{15-4}$$

Eliminating V_a from Eqs. 15-3 and 15-4 yields

$$M_{ab}^F = \left(\frac{w}{2}\right) \frac{\left(\int_0^l (x^2\,dx)/I\right)^2 - \int_0^l (x\,dx)/I \int_0^l (x^3\,dx)/I}{\int_0^l dx/I \int_0^l (x^2\,dx)/I - \left(\int_0^l (x\,dx)/I\right)^2} \tag{15-5}$$

The fixed-end moment at b can also be obtained from Eq. 15-5, by taking the integral origin at b and using reverse sign. For a member of varying I, M_{ab}^F and M_{ba}^F are usually not equal except in a symmetrical system.

For a member of uniform section Eq. 15-5 reduces to

$$M_{ab}^F = -M_{ba}^F = \left(\frac{w}{2}\right) \frac{\left(\int_0^l x^2\,dx\right)^2 - \int_0^l x\,dx \int_0^l x^3\,dx}{\int_0^l dx \int_0^l x^2\,dx - \left(\int_0^l x\,dx\right)^2} = -\frac{wl^2}{12}$$

To develop an expression for stiffness and carry-over factors for a member of varying I, recall the definition of *stiffness* for an end of a member as the end moment required to produce a unit rotation at this end (simple end) while the other end is fixed; and the definition of *carry-over factor* as the ratio of the induced moment at the other end (fixed) to the applied moment at this end. Consider the beam in Fig. 15-3 which is subjected to an end moment M_{ab} at

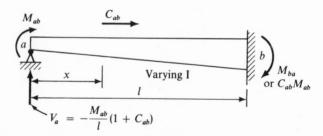

Fig. 15-3

end a (simple end). The induced moment at the other end b (fixed) is, by definition, $C_{ab}M_{ab}$, C_{ab} being the carry-over factor from a to b. Also indicated in Fig. 15-3 is the reaction or end shear at a, called V_a, expressed in terms of M_{ab} and C_{ab}. The moment at any section distance x from the left end is

$$M = M_{ab} + V_a x = M_{ab}\left[1 - \frac{x}{l}(1 + C_{ab})\right] \tag{15-6}$$

Since the vertical deflection at end a is zero,

$$\frac{\partial W}{\partial V_a} = \int_0^l \frac{M(\partial M/\partial V_a)\,dx}{I} = 0 \tag{15-7}$$

Substituting Eq. 15-6 and $\partial M/\partial V_a = x$ in Eq. 15-7, we have

$$M_{ab} \int_0^l \frac{[1 - (x/l)(1 + C_{ab})]x\,dx}{I} = 0$$

from which

$$C_{ab} = \frac{l\int_0^l (x\,dx)/I - \int_0^l (x^2\,dx)/I}{\int_0^l (x^2\,dx)/I} \tag{15-8}$$

The carry-over factor from b to a, called C_{ba}, can be obtained from the same expression by taking the integral origin at b. Note that for a member of nonuniform cross section C_{ab} and C_{ba} are usually not equal. For a member of constant cross section, Eq. 15-8 reduces to

$$C_{ab} = C_{ba} = \frac{l\int_0^l x\,dx - \int_0^l x^2\,dx}{\int_0^l x^2\,dx} = \frac{1}{2}$$

Refer to Fig. 15-3. If the end rotation at a, called θ_a, is made unity, then the applied end moment M_{ab} is, by definition, the stiffness of end a, denoted by S_{ab}. Using the least-work expression, we have

$$\theta_a = \frac{\partial W}{\partial M_{ab}} = \int_0^l \frac{M(\partial M/\partial M_{ab})\,dx}{EI} = 1 \tag{15-9}$$

Substituting
$$M = M_{ab}\left[1 - \frac{x}{l}(1 + C_{ab})\right]$$

$$= S_{ab}\left[1 - \frac{x}{l}(1 + C_{ab})\right] \tag{15-10}$$

in Eq. 15-9 yields
$$S_{ab}\int_0^l \frac{[1 - (x/l)(1 + C_{ab})]\,dx}{EI} = 1$$

or
$$S_{ab} = \frac{1}{\int_0^l [1 - (x/l)(1 + C_{ab})]\,dx/EI}$$

If E is taken as constant, the above expression becomes

$$S_{ab} = \frac{E}{\int_0^l [1 - (x/l)(1 + C_{ab})]\,dx/I}$$

Substituting C_{ab}, as expressed by Eq. 15-8, in the above equation gives

$$S_{ab} = \frac{E\int_0^l (x^2\,dx)/I}{\int_0^l dx/I \int_0^l (x^2\,dx)/I - \left(\int_0^l (x\,dx)/I\right)^2} \tag{15-11}$$

The stiffness of end b of member ab, called S_{ba}, can be obtained from the same equation by taking the integration origin at b. Note that for a member of nonuniform cross section S_{ba} is usually not equal to S_{ab}. For a member of constant cross section, Eq. 15-11 reduces to

$$S_{ab} = S_{ba} = \frac{EI \int_0^l x^2 \, dx}{\int_0^l dx \int_0^l x^2 \, dx - \left(\int_0^l x \, dx \right)^2} = \frac{4EI}{l}$$

Refer to Eqs. 15-5, 15-8, and 15-11. If the term I, a function of x, is given, then these expressions can be calculated either by regular integral procedure, which is often complicated, or by an approximate numerical approach; i.e., dividing the member into a number of segments and calculating each integral involved by replacing integration with summation. The accuracy of the results then depends upon the number of segments into which the member is divided.

15-3. FIXED-END MOMENTS, STIFFNESS, AND CARRY-OVER FACTORS SOLVED BY THE CONJUGATE-BEAM METHOD: NUMERICAL SOLUTION

A more practical way to determine fixed-end moments, stiffness, and carry-over factors for a member having a varying moment of inertia is by the conjugate-beam method, a numerical solution, which will be illustrated in the following examples.

Example 15-1. Given the tapered beam of constant width t, with the depth at the small end equal to d and at the large end $1.5d$, find the fixed-end moments due to a uniform load of intensity w per unit length distributed over the entire length l (see Fig. 15-4(a)).

Let the moment of inertia of the cross section at end a, the small end, be denoted by I_a.

$$I_a = \frac{1}{12} td^3$$

The moment of inertia at any section distance x from end a is given by

$$I_x = \frac{td_x^3}{12} = \frac{td^3}{12} \left(1 + \frac{x}{2l} \right)^3 = I_a \left(1 + \frac{x}{2l} \right)^3$$

If we divide the length of beam into ten equal divisions; i.e., $x = 0.1l, 0.2l, \ldots,$ etc., then the I value at each tenth point is as shown in Fig. 15-4(b).

The bending-moment diagram is drawn in three parts, each in terms of wl^2, M_{ab}^F, and M_{ba}^F as shown in Fig. 15-4(c). Note that the values in terms of

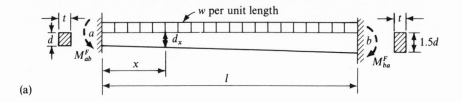

(a)

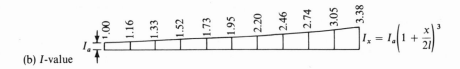

$$I_x = I_a\left(1 + \frac{x}{2l}\right)^3$$

(b) *I*-value

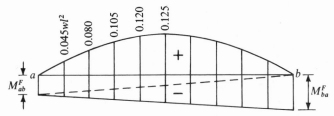

(c) *M*-diagram

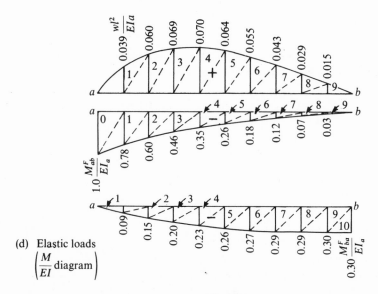

(d) Elastic loads
$\left(\dfrac{M}{EI}\text{ diagram}\right)$

Fig. 15-4

the fixed-end moments at the tenth points are not shown, since they are sufficiently indicated by the linear variation of moments and divisions of length.

The conjugate beam for the fixed-end beam is a floating beam maintaining equilibrium under the elastic load of the M/EI diagram as shown in Fig. 15-4(d).

The numerical arrangement corresponding to the divisions of the diagram of Fig. 15-4(d) for obtaining the total load and total moment about end a from which the unknowns M_{ab}^F and M_{ba}^F are determined is completely given in Table 15-1.

TABLE 15-1

Section	x	wl^2/EI_a		M_{ab}^F/EI_a		M_{ba}^F/EI_a	
		Area	Moment	Area	Moment	Area	Moment
0	$0.033l$	—	—	0.050	0.0017	—	—
1	$0.1l$	0.0039	0.0004	0.078	0.0078	0.009	0.0009
2	$0.2l$	0.0060	0.0012	0.060	0.0120	0.015	0.0030
3	$0.3l$	0.0069	0.0021	0.046	0.0138	0.020	0.0060
4	$0.4l$	0.0070	0.0028	0.035	0.0140	0.023	0.0092
5	$0.5l$	0.0064	0.0032	0.026	0.0130	0.026	0.0130
6	$0.6l$	0.0055	0.0033	0.018	0.0108	0.027	0.0162
7	$0.7l$	0.0043	0.0030	0.012	0.0084	0.029	0.0203
8	$0.8l$	0.0029	0.0023	0.007	0.0056	0.029	0.0232
9	$0.9l$	0.0015	0.0014	0.003	0.0027	0.030	0.0270
10	$0.967l$	—	—	—	—	0.015	0.0145
	Σ	0.0444	0.0197	0.335	0.0898	0.223	0.1333
		wl^3/EI_a	wl^4/EI_a	M_{ab}^Fl/EI_a	$M_{ab}^Fl^2/EI_a$	M_{ba}^Fl/EI_a	$M_{ba}^Fl^2/EI_a$

From

$$\sum F_y = 0 \qquad 0.0444wl^2 - 0.335M_{ab}^F - 0.223M_{ba}^F = 0 \qquad (15\text{-}12)$$

$$\sum M_a = 0 \qquad 0.0197wl^2 - 0.0898M_{ab}^F - 0.1333M_{ba}^F = 0 \qquad (15\text{-}13)$$

we solve for $\qquad M_{ab}^F = 0.061wl^2 \qquad$ (counter-clockwise)

$$M_{ba}^F = 0.107wl^2 \qquad \text{(clockwise)}$$

compared to the more exact values of $0.646wl^2$ and $0.105wl^2$ from published data.

Example 15-2. Determine the stiffness and carry-over factors for the tapered beam of the preceding example.

The set up in Fig. 15-5(a) is to determine the carry-over factor C_{ab}, i.e., the ratio of the moment induced at end b (fixed) to the applied moment at end a (simple end). The bending-moment diagram is as shown in Fig. 15-5(b). The M/EI diagram is plotted in two parts, as in Fig. 15-5(c), the ordinates to the curves being obtained by using the result of Fig. 15-4(d). With reference to Table 15-1, we see the total downward elastic load of $0.335M_{ab}l/EI_a$ acting at

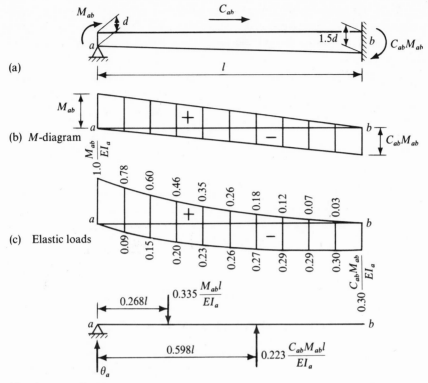

(a)

(b) M-diagram

(c) Elastic loads

(d) Conjugate beam

Fig. 15-5

a distance $0.0898l/0.335$ or $0.268l$ from end a and the total upward elastic load of $0.223C_{ab}M_{ab}l/EI_a$ acting at a distance $0.1333l/0.223$ or $0.598l$ from end a. The conjugate beam, together with these equivalent elastic loads and reaction, is shown in Fig. 15-5(d). Since the loaded conjugate beam is in equilibrium, we set $\sum M_a = 0$,

$$\left(\frac{0.335M_{ab}l}{EI_a}\right)(0.268l) - \left(\frac{0.223C_{ab}M_{ab}l}{EI_a}\right)(0.598l) = 0$$

from which $\qquad\qquad\qquad C_{ab} = 0.674$

compared with more exact value 0.675 from published data.

To determine the stiffness at end a, called S_{ab}, we need only set $\theta_a = 1$, $M_{ab} = S_{ab}$, and $C_{ab} = 0.674$ in Fig. 15-5(d). Using $\sum F_y = 0$, we find

$$\frac{0.335S_{ab}l}{EI_a} - \frac{(0.223)(0.674)S_{ab}l}{EI_a} - 1 = 0$$

to obtain $\qquad\qquad\qquad S_{ab} = 5.41\left(\frac{EI_a}{l}\right)$

compared with the more exact value $5.45EI_a/l$ from published data.

The carry-over factor from end b to end a and the stiffness at end b can be obtained similarly by referring to Fig. 15-6. Note that the conjugate beam loaded with the equivalent elastic weights shown in Fig. 15-6(b) is postured

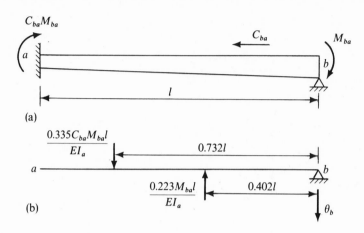

Fig. 15-6

from the result of Fig. 15-5(d). Thus from $\sum M_b = 0$, i.e.,

$$\left(\frac{0.223 M_{ba} l}{EI_a}\right)(0.402l) - \left(\frac{0.335 C_{ba} M_{ba} l}{EI_a}\right)(0.732l) = 0$$

we obtain $$C_{ba} = 0.365$$

compared with the more precise value 0.369 from published data. Setting $\theta_b = 1$, $C_{ba} = 0.365$, and $M_{ba} = S_{ba}$ in Fig. 15-6(b) and using $\sum F_y = 0$, i.e.,

$$\frac{0.223 S_{ba} l}{EI_a} - \frac{(0.335)(0.365) S_{ba} l}{EI_a} - 1 = 0$$

we obtain $$S_{ba} = 9.90\left(\frac{EI_a}{l}\right)$$

compared with the more precise value $10.1 EI_a/l$ from published data.

15-4. RELATIONSHIP OF STIFFNESS AND CARRY-OVER FACTORS

Suppose that a structural member ab has stiffnesses indicated by S_{ab} and S_{ba} at its two ends and carry-over factors of C_{ab} and C_{ba}. The general relationship between these values is given by

$$C_{ab} S_{ab} = C_{ba} S_{ba} \qquad (15\text{-}14)$$

which can be proved as follows:

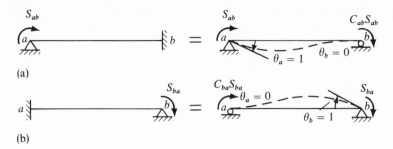

(a)

(b)

Fig. 15-7

Consider two systems of forces acting on the member together with the corresponding end distortions as shown in Figs. 15-7(a) and (b). By Betti's Law (Eq. 11-3)

$$W_{12} = W_{21}$$

It is readily seen that

$$(S_{ab})(0) + (C_{ab}S_{ab})(1) = (C_{ba}S_{ba})(1) + (S_{ba})(0)$$

or

$$C_{ab}S_{ab} = C_{ba}S_{ba}$$

This relationship is of importance for providing a check on separately computed values of the stiffness and carry-over factors.

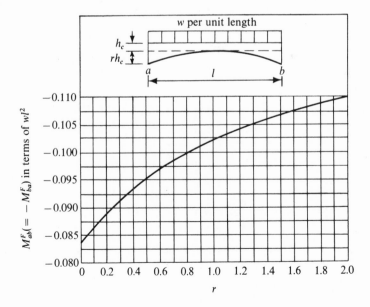

(Courtesy of the Portland Cement Association)
Fig. 15-8

15-5. CHARTS AND TABLES

The formulas for determining fixed-end moments, stiffness, and carry-over factors developed in Sec. 15-2 usually involve a large amount of computation, and the numerical solution for these factors, as described in Sec. 15-3, is also

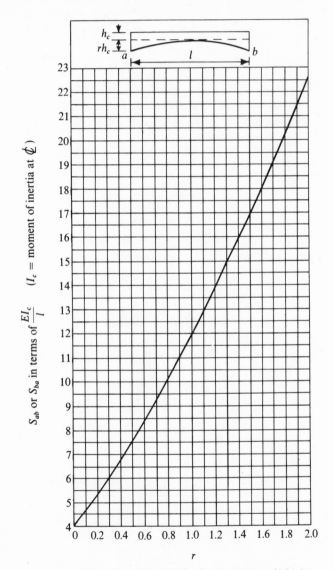

(Courtesy of the Portland Cement Association)

Fig. 15-9

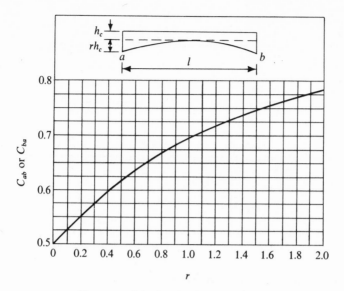

(Courtesy of the Portland Cement Association)

Fig. 15-10

laborious. Fortunately, the values of a considerable number of these factors for the more common types of nonprismatic members have been published for the convenience of structural engineers. One such source is the *Handbook of Frame Constants* published by the Portland Cement Association. Using the available data, we may, for instance, plot the curves for coefficients of fixed-end moments, stiffness, and carry-over factors for a symmetrical member with a soffit curve made for a parabola subjected to a uniform load, such as those shown in Figs. 15-8, 15-9, and 15-10.

15-6. MODIFIED STIFFNESS

The *modified stiffness* of this end of a member may be defined as the end moment required to produce this end (simple end) a unit rotation while the other end remains in the actual condition other than being fixed.

Case (1). The modified stiffness for end a of member ab, if end b is simply supported, is given by

$$S'_{ab} = S_{ab}(1 - C_{ab}C_{ba}) \tag{15-15}$$

in which S'_{ab} denotes the modified stiffness and S_{ab} is the stiffness found by the usual manner. The above equation can be proved as follows:

By definition the set up in Fig. 15-11(a) gives the configuration for finding

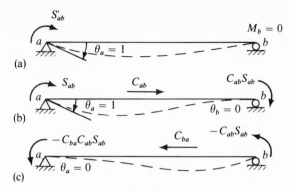

Fig. 15-11

the modified stiffness for end a of member ab, the other end being simply supported. To accomplish this, we may break member ab down into two separate steps as shown in Figs. 15-11(b) and (c). In Fig. 15-11(b) we temporarily lock end b against rotation ($\theta_b = 0$). A moment S_{ab} applied at end a will produce a unit rotation at a and induce a carry-over moment of $C_{ab}S_{ab}$ at b. In Fig. 15-11(c) we release end b to its actual condition of zero moment and at the same time lock end a against further rotation. A moment of $-C_{ab}S_{ab}$ must be developed at end b and, consequently, $-C_{ba}C_{ab}S_{ab}$ will be carried over to end a. The sum of above two steps for end a gives

$$S'_{ab} = S_{ab} - C_{ba}C_{ab}S_{ab}$$
$$= S_{ab}(1 - C_{ab}C_{ba})$$

as asserted.

For a prismatic member $C_{ab} = C_{ba} = 1/2$; therefore,

$$S'_{ab} = \tfrac{3}{4}S_{ab}$$

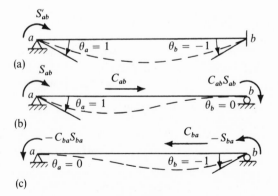

Fig. 15-12

Case (*2*). The modified stiffness for end a of member ab, if end b rotates an equal but opposite angle to that of end a as in the case of *symmetry*, is given by

$$S'_{ab} = S_{ab}(1 - C_{ab}) \qquad (15\text{-}16)$$

To prove this, we refer to Fig. 15-12(a), which is postulated according to the definition of modified stiffness for the present case.

As before, we break this into two separate steps as shown in Figs. 15-12(b) and (c). Figure 15-12(b) shows the usual way of determining S_{ab}. In Fig. 15-12(c) a moment of $-S_{ba}$ is applied at end b necessary to bring it back to its actual position. Consequently, a moment of $-C_{ba}S_{ba}$ is carried over to end a. The sum of the results of these two steps for end a gives

$$S'_{ab} = S_{ab} - C_{ba}S_{ba}$$

On substituting $C_{ab}S_{ab}$ for $C_{ba}S_{ba}$ (Eq. 15-14) in the above expression, we obtain

$$S'_{ab} = S_{ab}(1 - C_{ab})$$

as asserted.

For a prismatic member $C_{ab} = 1/2$; therefore,

$$S'_{ab} = \tfrac{1}{2}S_{ab}$$

Case (*3*). The modified stiffness of end a of member ab, if end b rotates an angle equal to that of end a as in the case of *antisymmetry*, is given by

$$S'_{ab} = S_{ab}(1 + C_{ab}) \qquad (15\text{-}17)$$

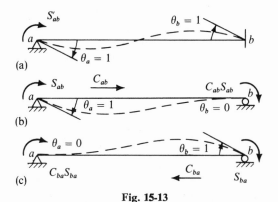

Fig. 15-13

To prove this, we refer to Fig. 15-13(a) for the set up of S'_{ab}. As before, this may be considered as the superposition of two separate cases, as shown in Figs. 15-13(b) and (c). Consequently,

$$S'_{ab} = S_{ab} + C_{ba}S_{ba}$$

On substituting $C_{ab}S_{ab}$ for $C_{ba}S_{ba}$, we obtain

$$S'_{ab} = S_{ab}(1 + C_{ab})$$

as asserted.

For a prismatic member $C_{ab} = 1/2$; therefore,

$$S'_{ab} = \tfrac{3}{2}S_{ab}$$

15-7. FIXED-END MOMENTS DUE TO JOINT TRANSLATION

We recall that the fixed-end moment developed at either end of a prismatic member because of relative end displacement Δ equals $-6EI\Delta/l^2$. This is no longer valid for a member of nonuniform cross section.

The moments developed at the ends because of the relative end translation Δ of a nonprismatic member ab (Fig. 15-14(a)) may be expressed in terms of its stiffness and carry-over factors. The reasoning is as follows:

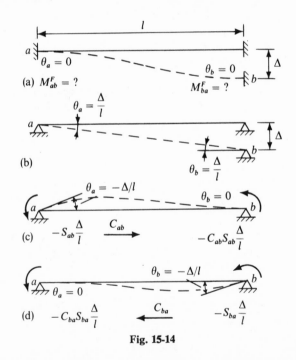

Fig. 15-14

1. Suppose that the fixity is removed from both ends and the member is displayed a relative end deflection Δ, as shown in Fig. 15-14(b). At this stage $\theta_a = \theta_b = \Delta/l$ for small deflection Δ.

2. Lock end b, and rotate end a until θ_a is restored to zero. This requires the set up shown in Fig. 15-14(c), i.e., the end moments of $-S_{ab}\Delta/l$ applied to a and $-C_{ab}S_{ab}\Delta/l$ carried over to b.

3. Lock end a against further rotation, and rotate end b until θ_b is restored to zero. This requires the set up shown in Fig. 15-14(d), i.e., the end moments of $-S_{ba}\Delta/l$ applied to b and $-C_{ba}S_{ba}\Delta/l$ carried over to a.

4. The above three steps will render the member to the same configuration as shown in Fig. 15-14(a). Therefore, the desired fixed-end moments can be obtained by the superposed effect

$$M_{ab}^F = -S_{ab}\frac{\Delta}{l} - C_{ba}S_{ba}\frac{\Delta}{l}$$

$$M_{ba}^F = -S_{ba}\frac{\Delta}{l} - C_{ab}S_{ab}\frac{\Delta}{l}$$

Using $C_{ab}S_{ab} = C_{ba}S_{ba}$ in each of above equations, we obtain

$$M_{ab}^F = -\frac{\Delta}{l}S_{ab}(1 + C_{ab}) \tag{15-18}$$

$$M_{ba}^F = -\frac{\Delta}{l}S_{ba}(1 + C_{ba}) \tag{15-19}$$

For member of uniform cross section, $S_{ab} = S_{ba} = 4EI/l$, $C_{ab} = C_{ba} = 1/2$, and the above expressions reduce to

$$M_{ab}^F = M_{ba}^F = -\frac{6EI\,\Delta}{l^2}$$

If end b is hinged, the modified fixed-end moment at a, called $M_{ab}^{F\prime}$ resulting from the relative end translation Δ, can be found by first assuming both ends fixed and subsequently restoring end b to its original hinged condition. Thus,

$$M_{ab}^{F\prime} = M_{ab}^F - C_{ba}M_{ba}^F$$

Using Eqs. 15-18, 15-19, and 15-14, we obtain

$$M_{ab}^{F\prime} = -\frac{\Delta}{l}S_{ab}(1 - C_{ab}C_{ba}) \tag{15-20}$$

For member of uniform cross section, the above expression reduces to

$$M_{ab}^{F\prime} = -\frac{3EI\,\Delta}{l^2}$$

as previously found.

15-8. NUMERICAL EXAMPLES

Example 15-3. Compute the moments at the interior supports of the three-span continuous beam in Fig. 15-15 by moment distribution. First do not use modified stiffnesses; then use modified stiffnesses.

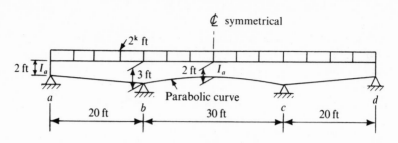

Fig. 15-15

With reference to Sec. 15-3, the fixed-end moments, stiffness, and carry-over factors for the tapered beams ab and cd of Fig. 15-15 are found to be

$$M_{ab}^F = -M_{dc}^F = -0.0646wl^2$$
$$= (-0.0646)(2)(20)^2 = -51.7 \text{ ft-kips}$$

$$M_{ba}^F = -M_{cd}^F = 0.105wl^2$$
$$= (0.105)(2)(20)^2 = 84 \text{ ft-kips}$$

$$S_{ab} = S_{dc} = 5.45\frac{EI_a}{l}$$

$$= 5.45\frac{EI_a}{20} = 0.273EI_a$$

$$S_{ba} = S_{cd} = 10.1\frac{EI_a}{l}$$

$$= 10.1\frac{EI_a}{20} = 0.505EI_a$$

$$C_{ab} = C_{dc} = 0.675$$
$$C_{ba} = C_{cd} = 0.369$$

Using coefficients from the charts of Sec. 15-5, we obtain the fixed-end moments, stiffness, and carry-over factors for the parabolic haunched beam bc of Fig. 15-15(a)

$$M_{bc}^F = -M_{cb}^F = -0.0954wl^2$$
$$= (-0.0954)(2)(30)^2 = -171.7 \text{ ft-kips}$$

$$S_{bc} = S_{cb} = 7.63\frac{EI_a}{l}$$

$$= 7.63\frac{EI_a}{30} = 0.254EI_a$$

$$C_{bc} = C_{cb} = 0.618$$

C.O.F	0.675 →	← 0.369	0.618 ↔	0.369 →	← 0.675	
$\dfrac{S}{\Sigma S}$	a 1	b 0.665	0.335	c 0.335	0.665	d 1
F.E.M.	−51.7	+84	−171.7	+171.7	−84	+51.7
D.M.	+51.7	+58.3	+29.4	−29.4	−58.3	−51.7
C.O.M.	+21.5	+35	−18.2	+18.2	−35	−21.5
D.M.	−21.5	−11.2	−5.6	+5.6	+11.2	+21.5
C.O.M.	−4.1	−14.5	+3.5	−3.5	+14.5	+4.1
D.M.	+4.1	+7.3	+3.7	−3.7	−7.3	−4.1
C.O.M.	+2.7	+2.8	−2.3	+2.3	−2.8	−2.7
D.M.	−2.7	−0.3	−0.2	+0.2	+0.3	+2.7
C.O.M.	−0.1	−1.8	+0.1	−0.1	+1.8	+0.1
D.M.	+0.1	+1.1	+0.6	−0.6	−1.1	−0.1
C.O.M.	+0.4	+0.1	−0.4	+0.4	−0.1	−0.4
D.M.	−0.4	+0.2	+0.1	−0.1	−0.2	+0.4
Σ (ft-kips)	0	+161	−161	+161	−161	0

Fig. 15-16

After these have been determined, the general technique we use to solve the beam of variable cross section is the same as that used for uniform cross section, as shown in Fig. 15-16, assuming that the working line of the beam is straight.

The above process can be simplified by using modified stiffnesses, developed in Sec. 15-6, since ends a and d are hinged and the whole system is symmetrical about the center line of member bc. Thus from Eq. 15-15

$$S'_{ba} = S_{ba}(1 - C_{ab}C_{ba})$$
$$= 0.505EI_a[1 - (0.675)(0.369)] = 0.380EI_a$$

and from Eq. 15-16

$$S'_{bc} = S_{bc}(1 - C_{bc})$$
$$= 0.254EI_a(1 - 0.618) = 0.097EI_a$$

The modified distribution factors for ba and bc are found to be

$$\frac{0.380}{0.380 + 0.097} \quad \text{or} \quad 0.795$$

$$\frac{0.097}{0.380 + 0.097} \quad \text{or} \quad 0.205$$

respectively. Using these and, because of symmetry, working with only half the structure, we complete the analysis by moment distribution, as shown in Fig. 15-17.

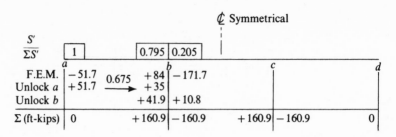

$\dfrac{S'}{\Sigma S'}$	1		0.795	0.205			
	a			*b*		*c*	*d*
F.E.M.	−51.7	0.675	+84	−171.7			
Unlock *a*	+51.7	⟶	+35				
Unlock *b*			+41.9	+10.8			
Σ (ft-kips)	0		+160.9	−160.9	+160.9	−160.9	0

Fig. 15-17

Example 15-4. Find the end moments for the frame in Fig. 15-18 because of a uniform horizontal load of 2 kips per ft from the left. The working line of the frame is shown by the dotted line. Also shown in Fig. 15-18 are the elements of frame for selecting coefficients.

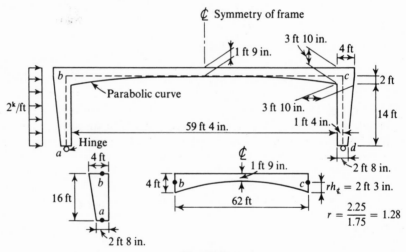

Fig. 15-18

Using the result of Sec. 15-3, we obtain

$$M_{ab}^F = (-0.0646)(2)(16)^2 = -33 \text{ ft-kips}$$

$$M_{ba}^F = (0.105)(2)(16)^2 = 53.7 \text{ ft-kips}$$

$$C_{ab} = C_{dc} = 0.675$$

$$C_{ba} = C_{cd} = 0.369$$

$$S_{ba} = S_{cd} = (10.1)\left(\frac{EI_a}{l_{ab}}\right) = (10.1)\frac{Et(2.67)^3/12}{16} = Et$$

for the columns. For other end *hinged*,

$$S'_{ba} = S'_{cd} = S_{ba}(1 - C_{ab}C_{ba})$$
$$= Et[1 - (0.675)(0.369)] = 0.751Et$$

Then, using the charts of Sec. 15-5, we obtain for the beam

$$C_{bc} = C_{cb} = 0.724$$

$$S_{bc} = S_{cb} = (14.89)\left(\frac{EI_t}{l_{bc}}\right) = (14.89)\frac{Et(1.75)^3/12}{62} = 0.107Et$$

In case of *antisymmetry*,

$$S'_{bc} = S'_{cb} = S_{bc}(1 + C_{bc})$$
$$= 0.107Et(1 + 0.724) = 0.185Et$$

With these values found, the analysis can be carried out by the usual moment-distribution procedures; i.e., by first providing restraint at the top of the column to prevent the frame from sideway and obtaining a moment-distribution solution, then removing the artificial effect by applying an equal and opposite force at the top of the column. The sum of the results of the two steps gives the final solution.

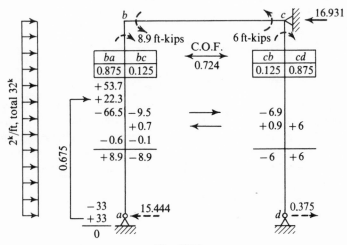

Fig. 15-19

The complete analysis of the first step is shown in Fig. 15-19 in which the modified stiffnesses for end b of ba and end c of cd are used. Thus the distribution factors at b or c are found to be

$$\frac{0.751}{0.751 + 0.107} = 0.875 \quad \text{(column)}$$

$$\frac{0.107}{0.751 + 0.107} = 0.125 \quad \text{(beam)}$$

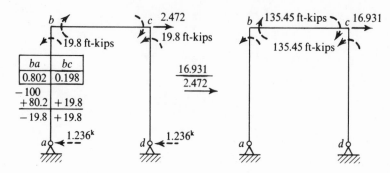

Fig. 15-20

It is found from the resulting end moments that a lateral force of 16.931 kips acting to the left at the top of the column is necessary to hold the frame from joint translation, as indicated in Fig. 15-19.

The complete analysis of the second step, i.e., the corrective measure to remove the effect due to the artificial force, is shown in Fig. 15-20. Advantage is taken of the antisymmetry of the system. The analysis is greatly simplified by using modified stiffnesses at b for both column ba and beam bc. Accordingly, the modified distribution factors at b are

$$\frac{0.751}{0.751 + 0.185} = 0.802 \qquad \text{(column)}$$

$$\frac{0.185}{0.751 + 0.185} = 0.198 \qquad \text{(beam)}$$

Refer to Fig. 15-20(a). We begin by assuming an arbitrary fixed-end moment due to joint translation at b of ba equal to -100. After distributing once, we halt the process since no carry over is needed. The resulting end moments are found to be consistent with a lateral force of 2.472 kips acting to the right at the top of the column. Proportioning the result of Fig. 15-20(a) with a ratio of $16.931/2.472$ gives the solution, as shown in Fig. 15-20(b).

Summing up the findings of the above two steps (Fig. 15-19 and Fig. 15-20(b)), we obtain the final solution as

$$M_{ab} = M_{dc} = 0$$
$$M_{ba} = -M_{bc} = 8.9 - 135.45 = -126.55 \text{ ft-kips}$$
$$M_{cb} = -M_{cd} = -6 + 135.45 = 129.45 \text{ ft kips}$$

15-9. THE GENERALIZED SLOPE-DEFLECTION EQUATIONS

The slope-deflection equations given in Sec. 12-2,

$$M_{ab} = 2E\frac{I}{l}\left[2\theta_a + \theta_b - 3\left(\frac{\Delta}{l}\right)\right] \pm M_{ab}^F$$

$$M_{ba} = 2E \frac{I}{l} \left[2\theta_b + \theta_a - 3\left(\frac{\Delta}{l}\right) \right] \pm M_{ba}^F$$

are developed for use in analyzing structures composed of *prismatic members*. In cases involving a variable moment of inertia, these equations are no longer valid, and some generalized slope-deflection equations must be formed to suit the purpose. To do this, we recall that the basic slope-deflection equations are derived from four separate elements of end moments:

1. The end moments due to θ_a
2. The end moments due to θ_b
3. The end moments due to a pure relative end translation Δ
4. The fixed-end moments due to loads on span ab

This reasoning, however, is not limited to the instances of uniform member. In the light of moment distribution developed in this chapter, we observe that

1. In order to produce θ_a while end b is fixed, an end moment $S_{ab}\theta_a$ is required at a and a carry-over moment $C_{ab}S_{ab}\theta_a$ (or $C_{ba}S_{ba}\theta_a$) is induced at b.

2. Similarly, to produce θ_b while end a is fixed, an end moment $S_{ba}\theta_b$ is required at b and $C_{ba}S_{ba}\theta_b$ (or $C_{ab}S_{ab}\theta_b$) is induced at a.

3. The end moments due to Δ are given by Eqs. 15-18 and 15-19,

$$M_{ab} = -\frac{\Delta}{l} S_{ab}(1 + C_{ab})$$

$$M_{ba} = -\frac{\Delta}{l} S_{ba}(1 + C_{ba})$$

4. The fixed-end moments M_{ab}^F and M_{ba}^F resulting from loads on the span have been discussed in Secs. 15-2, 15-3, and 15-5.

Adding these effects, we arrive at the generalized slope-deflection equations applicable to cases of variable moment of inertia.

$$M_{ab} = \left[S_{ab}\theta_a + C_{ba}S_{ba}\theta_b - \frac{\Delta}{l} S_{ab}(1 + C_{ab}) \right] \pm M_{ab}^F \qquad (15\text{-}21)$$

$$M_{ba} = \left[S_{ba}\theta_b + C_{ab}S_{ab}\theta_a - \frac{\Delta}{l} S_{ba}(1 + C_{ba}) \right] \pm M_{ba}^F \qquad (15\text{-}22)$$

or

$$M_{ab} = S_{ab}\left[\theta_a + C_{ab}\theta_b - \frac{\Delta}{l}(1 + C_{ab}) \right] \pm M_{ab}^F \qquad (15\text{-}23)$$

$$M_{ba} = S_{ba}\left[\theta_b + C_{ba}\theta_a - \frac{\Delta}{l}(1 + C_{ba}) \right] \pm M_{ba}^F \qquad (15\text{-}24)$$

For a member of uniform cross section, $S_{ab} = S_{ba} = 4EI/l$, $C_{ab} = C_{ba} = 1/2$, the above equations reduce to the basic form.

With the constants (stiffness, carry-over factors, and fixed-end moments) for each member determined, application of the above equations in analyzing frames composed of nonuniform members can be carried out using a procedure similar to that explained in Ch. 12.

PROBLEMS

15-1. Fig. 15-21 shows a beam of varying cross section with both ends fixed and subjected to the bending action of a concentrated load P. Find the integral expressions for the fixed-end moments.

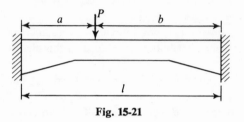

Fig. 15-21

15-2. For the haunched member of constant width shown in Fig. 15-22, find C_{ab}, C_{ba}, S_{ab}, and S_{ba} by the numerical method.

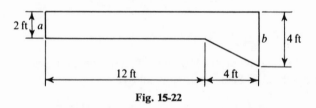

Fig. 15-22

15-3. Use the numerical method to determine the fixed-end moments for the haunched beam in Fig. 15-22 due to a uniform load of 3 kips per ft over the entire span.

15-4. Use the relevant calculations from Probs. 15-2 and 15-3 to find, by moment distribution, the end moments for the beam in Fig. 15-23 subjected to a uniform load of 3 kips per ft over the entire span.

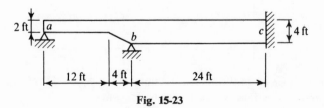

Fig. 15-23

15-5. Find, by moment distribution, the end moments for the beam in Fig. 15-23 due to a vertical support settlement of 1 in. at b. $E = 30,000$ kips per in². $I_a = 2,000$ in.⁴

15-6. Use the relevant calculations from Probs. 15-2 and 15-3 and the charts from Sec. 15-5 to determine, by moment distribution, the moment at support b of the continuous beam shown in Fig. 15-24.

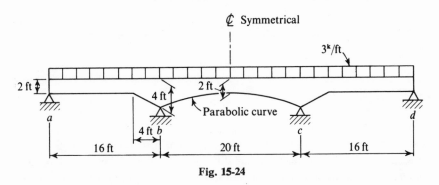

Fig. 15-24

15-7. Use the generalized slope-deflection equations to check your results for Prob. 15-6.

15-8. Use the relevant data from Secs. 15-3 and 15-5 to determine, by moment distribution, all the end moments of the frame in Fig. 15-25 due to (1) a uniform downward load of 1 kip per ft over the entire span from d to f; (2) a uniform horizontal load of 1 kip per ft acting to the right over the entire span of ad.

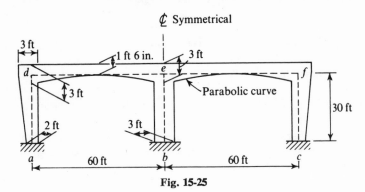

Fig. 15-25

16

MATRIX ALGEBRA FOR STRUCTURAL
ENGINEERS

16-1. INTRODUCTION

The usefulness of matrices in linear structural analysis lies in the following facts: first, the matrices furnish a very convenient mathematical means for expressing the theory; second, the solution expressing the theory can be best obtained by a sequence of matrix operations to which the high-speed computer is ideally adopted. Ease in thinking of structural theory in the matrix notion is, therefore, of great importance to the structural engineer.

The material covered in this chapter is intended for the reader who is not familiar with matrix albegra. It provides adequate background for understanding the subsequent two chapters on matrix analysis of structures but is not an exhausive treatment of matrix operations. Therefore the mathematics involved has been kept at as simple a level as possible.

16-2. MATRIX DEFINITIONS AND NOTATION

A *matrix* is defined as a rectangular array of elements assembled in rows and columns.

The *order* of a matrix refers to its *size*. A matrix containing m rows and n columns is said to be of order $(m \times n)$.

If no ambiguity arises, matrices are often represented by a single letter. More often, however, they are represented by displaying some or all of the constituent quantities between brackets. Thus,

$$A = [a_{ij}] = \begin{bmatrix} a_{11} & a_{12} \cdots a_{1n} \\ a_{21} & a_{22} \cdots a_{2n} \\ \vdots & \vdots \\ a_{m1} & a_{m2} \cdots a_{mn} \end{bmatrix} \tag{16-1}$$

In the above expression a_{ij} represents an element of matrix. Note particularly that the subscripts of the elements carry a position significance. That is, the first subscript represents the row position of the element and the second sub-script represents the column position.

The elements of a matrix may be numbers, vectors, algebraic functions, or any other quantities.

A matrix is not a determinant. A *determinant* is a square array arranged within vertical bars. It signifies a certain relationship among the elements. If the elements are numerical, the determinant can be evaluated. The matrix, on the other hand, merely represents the array and does not imply a relationship among the elements. Some special types of matrices are explained as follows:

Row Matrix. A row matrix is one which consists of a single row. It is of the order $(1 \times n)$. Thus,

$$A = [a_{11} \quad a_{12} \cdots a_{1n}]$$

Column Matrix. A column matrix is one which consists of a single column. It is of order $(m \times 1)$. Braces are usually used to denote a column matrix. Thus,

$$B = \begin{Bmatrix} b_{11} \\ b_{21} \\ \vdots \\ b_{m1} \end{Bmatrix}$$

Null Matrix. A null matrix is one in which all elements equal zero. For example,

$$0 = \begin{bmatrix} 0 & 0 & 0 \\ 0 & 0 & 0 \\ 0 & 0 & 0 \end{bmatrix}$$

A null matrix corresponds to zero in ordinary algebra.

Transposed Matrix. The transpose of matrix A is shown as A^T and is obtained by interchanging the rows and columns of A. Thus if

$$A = \begin{bmatrix} a & b & c \\ x & y & z \\ u & v & w \end{bmatrix}$$

then

$$A^T = \begin{bmatrix} a & x & u \\ b & y & v \\ c & z & w \end{bmatrix}$$

Apparently,

$$(A^T)^T = A \tag{16-2}$$

Any matrix may be transposed. Note that the transpose of a column matrix is a row matrix and vice versa.

Square Matrix. A matrix with the same number of rows and columns is called square matrix. Thus,

$$
A = \begin{bmatrix} a_{11} & a_{12} & \cdots & a_{1n} \\ a_{21} & a_{22} & \cdots & a_{2n} \\ \vdots & & & \vdots \\ a_{n1} & a_{n2} & \cdots & a_{nn} \end{bmatrix}
$$

The following definitions apply specifically to square matrices:

Diagonal Matrix. A square matrix in which all elements not on the main diagonal (from left down to right) are zero is called a diagonal matrix. For example,

$$
A = \begin{bmatrix} a_{11} & 0 & 0 & 0 \\ 0 & a_{22} & 0 & 0 \\ 0 & 0 & a_{33} & 0 \\ 0 & 0 & 0 & a_{44} \end{bmatrix} \quad \text{or} \quad B = \begin{bmatrix} 0 & 0 & 0 & 0 \\ 0 & 0 & 0 & 0 \\ 0 & 0 & b_{33} & 0 \\ 0 & 0 & 0 & 0 \end{bmatrix}
$$

Common notations for diagonal matrices are

$$
A = [\diagdown A \diagup] = [\diagdown a_{i \diagdown}]
$$

Unit Matrix. A diagonal matrix in which all elements on the main diagonal are equal to unity is called a unit matrix. A unit matrix is usually denoted by the symbol I. For example,

$$
I = \begin{bmatrix} 1 & 0 & 0 & 0 \\ 0 & 1 & 0 & 0 \\ 0 & 0 & 1 & 0 \\ 0 & 0 & 0 & 1 \end{bmatrix}
$$

Generally if A is a unit matrix, then

$$
a_{ij} = 1 \quad \text{if} \quad i = j
$$
$$
a_{ij} = 0 \quad \text{if} \quad i \neq j
$$

Note that a unit matrix serves the same function as unity does in ordinary algebra.

Symmetrical Matrix. A symmetrical matrix is a square matrix whose elements are symmetrical about its main diagonal. For example,

$$
A = \begin{bmatrix} a & b & c \\ b & d & e \\ c & e & f \end{bmatrix}
$$

We note that a symmetrical matrix is unchanged by the process of transposition, or

$$a_{ij} = a_{ji}$$

Antisymmetrical Matrix. A square matrix A is called an antisymmetrical matrix if

$$a_{ij} = -a_{ji}$$

For example,
$$A = \begin{bmatrix} 0 & y & x \\ -y & 0 & 1 \\ -x & -1 & 0 \end{bmatrix}$$

Note that in an antisymmetrical matrix the elements on the main diagonal must be zero (i.e., $a_{ii} = 0$) for only then will $a_{ij} = -a_{ji}$ be true for these elements.

Triangular Matrix. A square matrix in which all elements on one side of the main diagonal are zero is called a triangular matrix. For example,

$$A = \begin{bmatrix} 3 & 1 & 8 \\ 0 & 6 & 1 \\ 0 & 0 & 9 \end{bmatrix} \quad \text{or} \quad B = \begin{bmatrix} 1 & 0 & 0 \\ 2 & 1 & 0 \\ 3 & 4 & 1 \end{bmatrix}$$

16-3. EQUALITY, ADDITION, SUBTRACTION, AND
SCALAR MULTIPLICATION

Two matrices are said to be equal if and only if each has the same number of rows and the same number of columns, and all the corresponding elements are equal. Thus, given

$$A = \begin{bmatrix} a_{11} & a_{12} & a_{13} \\ a_{21} & a_{22} & a_{23} \\ a_{31} & a_{32} & a_{33} \end{bmatrix} \quad B = \begin{bmatrix} b_{11} & b_{12} & b_{13} \\ b_{21} & b_{22} & b_{23} \\ b_{31} & b_{32} & b_{33} \end{bmatrix}$$

and
$$a_{ij} = b_{ij}$$

for all values of i and j, we say that

$$A = B$$

The four simple algebraic equations

$$\alpha = a + 3$$
$$\beta = b - 7$$
$$\gamma = c + 2$$
$$\delta = d + 1$$

may be given in matrix form as

$$\begin{bmatrix} \alpha & \beta \\ \gamma & \delta \end{bmatrix} = \begin{bmatrix} a+3 & b-7 \\ c+2 & d+1 \end{bmatrix}$$

Two matrices having the same number of rows and the same number of columns can be added or subtracted from each other. Thus, if

$$A = [a_{ij}] = \begin{bmatrix} a_{11} & a_{12} & a_{13} \\ a_{21} & a_{22} & a_{23} \\ a_{31} & a_{32} & a_{33} \end{bmatrix} \qquad B = [b_{ij}] = \begin{bmatrix} b_{11} & b_{12} & b_{13} \\ b_{21} & b_{22} & b_{23} \\ b_{31} & b_{32} & b_{33} \end{bmatrix}$$

and $\quad C = [c_{ij}] = [a_{ij} + b_{ij}] = \begin{bmatrix} a_{11}+b_{11} & a_{12}+b_{12} & a_{13}+b_{13} \\ a_{21}+b_{21} & a_{22}+b_{22} & a_{23}+b_{23} \\ a_{31}+b_{31} & a_{32}+b_{32} & a_{33}+b_{33} \end{bmatrix}$

we say that
$$C = A + B$$

Similarly, if

$$D = [d_{ij}] = [a_{ij} - b_{ij}] = \begin{bmatrix} a_{11}-b_{11} & a_{12}-b_{12} & a_{13}-b_{13} \\ a_{21}-b_{21} & a_{22}-b_{22} & a_{23}-b_{23} \\ a_{31}-b_{31} & a_{32}-b_{32} & a_{33}-b_{33} \end{bmatrix}$$

we say that
$$D = A - B$$

It follows from the matrix definitions that addition (or subtraction) obeys the commutative and associative laws. That is,

$$A + B = B + A \tag{16-3}$$

$$A + (B + C) = (A + B) + C \tag{16-4}$$

Any square matrix may be given as the sum of a symmetrical and anti-symmetrical matrix; for, if A is a square matrix,

$$A = \frac{A + A^T}{2} + \frac{A - A^T}{2} \tag{16-5}$$

The first form on the right is a symmetrical matrix, and the second form is an antisymmetrical matrix. Since

$$\frac{A + A^T}{2} = \left[\frac{a_{ij} + a_{ji}}{2}\right]$$

$$\frac{A - A^T}{2} = \left[\frac{a_{ij} - a_{ji}}{2}\right]$$

the interchanging of i and j makes no change in the former expression but alters the sign of the latter expression. For example, if

$$A = \begin{bmatrix} a_{11} & a_{12} \\ a_{21} & a_{22} \end{bmatrix}$$

then
$$A^T = \begin{bmatrix} a_{11} & a_{21} \\ a_{12} & a_{22} \end{bmatrix}$$

so that
$$\frac{A + A^T}{2} = \begin{bmatrix} a_{11} & \dfrac{a_{12} + a_{21}}{2} \\ \dfrac{a_{21} + a_{12}}{2} & a_{22} \end{bmatrix}$$

which is a symmetrical matrix, and

$$\frac{A - A^T}{2} = \begin{bmatrix} 0 & \dfrac{a_{12} - a_{21}}{2} \\ \dfrac{a_{21} - a_{12}}{2} & 0 \end{bmatrix}$$

which is an antisymmetrical matrix.

The product of a matrix A and a scalar k is the matrix kA whose elements are the elements a_{ij} of A each multiplied by k. Thus

$$k \begin{bmatrix} a & b & c \\ x & y & z \\ u & v & w \end{bmatrix} = \begin{bmatrix} ka & kb & kc \\ kx & ky & kz \\ ku & kv & kw \end{bmatrix}$$

16-4. MATRIX MULTIPLICATION

Before giving the general definition of multiplication of two matrices, we must define the inner product of a row matrix and a column matrix and two matrices conformable for multiplication.

Consider a row matrix

$$R = [r_1 \quad r_2 \quad r_3 \cdots r_j]$$

and a column matrix
$$C = \begin{Bmatrix} c_1 \\ c_2 \\ c_3 \\ \vdots \\ c_i \end{Bmatrix}$$

The *inner product RC* is defined as the sum of the products of the corresponding components of the row matrix and the column matrix. That is,

$$RC = r_1 c_1 + r_2 c_2 + r_3 c_3 + \cdots + r_j c_i \tag{16-6}$$

It is obvious that i must be equal to j. A convenient way to do is to write the column horizontally above the row as shown below.

$$\{c_1 \quad c_2 \quad c_3 \cdots c_j\}$$
$$[r_1 \quad r_2 \quad r_3 \cdots r_j]$$

Two matrices A and B are said to be conformable in the order AB if the number of columns in A is equal to the number of rows in B. In other words, if A is an $(m \times n)$ matrix and B is a $(s \times t)$ matrix, A and B are conformable in the order AB if and only if $n = s$. Thus

$$A = \begin{bmatrix} a_{11} & a_{12} & a_{13} \\ a_{21} & a_{22} & a_{23} \\ a_{31} & a_{32} & a_{33} \end{bmatrix} \quad \text{and} \quad B = \begin{bmatrix} b_{11} & b_{12} \\ b_{21} & b_{22} \\ b_{31} & b_{32} \end{bmatrix}$$

$$(3 \times 3) \qquad\qquad (3 \times 2)$$

are conformable for multiplication.

We are now in a position to define the important concept of the *multiplication of two matrices* as:

If A is a $(p \times q)$ matrix and B is a $(q \times r)$ matrix, so that A and B are conformable in that order, the product $C = AB$ is a $(p \times r)$ matrix in which the element c_{ij} in the ith row and jth column of C is the inner product of ith row of A and jth column of B. Or

$$c_{ij} = \sum_{k=1}^{q} a_{ik} b_{kj} \tag{16-7}$$

Thus if matrices A and B are as given above, then

$$C = AB$$

is a (3×2) matrix expressed by

$$\begin{bmatrix} c_{11} & c_{12} \\ c_{21} & c_{22} \\ c_{31} & c_{32} \end{bmatrix} = \begin{bmatrix} a_{11} & a_{12} & a_{13} \\ a_{21} & a_{22} & a_{23} \\ a_{31} & a_{32} & a_{33} \end{bmatrix} \begin{bmatrix} b_{11} & b_{12} \\ b_{21} & b_{22} \\ b_{31} & b_{32} \end{bmatrix}$$
$$= \begin{bmatrix} a_{11}b_{11} + a_{12}b_{21} + a_{13}b_{31} & a_{11}b_{12} + a_{12}b_{22} + a_{13}b_{32} \\ a_{21}b_{11} + a_{22}b_{21} + a_{23}b_{31} & a_{21}b_{12} + a_{22}b_{22} + a_{23}b_{32} \\ a_{31}b_{11} + a_{32}b_{21} + a_{33}b_{31} & a_{31}b_{12} + a_{32}b_{22} + a_{33}b_{32} \end{bmatrix}$$

In the above expression the element c_{32}, for instance, is the inner product of the third row of A and the second column of B. That is,

$$\begin{bmatrix} a_{31} & a_{32} & a_{33} \end{bmatrix} \begin{Bmatrix} b_{12} \\ b_{22} \\ b_{32} \end{Bmatrix} \quad \text{or} \quad a_{31}b_{12} + a_{32}b_{22} + a_{33}b_{32}$$

Other elements are obtained in a similar way.

It follows from the law of multiplication that although two matrices may be conformable in the sequence AB, this may not be the case if they are interchanged. Even for matrices conformable in either order multiplication is not commutative; i.e., in general

$$AB \neq BA \tag{16-8}$$

which can be demonstrated by the following example.

$$A = \begin{bmatrix} 1 & 0 \\ 0 & 0 \end{bmatrix} \qquad B = \begin{bmatrix} 0 & 1 \\ 0 & 0 \end{bmatrix}$$

$$AB = \begin{bmatrix} 0 & 1 \\ 0 & 0 \end{bmatrix} \qquad BA = \begin{bmatrix} 0 & 0 \\ 0 & 0 \end{bmatrix}$$

In two cases, however, the multiplication of matrices is commutative. Unit matrices and null matrices commute with all suitably conformable matrices. More specifically,

$$AI = IA = A \tag{16-9}$$

$$A0 = 0A = 0 \tag{16-10}$$

In ordinary algebra the division law states that if $xy = 0$, then either $x = 0$ or $y = 0$. This is not necessarily true in matrix algebra. For example,

$$\begin{bmatrix} 8 & 6 \\ 4 & 3 \end{bmatrix} \begin{bmatrix} 3 & -3 \\ -4 & 4 \end{bmatrix} = \begin{bmatrix} 0 & 0 \\ 0 & 0 \end{bmatrix} = 0$$

The associative and distributive laws apply to multiplication of matrices, provided that the sequence of the terms is strictly maintained. Thus

$$A(BC) = (AB)C = ABC \tag{16-11}$$

$$A(B + C) = AB + AC \tag{16-12}$$

Matrix multiplication is often used to express the work quantities. Suppose that a force vector

$$P = \begin{Bmatrix} P_1 \\ P_2 \\ P_3 \end{Bmatrix}$$

in Cartesian coordinates is multiplied by a displacement vector

$$s = \begin{Bmatrix} s_1 \\ s_2 \\ s_3 \end{Bmatrix}$$

The work quantity may then be expressed by

$$W = P^T s = [P_1 \quad P_2 \quad P_3] \begin{Bmatrix} s_1 \\ s_2 \\ s_3 \end{Bmatrix} \tag{16-13}$$

$$= P_1 s_1 + P_2 s_2 + P_3 s_3$$

or

$$W = s^T P = [s_1 \quad s_2 \quad s_3] \begin{Bmatrix} P_1 \\ P_2 \\ P_3 \end{Bmatrix} \tag{16-14}$$

$$= s_1 P_1 + s_2 P_2 + s_3 P_3$$

16-5. MATRIX PARTITIONING

Often it is convenient to subdivide, or partition, a matrix into different groups called submatrices and then to regard the original matrix as a partitioned matrix. There are many ways to perform partitioning. For instance, among numerous other possibilities, we can write matrix A as

$$A = \begin{bmatrix} a_{11} & a_{12} & a_{13} & a_{14} \\ a_{21} & a_{22} & a_{23} & a_{24} \\ a_{31} & a_{32} & a_{33} & a_{34} \end{bmatrix} = \begin{bmatrix} A_{11} & A_{12} \\ A_{21} & A_{22} \end{bmatrix}$$

where

$$A_{11} = \begin{bmatrix} a_{11} & a_{12} \\ a_{21} & a_{22} \end{bmatrix} \qquad A_{12} = \begin{bmatrix} a_{13} & a_{14} \\ a_{23} & a_{24} \end{bmatrix}$$

$$A_{21} = [a_{31} \quad a_{32}] \qquad A_{22} = [a_{33} \quad a_{34}]$$

or equally well

$$A = \begin{bmatrix} a_{11} & a_{12} & a_{13} & a_{14} \\ a_{21} & a_{22} & a_{23} & a_{24} \\ a_{31} & a_{32} & a_{33} & a_{34} \end{bmatrix} = [A_{11} \quad A_{12} \quad A_{13} \quad A_{14}]$$

where now

$$A_{11} = \begin{Bmatrix} a_{11} \\ a_{21} \\ a_{31} \end{Bmatrix} \qquad A_{12} = \begin{Bmatrix} a_{12} \\ a_{22} \\ a_{32} \end{Bmatrix} \qquad A_{13} = \begin{Bmatrix} a_{13} \\ a_{23} \\ a_{33} \end{Bmatrix} \qquad A_{14} = \begin{Bmatrix} a_{14} \\ a_{24} \\ a_{34} \end{Bmatrix}$$

In constructing the product of two matrices A and B, we sometimes partition them and express the product in terms of the submatrices of A and B, provided that the given matrices are conformable and that the various submatrices that must be multiplied together are also conformable. This may be verified for the multiplication of a (3×3) matrix and a (3×2) matrix as follows:

Given

$$A = \begin{bmatrix} a_{11} & a_{12} & a_{13} \\ a_{21} & a_{22} & a_{23} \\ a_{31} & a_{32} & a_{33} \end{bmatrix} \qquad B = \begin{bmatrix} b_{11} & b_{12} \\ b_{21} & b_{22} \\ b_{31} & b_{32} \end{bmatrix}$$

$$(3 \times 3) \qquad\qquad (3 \times 2)$$

we may write

$$A = \begin{bmatrix} a_{11} & a_{12} & \vdots & a_{13} \\ a_{21} & a_{22} & \vdots & a_{23} \\ \cdots & \cdots & \cdots & \cdots \\ a_{31} & a_{32} & \vdots & a_{33} \end{bmatrix} = \begin{bmatrix} A_{11} & A_{12} \\ (2 \times 2) & (2 \times 1) \\ A_{21} & A_{22} \\ (1 \times 2) & (1 \times 1) \end{bmatrix}(2 \times 2)$$

$$B = \begin{bmatrix} b_{11} & b_{12} \\ b_{21} & b_{22} \\ \cdots & \cdots \\ b_{31} & b_{32} \end{bmatrix} = \begin{bmatrix} B_{11} \\ (2 \times 2) \\ B_{21} \\ (1 \times 2) \end{bmatrix}(2 \times 1)$$

so that

$$AB = \begin{bmatrix} A_{11}B_{11} + A_{12}B_{21} \\ A_{21}B_{11} + A_{22}B_{21} \end{bmatrix}$$

$$= \begin{bmatrix} \begin{bmatrix} a_{11}b_{11} + a_{12}b_{21} & a_{11}b_{12} + a_{12}b_{22} \\ a_{21}b_{11} + a_{22}b_{21} & a_{21}b_{12} + a_{22}b_{22} \end{bmatrix} + \begin{bmatrix} a_{13}b_{31} & a_{13}b_{32} \\ a_{23}b_{31} & a_{23}b_{32} \end{bmatrix} \\ [a_{31}b_{11} + a_{32}b_{21} \quad a_{31}b_{12} + a_{32}b_{22}] + [a_{33}b_{31} \quad a_{33}b_{32}] \end{bmatrix}$$

$$= \begin{bmatrix} a_{11}b_{11} + a_{12}b_{21} + a_{13}b_{31} & a_{11}b_{12} + a_{12}b_{22} + a_{13}b_{32} \\ a_{21}b_{11} + a_{22}b_{21} + a_{23}b_{31} & a_{21}b_{12} + a_{22}b_{22} + a_{23}b_{32} \\ a_{31}b_{11} + a_{32}b_{21} + a_{33}b_{31} & a_{31}b_{12} + a_{32}b_{22} + a_{33}b_{32} \end{bmatrix}$$

This is the same as the product found by direct multiplication in Sec. 16-4.

In order for multiplication of partitioned matrices to be possible, the partitioning of the columns of the first matrix A should correspond exactly to the partitioning of the rows of the second matrix B. Since A and B are conformable, the number of columns of A must be equal to the number of rows of B. As long as the conformability of the submatrices is kept intact, no restriction is imposed on the horizontal partitioning of A and the vertical partitioning of B, since there is no necessary relationship between the number of rows of A and the number of columns of B. Partitioning of matrices is often used to lessen the work in matrix multiplication especially when the matrices are large and some of the submatrices are unit matrices or null matrices.

16-6. TRANSPOSE OF PRODUCT

The transpose of the product of two conformable matrices is the product of the transposed matrices taken in the reverse order, i.e.,

$$(AB)^T = B^T A^T \tag{16-15}$$

The proof follows directly from the definition of matrix multiplication, Eq. 16-7. Suppose that A and B are of the order $(p \times q)$ and $(q \times r)$ respectively, so that they can form a product $C = AB$ of the order $(p \times r)$. An element in AB, c_{ij}, is given by

$$c_{ij} = \sum_{k=1}^{q} a_{ik} b_{kj}$$

Let $D = (AB)^T$. An element in $(AB)^T$ is therefore given by c_{ji}, i.e.,

$$d_{ij} = c_{ji} = \sum_{k=1}^{q} a_{jk} b_{ki} \qquad (16\text{-}16)$$

Note that $(AB)^T$ is of order $(r \times p)$.

Next consider B^T and A^T. Since B^T and A^T are of order $(r \times q)$ and $(q \times p)$ respectively, they can form a product $E = B^T A^T$ of the order $(r \times p)$ the same as that of $(AB)^T$. Let $F = B^T$ and $G = A^T$. An element of $B^T A^T$ is given by

$$e_{ij} = \sum_{k=1}^{q} f_{ik} g_{kj} = \sum_{k=1}^{q} b_{ki} a_{jk} \qquad (16\text{-}17)$$

Since $d_{ij} = e_{ij}$, the proof is complete.

As an example, consider

$$A = \begin{bmatrix} 1 & 2 & 3 \\ 3 & 2 & 1 \end{bmatrix} \qquad B = \begin{bmatrix} 1 & 4 \\ 2 & 5 \\ 3 & 6 \end{bmatrix}$$

$$AB = \begin{bmatrix} 14 & 32 \\ 10 & 28 \end{bmatrix} \qquad (AB)^T = \begin{bmatrix} 14 & 10 \\ 32 & 28 \end{bmatrix}$$

and

$$B^T A^T = \begin{bmatrix} 1 & 2 & 3 \\ 4 & 5 & 6 \end{bmatrix} \begin{bmatrix} 1 & 3 \\ 2 & 2 \\ 3 & 1 \end{bmatrix} = \begin{bmatrix} 14 & 10 \\ 32 & 28 \end{bmatrix}$$

The rule given in Eq. 16-15 can easily be extended to

$$(ABC)^T = C^T B^T A^T \qquad (16\text{-}18)$$

etc.

16-7. INVERSION OF MATRIX

There is no direct matrix division. The operation of division is performed by inversion. For example, given a square matrix A, we can find another square matrix B of the same order such that

$$AB = I$$

provided that A is *nonsingular*; i.e., the determinant $|A|$ containing the same elements as A is not zero.

The matrix B is then called the *inverse* of A and is denoted by A^{-1}.

To develop an expression for A^{-1}, we must define the following quantities and notations:

1. *Determinant A*

$$|A| = |a_{ij}| = \begin{vmatrix} a_{11} & a_{12} & \cdots & a_{1n} \\ a_{21} & a_{22} & \cdots & a_{2n} \\ \vdots & & & \\ a_{n1} & a_{n2} & \cdots & a_{nn} \end{vmatrix} \qquad (16\text{-}19)$$

2. *Minor of a_{ij}*

$$m_{ij} = \begin{vmatrix} A & \\ \hline & i \\ j & \end{vmatrix} = \begin{array}{l} \text{the determinant obtained} \\ \text{by erasing the } i\text{th row and} \\ j\text{th column of } |A| \end{array} \qquad (16\text{-}20)$$

3. *Cofactor of a_{ij}*

$$\bar{a}_{ij} = (-1)^{i+j} m_{ij} = \text{signed minor} \qquad (16\text{-}21)$$

4. *Cofactor matrix of A*

$$\text{co } A = \bar{A} = [\bar{a}_{ij}] = \begin{bmatrix} \bar{a}_{11} & \bar{a}_{12} & \cdots & \bar{a}_{1n} \\ \bar{a}_{21} & \bar{a}_{22} & \cdots & \bar{a}_{2n} \\ \vdots & & & \\ \bar{a}_{n1} & \bar{a}_{n2} & \cdots & \bar{a}_{nn} \end{bmatrix} \qquad (16\text{-}22)$$

5. *Adjoint of A*

$$\text{adj } A = (\bar{A})^T = \begin{bmatrix} \bar{a}_{11} & \bar{a}_{21} & \cdots & \bar{a}_{n1} \\ \bar{a}_{12} & \bar{a}_{22} & \cdots & \bar{a}_{n2} \\ \vdots & & & \\ \bar{a}_{1n} & \bar{a}_{2n} & \cdots & \bar{a}_{nn} \end{bmatrix} \qquad (16\text{-}23)$$

The value of the determinant is obtained by using all the cofactors associated with any one row (or any one column) of the determinant, known as *Laplace's expansion*. For instance, if the first row is used,

$$|A| = a_{11}\bar{a}_{11} + a_{12}\bar{a}_{12} + \cdots + a_{1n}\bar{a}_{1n}$$

Similarly,

$$|A| = a_{21}\bar{a}_{21} + a_{22}\bar{a}_{22} + \cdots + a_{2n}\bar{a}_{2n}$$
$$\vdots \qquad (16\text{-}24)$$
$$|A| = a_{n1}\bar{a}_{n1} + a_{n2}\bar{a}_{n2} + \cdots + a_{nn}\bar{a}_{nn}$$

Now consider the matrix product

$$C = A(\bar{A})^T = \begin{bmatrix} a_{11} & a_{12} & \cdots & a_{1n} \\ a_{21} & a_{22} & \cdots & a_{2n} \\ \vdots & & & \\ a_{n1} & a_{n2} & \cdots & a_{nn} \end{bmatrix} \begin{bmatrix} \bar{a}_{11} & \bar{a}_{21} & \cdots & \bar{a}_{n1} \\ \bar{a}_{12} & \bar{a}_{22} & \cdots & \bar{a}_{n2} \\ \vdots & & & \\ \bar{a}_{1n} & \bar{a}_{2n} & \cdots & \bar{a}_{nn} \end{bmatrix}$$

The (1, 1) element of C is

$$a_{11}\bar{a}_{11} + a_{12}\bar{a}_{12} + \cdots + a_{1n}\bar{a}_{1n} \tag{16-25}$$

which, according to Eq. 16-24, is $|A|$ or

$$\begin{vmatrix} a_{11} & a_{12} & \cdots & a_{1n} \\ a_{21} & a_{22} & \cdots & a_{2n} \\ \vdots & & & \\ a_{n1} & a_{n2} & \cdots & a_{nn} \end{vmatrix} \tag{16-26}$$

The same is true for the $(2, 2)$, $(3, 3)$, $\cdots$, (n, n) element on the main diagonal of C.

Next, consider the $(2, 1)$ element of C,

$$a_{21}\bar{a}_{11} + a_{22}\bar{a}_{12} + \cdots + a_{2n}\bar{a}_{1n} \tag{16-27}$$

Since this can be obtained by replacing a_{11} with a_{21}, a_{12} with a_{22}, and so on in Eq. 16-25 and, therefore, in Eq. 16-26, we have

$$a_{21}\bar{a}_{11} + a_{22}\bar{a}_{12} + \cdots + a_{2n}\bar{a}_{1n} = \begin{vmatrix} a_{21} & a_{22} & \cdots & a_{2n} \\ a_{21} & a_{22} & \cdots & a_{2n} \\ \vdots & & & \\ a_{n1} & a_{n2} & \cdots & a_{nn} \end{vmatrix} = 0$$

Note that a determinant with two rows identical is zero.

Generally, the elements of C off the main diagonal are zero. Thus

$$C = A(\bar{A})^T = \begin{bmatrix} |A| & 0 & \cdots & 0 \\ 0 & |A| & \cdots & 0 \\ \vdots & & \ddots & \\ 0 & 0 & & |A| \end{bmatrix} = |A|I$$

from which

$$A\frac{(\bar{A})^T}{|A|} = I$$

Therefore

$$A^{-1} = \frac{(\bar{A})^T}{|A|} = \frac{\text{adj } A}{|A|} \tag{16-28}$$

As an example, if

$$A = \begin{bmatrix} 1 & 5 & 2 \\ 1 & 1 & 7 \\ 0 & -3 & 4 \end{bmatrix}$$

then

$$|A| = \begin{vmatrix} 1 & 5 & 2 \\ 1 & 1 & 7 \\ 0 & -3 & 4 \end{vmatrix} = -1$$

$$\text{co } A = \bar{A} = \begin{bmatrix} \begin{vmatrix} 1 & 7 \\ -3 & 4 \end{vmatrix} & -\begin{vmatrix} 1 & 7 \\ 0 & 4 \end{vmatrix} & \begin{vmatrix} 1 & 1 \\ 0 & -3 \end{vmatrix} \\ -\begin{vmatrix} 5 & 2 \\ -3 & 4 \end{vmatrix} & \begin{vmatrix} 1 & 2 \\ 0 & 4 \end{vmatrix} & -\begin{vmatrix} 1 & 5 \\ 0 & -3 \end{vmatrix} \\ \begin{vmatrix} 5 & 2 \\ 1 & 7 \end{vmatrix} & -\begin{vmatrix} 1 & 2 \\ 1 & 7 \end{vmatrix} & \begin{vmatrix} 1 & 5 \\ 1 & 1 \end{vmatrix} \end{bmatrix}$$

$$= \begin{bmatrix} 25 & -4 & -3 \\ -26 & 4 & 3 \\ 33 & -5 & -4 \end{bmatrix}$$

$$\text{adj } A = (\bar{A})^T = \begin{bmatrix} 25 & -26 & 33 \\ -4 & 4 & -5 \\ -3 & 3 & -4 \end{bmatrix}$$

Therefore

$$A^{-1} = \frac{(\bar{A})^T}{|A|} = \frac{\begin{bmatrix} 25 & -26 & 33 \\ -4 & 4 & -5 \\ -3 & 3 & -4 \end{bmatrix}}{(-1)} = \begin{bmatrix} -25 & 26 & -33 \\ 4 & -4 & 5 \\ 3 & -3 & 4 \end{bmatrix}$$

It can be easily shown that

$$AA^{-1} = A^{-1}A = \begin{bmatrix} 1 & 0 & 0 \\ 0 & 1 & 0 \\ 0 & 0 & 1 \end{bmatrix} = I$$

Other theorems for matrix inversion are:

1. *The inverse of a matrix is unique.* Suppose that there exists another matrix X such that

$$AX = I$$

Then $$A^{-1}AX = A^{-1}I = A^{-1}$$

Therefore $$X = A^{-1}$$

2. *The inverse of the transpose of a matrix is equal to the transpose of the inverse of the matrix.* Since

$$(A^T)^{-1}A^T = I \tag{16-29}$$

and

$$(A^{-1})^T A^T = (AA^{-1})^T = I^T = I \tag{16-30}$$

Comparing Eqs. 16-29 and 16-30 gives

$$(A^T)^{-1} = (A^{-1})^T \tag{16-31}$$

3. *The inverse of the product of two matrices is equal to the product of the inverse of two matrices in reverse order.* Consider the product

$$(AB)(B^{-1}A^{-1}) = A(BB^{-1})A^{-1} = AA^{-1} = I$$

Hence

$$(AB)^{-1} = B^{-1}A^{-1} \tag{16-32}$$

This can be extended to

$$(ABC)^{-1} = C^{-1}B^{-1}A^{-1} \tag{16-33}$$

etc.

4. *The inverse of a symmetrical matrix is itself symmetrical.* Let A be symmetrical. Prove that A^{-1} is also symmetrical.

We begin with

$$AA^{-1}A^T = A^T \tag{16-34}$$

Taking the transpose of each side of Eq. 16-34 gives

$$A(A^{-1})^T A^T = A$$

Using the fact that $A = A^T$ if A is symmetrical, we have

$$A(A^{-1})^T A^T = A^T \tag{16-35}$$

Comparing Eqs. 16-34 and 16-35 gives

$$A^{-1} = (A^{-1})^T \tag{16-36}$$

Equation 16-36 can be true only if A^{-1} is symmetrical.

16-8. INVERSION BY SUCCESSIVE TRANSFORMATIONS

Consider the scheme of matrix A and a unit matrix of the same order:

$$A|I \tag{16-37}$$

If a series of operations are performed for matrix A such that A is transformed to a unit matrix, then by the same operations the unit matrix I will be transformed to A^{-1} as in the scheme shown below:

$$I|A^{-1} \tag{16-38}$$

The detailed procedure is similar to that of solving a set of simultaneous equations by elimination in ordinary algebra. It is illustrated in Table 16-1.

TABLE 16-1

Operations	Row	Matrix A			Matrix I		
	①	1	5	2	1	0	0
	②	1	1	7	0	1	0
	③	0	−3	4	0	0	1
①	④	1	5	2	1	0	0
② − ①	⑤	0	−4	5	−1	1	0
③	⑥	0	−3	4	0	0	1
④ − ⑧ × 5	⑦	1	0	$\frac{33}{4}$	$-\frac{1}{4}$	$\frac{5}{4}$	0
⑤ × $(-\frac{1}{4})$	⑧	0	1	$-\frac{5}{4}$	$\frac{1}{4}$	$-\frac{1}{4}$	0
⑥ + ⑧ × 3	⑨	0	0	$\frac{1}{4}$	$\frac{3}{4}$	$-\frac{3}{4}$	1
⑦ − ⑫ × $\frac{33}{4}$	⑩	1	0	0	−25	26	−33
⑧ + ⑫ × $\frac{5}{4}$	⑪	0	1	0	4	−4	5
⑨ × 4	⑫	0	0	1	3	−3	4

$$\text{The result} \qquad A^{-1} = \begin{bmatrix} -25 & 26 & -33 \\ 4 & -4 & 5 \\ 3 & -3 & 4 \end{bmatrix}$$

is checked with what is obtained in Sec. 16-7.

We have hitherto demonstrated the procedure by which the matrix A is successively transformed to a unit matrix by a suitable combination of the rows of A. A similar procedure can be developed for transforming A to I by a suitable combination of the columns of A.

Note that there are many other ways of computing A^{-1} available. Those described in this section and in Sec. 16-7 are suitabe for hand calculation if the order of A is small.

16-9. SOLUTION OF LINEAR SIMULTANEOUS
EQUATIONS

Many structural problems, either statically determinate or indeterminate, require solving simultaneous linear equations.

Consider the simultaneous linear equations:

$$\begin{aligned} a_{11}X_1 + a_{12}X_2 + \cdots + a_{1n}X_n &= b_1 \\ a_{21}X_1 + a_{22}X_2 + \cdots + a_{2n}X_n &= b_2 \\ &\vdots \\ a_{n1}X_1 + a_{n2}X_2 + \cdots + a_{nn}X_n &= b_n \end{aligned} \qquad (16\text{-}39)$$

These can be put in matrix form:

$$\begin{bmatrix} a_{11} & a_{12} \cdots a_{1n} \\ a_{21} & a_{22} \cdots a_{2n} \\ \vdots \\ a_{n1} & a_{n2} \cdots a_{nn} \end{bmatrix} \begin{Bmatrix} X_1 \\ X_2 \\ \vdots \\ X_n \end{Bmatrix} = \begin{Bmatrix} b_1 \\ b_2 \\ \vdots \\ b_n \end{Bmatrix} \tag{16-40}$$

or simply

$$AX = B \tag{16-41}$$

where A is a square matrix and X and B are column matrices. Provided that A is nonsingular, we can solve

$$X = A^{-1}B \tag{16-42}$$

since $\qquad\qquad A^{-1}AX = A^{-1}B$

and $\qquad\qquad A^{-1}A = I$

The problem is now to find A^{-1}. After A^{-1} is obtained, X is solved. For example,

$$X_1 + 5X_2 + 2X_3 = 9$$
$$X_1 + X_2 + 7X_3 = 6$$
$$-3X_2 + 4X_3 = -2$$

The matrix form will be

$$\begin{bmatrix} 1 & 5 & 2 \\ 1 & 1 & 7 \\ 0 & -3 & 4 \end{bmatrix} \begin{Bmatrix} X_1 \\ X_2 \\ X_3 \end{Bmatrix} = \begin{Bmatrix} 9 \\ 6 \\ -2 \end{Bmatrix}$$

or $\qquad\qquad AX = B$

If $\qquad\qquad A = \begin{bmatrix} 1 & 5 & 2 \\ 1 & 1 & 7 \\ 0 & -3 & 4 \end{bmatrix}$

then $\qquad\qquad A^{-1} = \begin{bmatrix} -25 & 26 & -33 \\ 4 & -4 & 5 \\ 3 & -3 & 4 \end{bmatrix}$

as obtained previously in Sec. 16-7 or 16-8. Thus

$$X = A^{-1}B = \begin{bmatrix} -25 & 26 & -33 \\ 4 & -4 & 5 \\ 3 & -3 & 4 \end{bmatrix} \begin{Bmatrix} 9 \\ 6 \\ -2 \end{Bmatrix} = \begin{Bmatrix} -3 \\ 2 \\ 1 \end{Bmatrix}$$

i.e., $\qquad\qquad X_1 = -3 \qquad X_2 = 2 \qquad X_3 = 1$

If we have several systems of equations, all with the same coefficient matrix A, it is advantageous to find A^{-1} since the problem of determining the solutions for these systems would then reduce to mere matrix multiplications.

PROBLEMS

16-1. Given

$$A = \begin{bmatrix} 2 & 8 & -4 \\ 3 & 4 & 2 \\ 1 & 0 & 3 \end{bmatrix} \quad B = \begin{bmatrix} 2 & -1 & 0 \\ 4 & 0 & -1 \\ -3 & 3 & 2 \end{bmatrix} \quad C = \begin{bmatrix} 1 & 0 \\ 0 & 1 \\ 5 & 4 \end{bmatrix}$$

determine $A + B$, $A - B$, AB, AC, BC, and $B^T AB$.

16-2. Use partitioning to find the products of AC and BC in Prob. 16-1.

16-3. Split matrix A in Prob. 16-1 into a symmetrical and an antisymmetrical matrix.

16-4. If A and B are symmetrical matrices of the same order, prove that AB is symmetrical if and only if $AB = BA$.

16-5. What is the adjoint of the matrix A in Prob. 16-1? What is the inverse of A?

16-6. Verify the result for A^{-1} in Prob. 16-5 by the method of successive transformation.

16-7. Given

$$A = \begin{bmatrix} a_{11} & a_{12} \\ a_{21} & a_{22} \end{bmatrix} \quad \text{and} \quad B = \begin{bmatrix} b_{11} & b_{12} \\ b_{21} & b_{22} \end{bmatrix}$$

If $B = A^{-1}$, find each element of B in terms of the elements of A by direct application to the equation

$$AA^{-1} = I$$

17

MATRIX ANALYSIS OF STRUCTURES
BY THE FINITE ELEMENT METHOD
PART I: THE FORCE METHOD

17-1. GENERAL

During the past decade, the rapid development of computers and the growing demand for better methods of analysis for complex and light-weight structures led to the development of methods for matrix analysis of structures. The use of matrix notation in expressing structural theory is in itself simple and elegant, but the practical value of matrix analysis would not have become clear without the invention of the high-speed digital computer.

It is true that the classical methods of structural analysis, such as the method of consistent deformations, Castigliano's theorems, or the slope-deflection method, which had only limited use in the past because of operational difficulties, have now regained their strength because of the invention of the digital computer. Indeed, solving a set of one hundred simultaneous equations with a modern computer would hardly take a minute, and the solution of simultaneous equations is equivalent to inverting a matrix. However, the matrix method, to be discussed in this and the subsequent chapter, has its unique theoretical basis and particular procedures. According to its basic principle, it is sometimes better known as the *finite element* method, because a structure, instead of being a continuation of differential elements, is idealized as a composition of a number of finite pieces. This idea enables the step-by-step build up of the force-displacement relationship of a structure from those basic elements of which the structure is composed.

Many complicated problems in various fields are thus solved by this new computational technique including:

1. Trusses, beams, and rigid frames,
2. Plates and shells of arbitrary shape and loading,
3. Composite structures,
4. Pressure vessels,
5. Torsion in members with irregular section,

6. Dynamic analysis of framework,

etc.

Since this book is an elementary treatment, we will confine the discussion to plane structures of the type listed as item (1).

The principal matrix procedures based on a finite element representation commonly fall into two categories—the force method and the displacement method. The force method treats the member forces as the basic unknowns; whereas the displacement method regards the nodal displacements as the basic unknowns. As will be seen, a duality exists between the two approaches. We shall discuss the force method in this chapter and the displacement method in the next chapter.

17-2. BASIC CONCEPTS OF STRUCTURES

Structures, such as trusses, beams, and rigid frames, are defined as an assemblage of structural elements jointed together at a finite number of discrete points, called *nodes* or *nodal points*, and loaded only at these points. The term node is used instead of joint because often a concentratedly loaded point, not the conventional joint, is taken as a node.

Any complicated structure can be cut into simpler components. For instances, a truss may be considered as composed of many two-force members pin connected at their ends. A rigid frame may be taken as a composition of a number of three-force members or other convenient units. The behavior of the subdivided elements, as well as the whole structure, must satisfy the following basic conditions:

1. Equilibrium of forces;
2. Compatibility of displacements;
3. Force-displacement relationship specified by the geometric and elastic properties of the elements.

These conditions are generally required by a linear structure no matter what method is used.

Frequently involved in the subsequent discussion is the principle of virtual work, which serves in many cases as an effective substitute for the equations of equilibrium or compatibility. The principle of virtual work states:

If a system in equilibrium under the action of a set of external forces is given a small virtual displacement compatible with the constraint imposed on the system, then the work done by the external forces equals the increase in strain energy stored in the system.

Before we elaborate, we must introduce the notation used for force and displacement in this and the next chapter. Consider the frame in Fig. 17-1(a), which is composed of three elements (members a, b, and c) subjected to external loads (nodal forces) denoted by R_1, R_2, and R_3. Neglecting the small shear deformations, we denote the nodal displacements caused by these loads by r_1, r_2, and r_3 respectively.

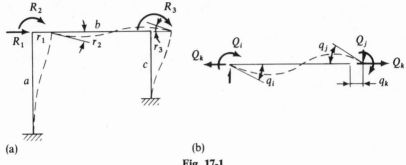

(a)   (b)

Fig. 17-1

A typical member element is shown in Fig. 17-1(b) and is generally subjected to the internal forces of the end moments, denoted by Q_i and Q_j, and the axial forces, denoted by Q_k. The member is also subjected to end shears; however, the end shears can be expressed in terms of the end moments and therefore are not considered independent forces. Associated with the end moments and axial forces are the end rotations, q_i and q_j, and the axial elongation q_k. The signs shown in Fig. 17-1 are considered as positive. A superscript is used for these quantities to identify their belonging to a particular member. For instance, Q_i^a, Q_j^a, and Q_k^a indicate the internal forces for member a, and q_i^a, q_j^a and q_k^a their corresponding deformations.

Thus in a structure, if we use a column matrix R

$$R = \begin{Bmatrix} R_1 \\ R_2 \\ R_3 \\ \vdots \end{Bmatrix}$$

to denote all the nodal forces, and a column matrix r

$$r = \begin{Bmatrix} r_1 \\ r_2 \\ r_3 \\ \vdots \end{Bmatrix}$$

to denote the corresponding nodal displacements, then the external work done may be expressed by

$$W_E = R^T r = r^T R \tag{17-1}$$

Similarly, if we use a column matrix

$$Q = \left\{ \begin{matrix} Q_i^a \\ Q_j^a \\ \vdots \\ Q_i^b \\ Q_j^b \\ \vdots \end{matrix} \right\}$$

to denote all the internal end forces for members a, b, $\cdots$, etc., and a column matrix q

$$q = \left\{ \begin{matrix} q_i^a \\ q_j^a \\ \vdots \\ q_i^b \\ q_j^b \\ \vdots \end{matrix} \right\}$$

to denote the corresponding internal displacements, then the internal work or strain energy restored may be expressed by

$$W_I = Q^T q = q^T Q \tag{17-2}$$

Equating W_E and W_I, we have

$$R^T r = Q^T q \tag{17-3}$$

or

$$r^T R = q^T Q \tag{17-4}$$

Equation 17-3 or 17-4 is valid if R and Q are in equilibrium or r and q are compatible. The virtual work may be the result of either a virtual displacement or a virtual force. A bar is added to the top of each of the virtual elements in Eqs. 17-3 and 17-4 to distinguish them from the real elements. Thus if virtual displacements are used, we have from Eq. 17-4:

$$\bar{r}^T R = \bar{q}^T Q \tag{17-5}$$

On the other hand, if virtual forces are used, then from Eq. 17-3

$$\bar{R}^T r = \bar{Q}^T q \tag{17-6}$$

17-3. EQUILIBRIUM, FORCE TRANSFORMATION
MATRIX

For a statically determinate structure each of the member forces may be expressed in terms of the external nodal loads by using the equilibrium conditions of the system alone. Thus,

$$Q_1 = b_{11}R_1 + b_{12}R_2 + \cdots + b_{1n}R_n$$
$$Q_2 = b_{21}R_1 + b_{22}R_2 + \cdots + b_{2n}R_n$$
$$\vdots$$
$$Q_m = b_{m1}R_1 + b_{m2}R_2 + \cdots + b_{mn}R_n$$

(17-7)

in which $Q_1 = Q_i^a$, $Q_2 = Q_j^a, \cdots$, etc. Observe that $R_1, R_2, \cdots, R_n$ represent the total set of applied loads and $Q_1, Q_2, \cdots, Q_m$ the total set of member forces. No connection between the subscripts on R and Q is implied.

The matrix form for Eq. 17-7 is

$$Q = bR$$

(17-8)

where

$$b = \begin{bmatrix} b_{11} & b_{12} & \cdots & b_{1n} \\ b_{21} & b_{22} & \cdots & b_{2n} \\ \vdots & & & \\ b_{m1} & b_{m2} & \cdots & b_{mn} \end{bmatrix}$$

(17-8a)

called the *force transformation matrix* which relates the internal forces to the external forces. Matrix b is usually a rectangular matrix in which the typical element b_{ij} is the value of the internal force component Q_i caused by a unit value of external load R_j. Note that b is merely an expression of equilibrium for the system.

As for a statically indeterminate structure, the internal member forces cannot be expressed in terms of the external loads by equilibrium alone. However, as previously stated (see Sec. 9-1), a statically indeterminate structure can be made determinate by removing the redundant elements. The statically determinate and stable structure that remains after the removal of the extra restraints is called a *primary structure*. We then consider the original structure as equivalent to the primary structure subjected to the combined influences of the applied loads and the unknown redundant forces, thereby treating the redundants as a part of the external loads of unknown magnitude. In this way, we can express member forces in terms of the original applied loads R and the redundant forces X as

$$Q = b_R R + b_X X$$

(17-9)

or

$$Q = [b_R \mid b_X] \left\{ \frac{R}{X} \right\}$$

(17-10)

where b_R and b_X are force transformation matrices representing the separate influences of the known applied loads R and the unknown redundants X on the member forces. They are generally rectangular matrices.

17-4. COMPATIBILITY

Compatibility is a continuity condition on the displacements of the structure after the external loads are applied to the structure. Compatibility must be brought into the analysis of statically indeterminate structures since the equilibrium equations alone do not suffice to solve the problem.

The compatibility conditions used in the force method for solving static structures mounted on rigid supports are that the displacements at all the cuts, or releases of redundant points caused by the original applied loads, and the redundant forces must be made to vanish in order that the continuity of the structures can be maintained. That is, the gap in the displacements resulting from the applied loads is precisely removed by the redundant forces.

If we use r_x to denote the displacement matrix (a column matrix) at the releases of the redundant points of an indeterminate structure, we have

$$r_x = 0 \qquad (17\text{-}11)$$

17-5. FORCE-DISPLACEMENT RELATIONSHIP,
FLEXIBILITY COEFFICIENT, FLEXIBILITY MATRIX

A *flexibility coefficient* f_{ij} is the displacement at point i due to a unit action at point j, all other points being unloaded. Apparently the flexibility coefficient constitutes a relationship between deformation and force. Applying the principle of superposition, we may express the deformation at any point of a system caused by a set of forces in terms of the flexibility coefficients.

Our intention is, first of all, to establish the relationship between the member displacements and the member forces of a structure. Consider a typical member a taken from a plane structure as shown in Fig. 17-2. As before, the member forces are represented by a column matrix Q^a,

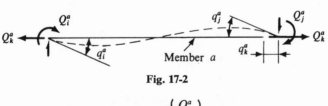

Fig. 17-2

$$Q^a = \left\{ \begin{array}{c} Q_i^a \\ Q_j^a \\ Q_k^a \end{array} \right\}$$

and the corresponding member deformations are represented by a column matrix q^a,

$$q^a = \left\{ \begin{array}{c} q_i^a \\ q_j^a \\ q_k^a \end{array} \right\}$$

Note that the clockwise end moments and rotations and the tensile axial forces and elongation are considered as positive.

Using the flexibility coefficient f_{ij}^a, we may express each of the member deformations in terms of the separate influences of the whole set of member forces

$$q_i^a = f_{ii}^a Q_i^a + f_{ij}^a Q_j^a + f_{ik}^a Q_k^a$$
$$q_j^a = f_{ji}^a Q_i^a + f_{jj}^a Q_j^a + f_{jk}^a Q_k^a \qquad (17\text{-}12)$$
$$q_k^a = f_{ki}^a Q_i^a + f_{kj}^a Q_j^a + f_{kk}^a Q_k^a$$

or in matrix form

$$q^a = f^a Q^a \qquad (17\text{-}13)$$

in which

$$f^a = \begin{bmatrix} f_{ii}^a & f_{ij}^a & f_{ik}^a \\ f_{ji}^a & f_{jj}^a & f_{jk}^a \\ f_{ki}^a & f_{kj}^a & f_{kk}^a \end{bmatrix} \qquad (17\text{-}13a)$$

defined as the *element flexibility matrix*. Clearly the coefficient, for instance, f_{ii}^a is given by

$$f_{ii}^a = q_i^a \quad \text{as} \quad Q_i^a = 1 \qquad Q_j^a = Q_k^a = 0$$

The rest can similarly be defined.

The above description is referred to an individual element. For a structure consisting of $a, b, \cdots$ elements, we have

$$q^a = f^a Q^a$$
$$q^b = f^b Q^b$$
$$\vdots$$

etc.

Let

$$q = \left\{ \begin{array}{c} q^a \\ q^b \\ \vdots \end{array} \right\} \quad \text{and} \quad Q = \left\{ \begin{array}{c} Q^a \\ Q^b \\ \vdots \end{array} \right\}$$

The above equations can be put in matrix form as

$$q = fQ \qquad (17\text{-}14)$$

where

$$f = \begin{bmatrix} f^a & & \\ & f^b & \\ & & \ddots \end{bmatrix} \qquad (17\text{-}14a)$$

which is a diagonal matrix with element flexibility matrices as its constituents.

Since the flexibility coefficients of Eq. 17-13a serve to relate the member deformations to the member forces, they are certainly governed by the geometric and material properties of the member. Suppose that the member is prismatic with length L, cross-sectional area A, moment of inertia I, and modulus of elasticity E. The elements in the first column of f^a are, by definition, the member deformation resulting from $Q_i^a = 1$. These are found to be

$$f_{ii}^a = \text{rotation of the left end} \quad = \frac{L}{3EI}$$

$$f_{ji}^a = \text{rotation at the right end} \quad = -\frac{L}{6EI}$$

$$f_{ki}^a = \text{elongation of the member} = 0$$

Note that f_{ii}^a and f_{ji}^a can easily be determined by the conjugate-beam method and that $f_{ki}^a = 0$ is apparent. All the other elements can be obtained similarly. Thus the member flexibility matrix is given by

$$f^a = \begin{bmatrix} \dfrac{L}{3EI} & -\dfrac{L}{6EI} & 0 \\[2ex] -\dfrac{L}{6EI} & \dfrac{L}{3EI} & 0 \\[2ex] 0 & 0 & \dfrac{L}{AE} \end{bmatrix} \qquad (17\text{-}15)$$

Note that the member flexibility matrix is symmetric because of reciprocity.

If the effect of axial forces in the member is neglected, as is usually done in rigid-frame analysis, then

$$f^a = \frac{L}{6EI} \begin{bmatrix} 2 & -1 \\ -1 & 2 \end{bmatrix} \qquad (17\text{-}16)$$

For truss member subjected to axial forces only

$$f^a = \begin{bmatrix} \dfrac{L}{AE} \end{bmatrix} \qquad (17\text{-}17)$$

We have already established the relationship between the member deformations and the member forces (the internal relationship). Next we must establish the relationship between the nodal displacements r and the nodal forces R (the external relationship). This may be accomplished by using the technique of virtual work. We start by Eq. 17-6

$$\bar{R}^T r = \bar{Q}^T q \qquad (17\text{-}18)$$

On account of Eqs. 17-8 and 17-14, namely,

$$Q = bR \quad \text{and} \quad q = fQ$$

we have
$$\bar{Q} = b\bar{R}$$

or

$$\bar{Q}^T = \bar{R}^T b^T \qquad (17\text{-}19)$$

and

$$q = fbR \qquad (17\text{-}20)$$

Substituting Eqs. 17-19 and 17-20 in Eq. 17-18 yields

$$\bar{R}^T r = \bar{R}^T b^T fbR$$

from which

$$r = b^T fbR \qquad (17\text{-}21)$$

If we let

$$F = b^T fb \qquad (17\text{-}22)$$

F being called the *total flexibility matrix* or *flexibility matrix of the structure*, then

$$r = FR \qquad (17\text{-}23)$$

For a statically determinate structure Eq. 17-23 gives a direct solution of all the nodal displacements in terms of the external nodal forces.

If the structure is statically indeterminate, then the external work must include the work caused by the redundant forces; also Eq. 17-9 must be used instead of Eq. 17-8. To relate r to R, we begin the derivation by

$$\bar{R}^T r + \bar{X}^T r_X = \bar{Q}^T q \qquad (17\text{-}24)$$

Using equilibrium Eq. 17-9 and virtual force, we have

$$\bar{Q} = b_R \bar{R} + b_X \bar{X}$$

so that

$$\bar{Q}^T = \bar{R}^T b_R^T + \bar{X}^T b_X^T \qquad (17\text{-}25)$$

Also because of Eqs. 17-14 and 17-9,

$$q = fQ = fb_R R + fb_X X \qquad (17\text{-}26)$$

Substituting Eqs. 17-25 and 17-26 in Eq. 17-24 yields

$$\bar{R}^T r + \bar{X}^T r_X = (\bar{R}^T b_R^T + \bar{X}^T b_X^T)(fb_R R + fb_X X)$$

or $\bar{R}^T r + \bar{X}^T r_X = \bar{R}^T (b_R^T fb_R R + b_R^T fb_X X) + \bar{X}^T (b_X^T fb_R R + b_X^T fb_X X)$

Comparing the virtual forces on the left and the right sides of the above equation, we have

$$r = b_R^T fb_R R + b_R^T fb_X X$$

$$r_X = b_X^T fb_R R + b_X^T fb_X X$$

These may be arranged as

$$r = F_{RR} R + F_{RX} X \qquad (17\text{-}27)$$

$$r_X = F_{XR} R + F_{XX} X \qquad (17\text{-}28)$$

if we let
$$F_{RR} = b_R^T f b_R \qquad F_{RX} = b_R^T f b_X$$
$$F_{XR} = b_X^T f b_R \qquad F_{XX} = b_X^T f b_X \tag{17-29}$$

Equations 17-27 and 17-28 may be put in matrix form as

$$\left\{ \frac{r}{r_X} \right\} = \left[\begin{array}{c|c} F_{RR} & F_{RX} \\ \hline F_{XR} & F_{XX} \end{array} \right] \left\{ \frac{R}{X} \right\} \tag{17-30}$$

For structures on rigid supports Eq. 17-30 becomes

$$\left\{ \frac{r}{r_X = 0} \right\} = \left[\begin{array}{c|c} F_{RR} & F_{RX} \\ \hline F_{XR} & F_{XX} \end{array} \right] \left\{ \frac{R}{X} \right\} \tag{17-31}$$

The compatibility condition is, therefore,

$$F_{XR}R + F_{XX}X = 0 \tag{17-32}$$

from which

$$X = - F_{XX}^{-1} F_{XR} R \tag{17-33}$$

Equation 17-33 expresses the solution for the redundants.

Substituting Eq. 17-33 in Eq. 17-27, we finally relate the nodal displacements to the nodal forces,

$$r = (F_{RR} - F_{RX} F_{XX}^{-1} F_{XR}) R$$

Or simply

$$r = F'R \tag{17-34}$$

if we let

$$F' = F_{RR} - F_{RX} F_{XX}^{-1} F_{XR} \tag{17-34a}$$

F' is called the *flexibility matrix of the indeterminate structure*, which relates directly the nodal displacements to the nodal loads covering the effects of the redundants.

With the redundants X found to be $- F_{XX}^{-1} F_{XR} R$, the member forces are then solved by equilibrium;

$$Q = b_R R + b_X X = (b_R - b_X F_{XX}^{-1} F_{XR}) R$$

That is

$$Q = b'R \tag{17-35}$$

if we let

$$b' = b_R - b_X F_{XX}^{-1} F_{XR} \tag{17-35a}$$

b' being the *force transformation matrix of the indeterminate structure*, relating directly the member forces to the applied nodal loads covering the effects of the redundants.

An alternative form of F' may be obtained in terms of b'; i.e.,

$$F' = b_R^T f b' \tag{17-36}$$

since $\qquad F' = F_{RR} - F_{RX}F_{XX}^{-1}F_{XR}$

$$= b_R^T f b_R - b_R^T f b_X F_{XX}^{-1} F_{XR} = b_R^T f(b_R - b_X F_{XX}^{-1} F_{XR})$$

which yields Eq. 17-36. Note that Eq. 17-34a for finding F' is quite general but that the alternative form given by Eq. 17-36 is more convenient if b' is first determined.

The following identity is useful for checking results:

$$b_X^T f b' = 0 \qquad\qquad (17\text{-}37)$$

This can easily be proved as follows:

$$b_X^T f b' = b_X^T f(b_R - b_X F_{XX}^{-1} F_{XR}) = F_{XR} - F_{XX}F_{XX}^{-1}F_{XR}$$

$$= F_{XR} - F_{XR} = 0$$

17-6. ANALYSIS OF STATICALLY DETERMINATE STRUCTURES BY THE FORCE METHOD

As developed in Sec. 17-3, for a statically determinate structure, the internal forces Q can be solved by equilibrium alone; i.e.,

$$Q = bR$$

See Eq. 17-8.

Also the nodal displacements r can be solved by

$$r = b^T f b R$$

See Eq. 17-21.

Assume that the purely statical task of evaluating the force transformation matrix b is not difficult, although this phase of analysis may cost a considerable amount of labour in complicated problems.

The procedures for analyzing a statically determinate structure by the force method are contained in the following steps:

1. Define the external nodal loads R.
2. Define the internal member forces Q.
3. Determine the force transformation matrix b.

Consider the elements of the first column of b. If we let

$$R_1 = 1 \qquad R_2 = R_3 = \cdots = R_n = 0$$

it is readily seen from Eq. 17-7 that $Q_1, Q_2, \cdots, Q_m$ are the elements of the first column. The rest can be obtained similarly.

4. The internal member forces Q are then solved by

$$Q = bR$$

5. Determine individual element flexibility matrices f^a, f^b, $\cdots$ according to Eq. 17-16 or Eq. 17-17, and assemble them as a diagonal matrix

$$f = \begin{bmatrix} f^a & & \\ & f^b & \\ & & \ddots \end{bmatrix}$$

6. Compute the flexibility matrix of the structure

$$F = b^T f b$$

7. Find the nodal displacements r,

$$r = FR$$

Example 17-1. Find the bar forces of the truss shown in Fig. 17-3. Find also the deflections corresponding to the applied loads R_1 and R_2. Assume that $L/A = 1$ for all members.

The load matrix is

$$R = \begin{Bmatrix} R_1 \\ R_2 \end{Bmatrix}$$

The truss has five bars designated by a, b, c, d, and e. The member-force matrix is

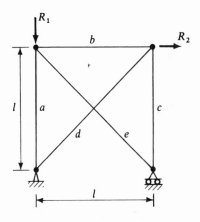

Fig. 17-3

$$Q = \begin{Bmatrix} Q^a \\ Q^b \\ Q^c \\ Q^d \\ Q^e \end{Bmatrix}$$

The force transformation matrix b is given by

$$b = \begin{bmatrix} -1 & 0 \\ 0 & 0 \\ 0 & -1 \\ 0 & \sqrt{2} \\ 0 & 0 \end{bmatrix}$$

in which the first column contains the bar forces of the truss in Fig. 17-3 in the order a, b, c, d, e, resulting from $R_1 = 1$, $R_2 = 0$. The second column contains the corresponding bar forces resulting from $R_2 = 1$, $R_1 = 0$. From equilibrium

$$\begin{Bmatrix} Q^a \\ Q^b \\ Q^c \\ Q^d \\ Q^e \end{Bmatrix} = \begin{bmatrix} -1 & 0 \\ 0 & 0 \\ 0 & -1 \\ 0 & \sqrt{2} \\ 0 & 0 \end{bmatrix} \begin{Bmatrix} R_1 \\ R_2 \end{Bmatrix}$$

For individual members the flexibility matrices are found to be

$$f^a = f^b = f^c = f^d = f^e = \frac{1}{E}$$

since $L/A = 1$ for all members. Thus the diagonal matrix

$$f = \begin{bmatrix} f^a & & & & \\ & f^b & & & \\ & & f^c & & \\ & & & f^d & \\ & & & & f^e \end{bmatrix} = \frac{1}{E} \begin{bmatrix} 1 & & & & \\ & 1 & & & \\ & & 1 & & \\ & & & 1 & \\ & & & & 1 \end{bmatrix}$$

The total flexibility matrix is then determined;

$$F = b^T f b$$

$$= \begin{bmatrix} -1 & 0 & 0 & 0 & 0 \\ 0 & 0 & -1 & \sqrt{2} & 0 \end{bmatrix} \frac{1}{E} \begin{bmatrix} 1 & & & & \\ & 1 & & & \\ & & 1 & & \\ & & & 1 & \\ & & & & 1 \end{bmatrix} \begin{bmatrix} -1 & 0 \\ 0 & 0 \\ 0 & -1 \\ 0 & \sqrt{2} \\ 0 & 0 \end{bmatrix}$$

$$= \frac{1}{E} \begin{bmatrix} 1 & 0 \\ 0 & 3 \end{bmatrix}$$

The nodal displacements r are solved by

$$r = FR$$

$$\begin{Bmatrix} r_1 \\ r_2 \end{Bmatrix} = \frac{1}{E} \begin{bmatrix} 1 & 0 \\ 0 & 3 \end{bmatrix} \begin{Bmatrix} R_1 \\ R_2 \end{Bmatrix}$$

or

$$r_1 = \frac{R_1}{E} \qquad r_2 = \frac{3R_2}{E}$$

Example 17-2. Find the deflections corresponding to the applied loads for the cantilever beam shown in Fig. 17-4(a). Assume constant EI.

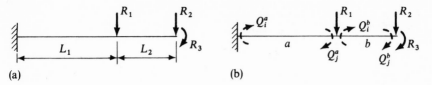

Fig. 17-4

Since the loaded point of R_1 must be considered as a nodal point, it divides the beam into two segments, designated as member a and member b in Fig. 17-4 (b). The internal member forces are shown by dotted lines. From equilibrium

$$R_1 = 1 \qquad R_2 = 1 \qquad R_3 = 1$$

$$Q = bR \qquad \begin{Bmatrix} Q_i^a \\ Q_j^a \\ Q_i^b \\ Q_j^b \end{Bmatrix} = \begin{bmatrix} -L_1 & -(L_1 + L_2) & -1 \\ 0 & L_2 & 1 \\ 0 & -L_2 & -1 \\ 0 & 0 & 1 \end{bmatrix} \begin{Bmatrix} R_1 \\ R_2 \\ R_3 \end{Bmatrix}$$

Note that the elements of the first column of matrix b are the member forces caused by $R_1 = 1$, $R_2 = R_3 = 0$ for the beam shown in Fig. 17-4(b). This gives

$$Q_i^a = -L_1 \qquad Q_j^a = Q_i^b = Q_j^b = 0$$

The second column of matrix b contains the member forces resulting from $R_2 = 1$, $R_1 = R_3 = 0$. Thus

$$Q_i^a = -(L_1 + L_2) \qquad Q_j^a = L_2 \qquad Q_i^b = -L_2 \qquad Q_j^b = 0$$

And the third column of matrix b contains the member forces due to a unit couple applied only at the free end of the beam, i.e., $R_3 = 1$, $R_1 = R_2 = 0$. This gives

$$Q_i^a = -1 \qquad Q_j^a = 1 \qquad Q_i^b = -1 \qquad Q_j^b = 1$$

The individual member flexibility matrices are

$$f^a = \frac{1}{6EI} \begin{bmatrix} 2L_1 & -L_1 \\ -L_1 & 2L_1 \end{bmatrix} \qquad f^b = \frac{1}{6EI} \begin{bmatrix} 2L_2 & -L_2 \\ -L_2 & 2L_2 \end{bmatrix}$$

from which $\qquad f = \dfrac{1}{6EI} \begin{bmatrix} 2L_1 & -L_1 & 0 & 0 \\ -L_1 & 2L_1 & 0 & 0 \\ 0 & 0 & 2L_2 & -L_2 \\ 0 & 0 & -L_2 & 2L_2 \end{bmatrix}$

The total flexibility matrix F is obtained from

$$F = b^T f b = \begin{bmatrix} -L_1 & 0 & 0 & 0 \\ -(L_1 + L_2) & L_2 & -L_2 & 0 \\ -1 & 1 & -1 & 1 \end{bmatrix} \left(\frac{1}{6EI}\right)$$

$$\cdot \begin{bmatrix} 2L_1 & -L_1 & 0 & 0 \\ -L_1 & 2L_1 & 0 & 0 \\ 0 & 0 & 2L_2 & -L_2 \\ 0 & 0 & -L_2 & 2L_2 \end{bmatrix} \begin{bmatrix} -L_1 & -(L_1 + L_2) & -1 \\ 0 & L_2 & 1 \\ 0 & -L_2 & -1 \\ 0 & 0 & 1 \end{bmatrix}$$

$$= \begin{bmatrix} \dfrac{L_1^3}{3EI} & \dfrac{2L_1^3 + 3L_1^2 L_2}{6EI} & \dfrac{L_1^2}{2EI} \\ \dfrac{2L_1^3 + 3L_1^2 L_2}{6EI} & \dfrac{(L_1 + L_2)^3}{3EI} & \dfrac{(L_1 + L_2)^2}{2EI} \\ \dfrac{L_1^2}{2EI} & \dfrac{(L_1 + L_2)^2}{2EI} & \dfrac{L_1 + L_2}{EI} \end{bmatrix}$$

Thus, $$\begin{Bmatrix} r_1 \\ r_2 \\ r_3 \end{Bmatrix} = \begin{bmatrix} \dfrac{L_1^3}{3EI} & \dfrac{2L_1^3 + 3L_1^2 L_2}{6EI} & \dfrac{L_1^2}{2EI} \\ \dfrac{2L_1^3 + 3L_1^2 L_2}{6EI} & \dfrac{(L_1 + L_2)^3}{3EI} & \dfrac{(L_1 + L_2)^2}{2EI} \\ \dfrac{L_1^2}{2EI} & \dfrac{(L_1 + L_2)^2}{2EI} & \dfrac{L_1 + L_2}{EI} \end{bmatrix} \begin{Bmatrix} R_1 \\ R_2 \\ R_3 \end{Bmatrix}$$

or

$$r_1 = \frac{R_1 L_1^3}{3EI} + \frac{R_2(2L_1^3 + 3L_1^2 L_2)}{6EI} + \frac{R_3 L_1^2}{2EI} \tag{17-38}$$

$$r_2 = \frac{R_1(2L_1^3 + 3L_1^2 L_2)}{6EI} + \frac{R_2(L_1 + L_2)^3}{3EI} + \frac{R_3(L_1 + L_2)^2}{2EI} \tag{17-39}$$

$$r_3 = \frac{R_1 L_1^2}{2EI} + \frac{R_2(L_1 + L_2)^2}{2EI} + \frac{R_3(L_1 + L_2)}{EI} \tag{17-40}$$

As a particular problem, find the vertical deflection and the rotation at the free end of the loaded cantilever beam shown in Fig. 17-5.

To do this, we set $R_1 = P$, $R_2 = R_3 = 0$ in Eq. 17-39 to obtain

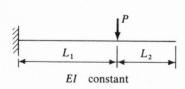

EI constant

Fig. 17-5

$$r_2 = \frac{P(2L_1^3 + 3L_1^2 L_2)}{6EI}$$

which is the resulting vertical deflection of the end of the beam, and we set $R_1 = P$, $R_2 = R_3 = 0$ in Eq. 17-40 to obtain

$$r_3 = \frac{PL_1^2}{2EI}$$

which is the resulting rotation of the end of the beam.

17-7. ANALYSIS OF STATICALLY INDETERMINATE STRUCTURES BY THE FORCE METHOD

As developed in Secs. 17-3, 17-4, and 17-5, the procedures for analyzing a statically indeterminate structure by the force method are given as follows:

1. Define the external loads R.
2. Define the internal member forces Q, and specify the redundants X.
3. Calculate the force transformation matrices b_R and b_X from equilibrium;

$$Q = [b_R \mid b_X] \left\{ \frac{R}{X} \right\}$$

4. Determine the individual element flexibility matrices f^a, f^b, $\cdots$, and assemble them to obtain f;

$$f = \begin{bmatrix} f^a & & \\ & f^b & \\ & & \ddots \end{bmatrix}$$

5. Calculate F_{XR};

$$F_{XR} = b_X^T f b_R$$

6. Calculate F_{XX};

$$F_{XX} = b_X^T f b_X$$

7. Find the inverse of F_{XX}.
8. Solve the redundants X by

$$X = - F_{XX}^{-1} F_{XR} R$$

and substitute X in the equilibrium equation to obtain the member forces Q.
9. Alternatively, we may find b' by

$$b' = b_R - b_X F_{XX}^{-1} F_{XR}$$

and obtain the member forces Q by

$$Q = b'R$$

10. If the nodal displacements are desired, calculate F' by

$$F' = b_R^T f b'$$

and find r by
$$r = F'R$$

As seen in the latter part of Example 17-2, if the points where the displacements are desired are not actually loaded, then we must apply fictitious loads of zero value at these points in order to carry out the above procedures.

Example 17-3. Find the bar forces of the truss in Fig. 17-6(a) by the force method. Also find the nodal displacement corresponding to the applied load. Assume that $E = 30,000$ kips per in². and $L(\text{ft})/A(\text{in}^2.) = 1$ for all members.

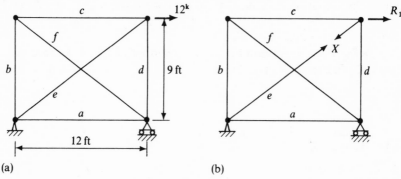

Fig. 17-6

The truss shown in Fig. 17-6(a) is statically indeterminate to the first degree. Let us select bar e as the redundant and denote the external load of 12 kips by R_1 as shown in Fig. 17-6(b). The bar forces are denoted by $Q^a, Q^b, \ldots, Q^f$. From equilibrium based on the primary structure of Fig. 17-6(b),

$$R_1 = 1 \quad X = 1$$

$$
\begin{Bmatrix} Q^a \\ Q^b \\ Q^c \\ Q^d \\ Q^e \\ Q^f \end{Bmatrix} =
\begin{bmatrix} 1 & -\frac{4}{5} \\ \frac{3}{4} & -\frac{3}{5} \\ 1 & -\frac{4}{5} \\ 0 & -\frac{3}{5} \\ 0 & 1 \\ -\frac{5}{4} & 1 \end{bmatrix}
\begin{Bmatrix} R_1 \\ X \end{Bmatrix}
$$

$$\qquad\qquad b_R \qquad b_X$$

since $L/A = 1$ for all members

$$
f = \frac{1}{E}
\begin{bmatrix}
1 & & & & & \\
& 1 & & & & \\
& & 1 & & & \\
& & & 1 & & \\
& & & & 1 & \\
& & & & & 1
\end{bmatrix}
$$

Thus

$$F_{XR} = b_X^T f b_R$$

$$= [-\tfrac{4}{5} \quad -\tfrac{3}{5} \quad -\tfrac{4}{5} \quad -\tfrac{3}{5} \quad 1 \quad 1]\left(\frac{1}{E}\right)\left\{\begin{array}{c} 1 \\ \tfrac{3}{4} \\ 1 \\ 0 \\ 0 \\ -\tfrac{5}{4} \end{array}\right\} = -\frac{3.3}{E}$$

$$F_{XX} = b_X^T f b_X$$

$$= [-\tfrac{4}{5} \quad -\tfrac{3}{5} \quad -\tfrac{4}{5} \quad -\tfrac{3}{5} \quad 1 \quad 1]\left(\frac{1}{E}\right)\left\{\begin{array}{c} -\tfrac{4}{5} \\ -\tfrac{3}{5} \\ -\tfrac{4}{5} \\ -\tfrac{3}{5} \\ 1 \\ 1 \end{array}\right\} = \frac{4}{E}$$

$$F_{XX}^{-1} = \frac{E}{4}$$

The redundant force X is then solved by

$$X = -F_{XX}^{-1}F_{XR}R$$

$$= -\left(\frac{E}{4}\right)\left(-\frac{3.3}{E}\right)(12) = 9.9 \text{ kips}$$

Substituting in the equilibrium equation, we obtain

$$\left\{\begin{array}{c} Q^a \\ Q^b \\ Q^c \\ Q^d \\ Q^e \\ Q^f \end{array}\right\} = \left[\begin{array}{cc} 1 & -\tfrac{4}{5} \\ \tfrac{3}{4} & -\tfrac{3}{5} \\ 1 & -\tfrac{4}{5} \\ 0 & -\tfrac{3}{5} \\ 0 & 1 \\ -\tfrac{5}{4} & 1 \end{array}\right]\left\{\begin{array}{c} 12 \\ 9.9 \end{array}\right\} = \left\{\begin{array}{c} 4.08 \\ 3.06 \\ 4.08 \\ -5.94 \\ 9.90 \\ 5.10 \end{array}\right\} \text{ kips}$$

Alternatively, we find

$$b' = b_R - b_X F_{XX}^{-1} F_{XR}$$

$$= \left\{\begin{array}{c} 1 \\ \tfrac{3}{4} \\ 1 \\ 0 \\ 0 \\ -\tfrac{5}{4} \end{array}\right\} - \left\{\begin{array}{c} -\tfrac{4}{5} \\ -\tfrac{3}{5} \\ -\tfrac{4}{5} \\ -\tfrac{3}{5} \\ 1 \\ 1 \end{array}\right\}\left(\frac{E}{4}\right)\left(-\frac{3.3}{E}\right) = \left\{\begin{array}{c} 0.340 \\ 0.255 \\ 0.340 \\ -0.495 \\ 0.825 \\ -0.425 \end{array}\right\}$$

and obtain Q by $Q = b'R$; i.e.,

$$
\begin{Bmatrix} Q^a \\ Q^b \\ Q^c \\ Q^d \\ Q^e \\ Q^f \end{Bmatrix} = \begin{Bmatrix} 0.340 \\ 0.255 \\ 0.340 \\ -0.495 \\ 0.825 \\ -0.425 \end{Bmatrix} (12) = \begin{Bmatrix} 4.08 \\ 3.06 \\ 4.08 \\ -5.94 \\ 9.90 \\ 5.10 \end{Bmatrix} \text{kips}
$$

To find r_1, we first calculate the flexibility matrix of structure F';

$$F' = b_R^T f b'$$

$$
= [1 \quad \tfrac{3}{4} \quad 1 \quad 0 \quad 0 \quad -\tfrac{5}{4}]\left(\frac{1}{E}\right) \begin{Bmatrix} 0.340 \\ 0.255 \\ 0.340 \\ -0.495 \\ 0.825 \\ -0.425 \end{Bmatrix} = \frac{0.340}{E}
$$

The displacement r_1 is then solved as

$$r_1 = F' R_1$$

$$
= \left(\frac{0.340}{E}\right)(12) = \frac{(0.340)(12)}{30,000} = 0.000136 \text{ ft}
$$

in the direction of the applied load.

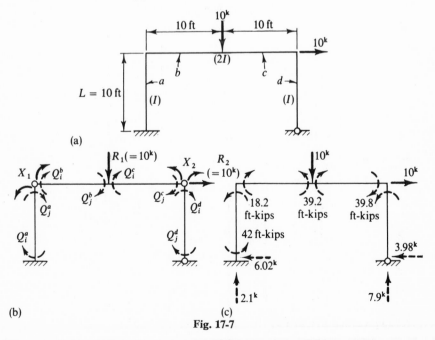

(a)

(b)   (c)

Fig. 17-7

Example 17-4. Find the member forces (end moments) of the rigid frame in Fig. 17-7(a) by the force method. E is constant.

The frame shown in Fig. 17-7(a) is statically indeterminate to the second degree. It may be made determinate by inserting two pins as in Fig. 17-7(b). Then the structure is subjected to the original applied loads denoted by R_1 and R_2 together with the redundant couples X_1 and X_2. The member forces (end moments) in Fig. 17-7(b), Q_i^a, $Q_j^a \cdots$ etc, are shown by dotted lines.

The force transformation matrix is obtained by considering the influences of $R_1 = 1$, $R_2 = 1$, $X_1 = 1$, and $X_2 = 1$ successively and separately, as shown in Fig. 17-8.

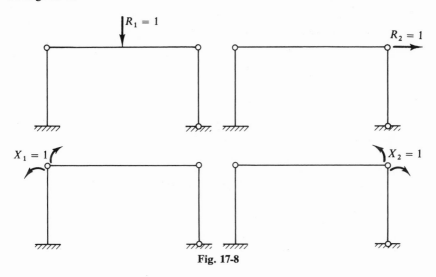

Fig. 17-8

$$R_1 = 1 \quad R_2 = 1 \quad X_1 = 1 \quad X_2 = 1$$

$$\begin{Bmatrix} Q_i^a \\ Q_j^a \\ Q_i^b \\ Q_j^b \\ Q_i^c \\ Q_j^c \\ Q_i^d \\ Q_j^d \end{Bmatrix} = \begin{bmatrix} 0 & -L & \bigm| & 1 & 1 \\ 0 & 0 & \bigm| & -1 & 0 \\ 0 & 0 & \bigm| & 1 & 0 \\ -\dfrac{L}{2} & 0 & \bigm| & -\dfrac{1}{2} & \dfrac{1}{2} \\ \dfrac{L}{2} & 0 & \bigm| & \dfrac{1}{2} & -\dfrac{1}{2} \\ 0 & 0 & \bigm| & 0 & 1 \\ 0 & 0 & \bigm| & 0 & -1 \\ 0 & 0 & \bigm| & 0 & 0 \end{bmatrix} \begin{Bmatrix} R_1 \\ R_2 \\ \overline{} \\ X_1 \\ X_2 \end{Bmatrix}$$

$$\underbrace{}_{b_R} \quad \underbrace{}_{b_X}$$

From individual member flexibility matrices, we form

$$f = \frac{L}{6EI} \begin{bmatrix} 2 & -1 & & & & & & \\ -1 & 2 & & & & & & \\ & & 1 & -\frac{1}{2} & & & & \\ & & -\frac{1}{2} & 1 & & & & \\ & & & & 1 & -\frac{1}{2} & & \\ & & & & -\frac{1}{2} & 1 & & \\ & & & & & & 2 & -1 \\ & & & & & & -1 & 2 \end{bmatrix}$$

Using b_R, b_X, and f found previously, we obtain

$$F_{XR} = b_X^T f b_R = \frac{L^2}{6EI} \begin{bmatrix} \frac{3}{4} & -3 \\ -\frac{3}{4} & -2 \end{bmatrix}$$

$$F_{XX} = b_X^T f b_X = \frac{L}{6EI} \begin{bmatrix} 8 & 2 \\ 2 & 6 \end{bmatrix}$$

and

$$F_{XX}^{-1} = \frac{6EI}{L} \frac{\begin{bmatrix} 6 & -2 \\ -2 & 8 \end{bmatrix}^T}{\begin{bmatrix} 8 & 2 \\ 2 & 6 \end{bmatrix}} = \left(\frac{6EI}{L}\right)\left(\frac{1}{44}\right) \begin{bmatrix} 6 & -2 \\ -2 & 8 \end{bmatrix}$$

The force transformation matrix of the indeterminate structure is

$$b' = b_R - b_X F_{XX}^{-1} F_{XR}$$

$$= \begin{bmatrix} 0 & -L \\ 0 & 0 \\ 0 & 0 \\ -\dfrac{L}{2} & 0 \\ \dfrac{L}{2} & 0 \\ 0 & 0 \\ 0 & 0 \\ 0 & 0 \end{bmatrix} - \begin{bmatrix} 1 & 1 \\ -1 & 0 \\ 1 & 0 \\ -\dfrac{1}{2} & \dfrac{1}{2} \\ \dfrac{1}{2} & -\dfrac{1}{2} \\ 0 & 1 \\ 0 & -1 \\ 0 & 0 \end{bmatrix}$$

$$\cdot \left(\frac{6EI}{L}\right)\left(\frac{1}{44}\right) \begin{bmatrix} 6 & -2 \\ -2 & 8 \end{bmatrix} \left(\frac{L^2}{6EI}\right) \begin{bmatrix} \frac{3}{4} & -3 \\ -\frac{3}{4} & -2 \end{bmatrix}$$

$$= \frac{L}{88} \begin{bmatrix} 0 & -88 \\ 0 & 0 \\ 0 & 0 \\ -44 & 0 \\ 44 & 0 \\ 0 & 0 \\ 0 & 0 \\ 0 & 0 \end{bmatrix} - \frac{L}{88} \begin{bmatrix} -3 & -48 \\ -12 & 28 \\ 12 & -28 \\ -13.5 & 4 \\ 13.5 & -4 \\ -15 & -20 \\ 15 & 20 \\ 0 & 0 \end{bmatrix} = \frac{L}{88} \begin{bmatrix} 3 & -40 \\ 12 & -28 \\ -12 & 28 \\ -30.5 & -4 \\ 30.5 & 4 \\ 15 & 20 \\ -15 & -20 \\ 0 & 0 \end{bmatrix}$$

The end moments are then solved by $Q = b'R$; i.e.,

$$\begin{Bmatrix} Q_i^a \\ Q_j^a \\ Q_i^b \\ Q_j^b \\ Q_i^c \\ Q_j^c \\ Q_i^d \\ Q_j^d \end{Bmatrix} = \frac{L}{88} \begin{bmatrix} 3 & -40 \\ 12 & -28 \\ -12 & 28 \\ -30.5 & -4 \\ 30.5 & 4 \\ 15 & 20 \\ -15 & -20 \\ 0 & 0 \end{bmatrix} \begin{Bmatrix} R_1 \\ R_2 \end{Bmatrix}$$

Using $L = 10$ ft and $R_1 = R_2 = 10$ kips, we obtain

$$Q_i^a = -42 \text{ ft-kips}$$
$$Q_j^a = -Q_i^b = -18.2 \text{ ft-kips}$$
$$Q_j^b = -Q_i^c = -39.2 \text{ ft-kips}$$
$$Q_j^c = -Q_i^d = 39.8 \text{ ft-kips}$$
$$Q_j^d = 0$$

To check, we find that the identity

$$b_x f b' = 0$$

is satisfied by substituting in the values of b_x, f, and b' previously found. The answer diagram for the end moments together with the reactions at the supports found by statics is shown by the dotted line in Fig. 17-7(c).

17-8. THE TREATMENT OF DISTRIBUTED LOAD

As stated in Sec. 17-2, in matrix analysis only *nodal loads* (concentrated loads or moments) are considered; i.e., external loads are only applied to nodes. Because of this restriction, distributed loads cannot be handled directly. When distributed loads are involved, one way to handle them is to divide the distributed loads into a series of closely spaced concentrated loads and to consider each concentrated load point as a node. The disadvantage of this procedure is obvious. It increases the number of elements of the structure and, therefore, the size of the matrices that must be used in the analysis. The alternative is to fix the loaded beam and to apply to its two ends (nodes) the reverse of the fixed-end moments and shears, as illustrated in Fig. 17-9. The final moments and shears in the loaded member must be obtained by adding ι̭e internal forces of the fixed-end beam to those resulting from the nodal force analysis. This latter procedure is illustrated in the next example.

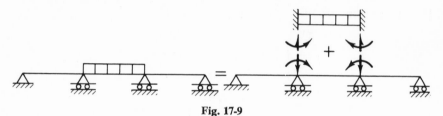

Fig. 17-9

Example 17-5. Find the end moments for the rigid frame shown in Fig. 17-10(a) by the force method. Assume constant EI.

The equivalent form of the given loaded frame is shown in Fig. 17-10(b). Because of symmetry, the vertical reaction at each support of the frame is known to be 6 kips acting upward, as indicated. If only flexural deformation is considered, then the nodal axial forces, shown in the frame in Fig. 17-10(b), only

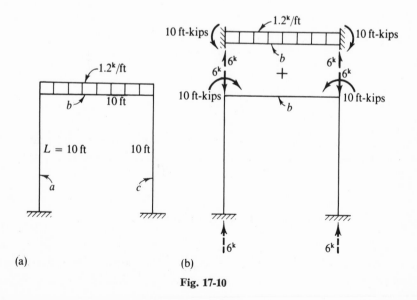

(a)   (b)

Fig. 17-10

increase the compression in the two columns but cause no effect on the end moments of the frame and can, therefore, be neglected in the nodal force analysis for obtaining end moments. The primary structure may be chosen as the one shown in Fig. 17-11, subjected to nodal moments R_1 and R_2 and

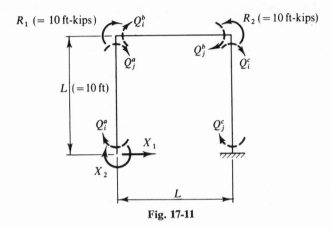

Fig. 17-11

redundant reaction components of the left support, denoted by X_1 and X_2. Those shown by dotted lines are member end moments Q_i^a, Q_j^a, $\cdots$, etc. They can be expressed in terms of R and X as

$$R_1 = 1 \quad R_2 = 1 \quad X_1 = 1 \quad X_2 = 1$$

$$\begin{Bmatrix} Q_i^a \\ Q_j^a \\ Q_i^b \\ Q_j^b \\ Q_i^c \\ Q_j^c \end{Bmatrix} = \left[\begin{array}{cc|cc} 0 & 0 & 0 & 1 \\ 0 & 0 & L & -1 \\ 1 & 0 & -L & 1 \\ -1 & 0 & L & -1 \\ 1 & -1 & -L & 1 \\ -1 & 1 & 0 & -1 \end{array} \right] \begin{Bmatrix} R_1 \\ R_2 \\ \hline X_1 \\ X_2 \end{Bmatrix}$$

$$\qquad\qquad\quad b_R \qquad\qquad b_X$$

From the member flexibility matrices, we form

$$f = \frac{L}{6EI} \begin{bmatrix} 2 & -1 & & & & \\ -1 & 2 & & & & \\ & & 2 & -1 & & \\ & & -1 & 2 & & \\ & & & & 2 & -1 \\ & & & & -1 & 2 \end{bmatrix}$$

Thus,

$$F_{XR} = b_X^T f b_R$$

$$= \begin{bmatrix} 0 & L & -L & L & -L & 0 \\ 1 & -1 & 1 & -1 & 1 & -1 \end{bmatrix} \left(\frac{L}{6EI}\right)$$

$$\cdot \begin{bmatrix} 2 & -1 & & & & \\ -1 & 2 & & & & \\ & & 2 & -1 & & \\ & & -1 & 2 & & \\ & & & & 2 & -1 \\ & & & & -1 & 2 \end{bmatrix} \begin{bmatrix} 0 & 0 \\ 0 & 0 \\ 1 & 0 \\ -1 & 0 \\ 1 & -1 \\ -1 & 1 \end{bmatrix}$$

$$= \frac{L}{6EI} \begin{bmatrix} -9L & 3L \\ 12 & -6 \end{bmatrix}$$

Similarly, $$F_{XX} = b_X^T f b_X$$

$$= \frac{L}{6EI} \begin{bmatrix} 10L^2 & -12L \\ -12L & 18 \end{bmatrix}$$

from which $$F_{XX}^{-1} = \frac{6EI}{L} \frac{\begin{bmatrix} 18 & 12L \\ 12L & 10L^2 \end{bmatrix}^T}{(180L^2 - 144L^2)} = \frac{6EI}{L} \frac{\begin{bmatrix} 18 & 12L \\ 12L & 10L^2 \end{bmatrix}}{36L^2}$$

The force transformation matrix of the indeterminate structure is

$$b' = b_R - b_X F_{XX}^{-1} F_{XR}$$

$$= \begin{bmatrix} 0 & 0 \\ 0 & 0 \\ 1 & 0 \\ -1 & 0 \\ 1 & -1 \\ -1 & 1 \end{bmatrix} - \begin{bmatrix} 0 & 1 \\ L & -1 \\ -L & 1 \\ L & -1 \\ -L & 1 \\ 0 & -1 \end{bmatrix} \left(\frac{1}{36L^2}\right) \begin{bmatrix} 18 & 12L \\ 12L & 10L^2 \end{bmatrix} \begin{bmatrix} -9L & 3L \\ 12 & -6 \end{bmatrix}$$

$$= \begin{bmatrix} 0 & 0 \\ 0 & 0 \\ 1 & 0 \\ -1 & 0 \\ 1 & -1 \\ -1 & 1 \end{bmatrix} - \begin{bmatrix} \frac{1}{3} & -\frac{2}{3} \\ -\frac{5}{6} & \frac{1}{6} \\ \frac{5}{6} & -\frac{1}{6} \\ -\frac{5}{6} & \frac{1}{6} \\ \frac{5}{6} & -\frac{1}{6} \\ -\frac{1}{3} & \frac{2}{3} \end{bmatrix} = \begin{bmatrix} -\frac{1}{3} & \frac{2}{3} \\ \frac{5}{6} & -\frac{1}{6} \\ \frac{1}{6} & \frac{1}{6} \\ -\frac{1}{6} & -\frac{1}{6} \\ \frac{1}{6} & -\frac{5}{6} \\ -\frac{2}{3} & \frac{1}{3} \end{bmatrix}$$

The end moments based on the nodal-force analysis are then solved by $Q = b'R$; i.e.,

$$\begin{Bmatrix} Q_i^a \\ Q_j^a \\ Q_i^b \\ Q_j^b \\ Q_i^c \\ Q_j^c \end{Bmatrix} = \begin{bmatrix} -\frac{1}{3} & \frac{2}{3} \\ \frac{5}{6} & -\frac{1}{6} \\ \frac{1}{6} & \frac{1}{6} \\ -\frac{1}{6} & -\frac{1}{6} \\ \frac{1}{6} & -\frac{5}{6} \\ -\frac{2}{3} & \frac{1}{3} \end{bmatrix} \begin{Bmatrix} 10 \\ 10 \end{Bmatrix} = \begin{Bmatrix} \frac{10}{3} \\ \frac{20}{3} \\ \frac{10}{3} \\ -\frac{10}{3} \\ -\frac{20}{3} \\ -\frac{10}{3} \end{Bmatrix} \text{ft-kips}$$

The final result is obtained by adding the fixed-end moments (see upper part of Fig. 17-10(b)) to the end moments of member b. Thus,

$$\begin{Bmatrix} Q_i^a \\ Q_j^a \\ Q_i^b \\ Q_j^b \\ Q_i^c \\ Q_j^c \end{Bmatrix} = \begin{Bmatrix} \frac{10}{3} \\ \frac{20}{3} \\ \frac{10}{3} - 10 \\ -\frac{10}{3} + 10 \\ -\frac{20}{3} \\ -\frac{10}{3} \end{Bmatrix} = \begin{Bmatrix} \frac{10}{3} \\ \frac{20}{3} \\ -\frac{20}{3} \\ \frac{20}{3} \\ -\frac{20}{3} \\ -\frac{10}{3} \end{Bmatrix} \text{ft-kips}$$

The above procedure for fixing a loaded beam is not limited to the case of distributed loads. The procedure can also be applied to members subjected to a set of concentrated loads, if reducing the number of nodes is desirable.

17-9. ON THE NOTION OF PRIMARY STRUCTURE

The procedures for the analysis of statically indeterminate structures by the force method already discussed are based on the concept of *primary structure* previously developed in the method of consistent deformations. The notion of primary structure serves a convenient means of setting up equilibrium equation. However, if we, without considering the notion of primary structure, examine the basic equation

$$Q = b_R R + b_X X$$

we observe that it merely states: Q is linearly related to a set of applied forces R and a set of unknown forces X. The equation itself does not necessarily suggest a primary structure. As a result, we may separate these two sets of influences from the two independent force systems imposed on the original structure. So doing does not violate the truth of the above equation but certainly broadens our view of handling the problem.

Now b_R represents an array of member forces in equilibrium with unit applied loads based on the original structure. More specifically, each column of b_R represents member forces in equilibrium with a certain unit load applied to the original structure. Since the original structure is statically indeterminate, many equilibrating systems may be chosen from to establish each column of b_R.

Likewise, each column of b_X can be thought of as independent self-equilibrating internal force system for the original structure. For a structure indeterminate to the nth degree, b_X will represent any group of n independent self-equilibrating member force systems, one for each redundant.

If it is convenient, these member forces may be determined by introducing a primary structure. However, in a larger sense, the traditional notion of primary structure is not essential to the analysis of a statically indeterminate structure; rather, it introduces unnecessary restrictions to the analysis.

Example 17-6. Solve the bar forces of the truss in Fig. 17-6(a) (Example 17-3) by the preceding generalized procedures.

Solution (1). Disregarding the notion of a primary structure, we may choose a set of member forces in equilibrium with external load $R_1 = 1$, as shown in Fig. 17-12(a), and a set of self-equilibrating internal forces, as shown in Fig. 17-12(b). Thus

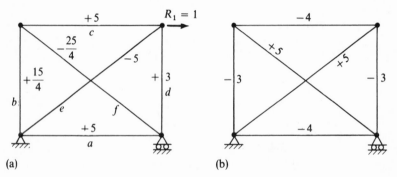

Fig. 17-12

$$b_R = \begin{Bmatrix} 5 \\ \frac{15}{4} \\ 5 \\ 3 \\ -5 \\ -\frac{25}{4} \end{Bmatrix} \qquad b_X = \begin{Bmatrix} -4 \\ -3 \\ -4 \\ -3 \\ 5 \\ 5 \end{Bmatrix}$$

There are, of course, many other choices which might be made.

Using b_R and b_X and f, obtained in Example 17-3, we have

$$F_{XR} = b_X^T f b_R$$

$$= [-4 \quad -3 \quad -4 \quad -3 \quad 5 \quad 5] \left(\frac{1}{E}\right) \left\{ \begin{array}{c} 5 \\ \frac{15}{4} \\ 5 \\ 3 \\ -5 \\ -\frac{25}{4} \end{array} \right\} = -\frac{116.5}{E}$$

$$F_{XX} = b_X^T f b_X$$

$$= [-4 \quad -3 \quad -4 \quad -3 \quad 5 \quad 5] \left(\frac{1}{E}\right) \left\{ \begin{array}{c} -4 \\ -3 \\ -4 \\ -3 \\ 5 \\ 5 \end{array} \right\} = \frac{100}{E}$$

$$F_{XX}^{-1} = \frac{E}{100}$$

The force transformation matrix is then determined;

$$b' = b_R - b_X F_{XX}^{-1} F_{XR}$$

$$= \left\{ \begin{array}{c} 5 \\ \frac{15}{4} \\ 5 \\ 3 \\ -5 \\ -\frac{25}{4} \end{array} \right\} - \left\{ \begin{array}{c} -4 \\ -3 \\ -4 \\ -3 \\ 5 \\ 5 \end{array} \right\} \left(\frac{E}{100}\right)\left(\frac{-116.5}{E}\right) = \left\{ \begin{array}{c} 5 \\ 3.75 \\ 5 \\ 3 \\ -5 \\ -6.25 \end{array} \right\} - \left\{ \begin{array}{c} 4.660 \\ 3.495 \\ 4.660 \\ 3.495 \\ -5.825 \\ -5.825 \end{array} \right\} = \left\{ \begin{array}{c} 0.340 \\ 0.255 \\ 0.340 \\ -0.495 \\ 0.825 \\ -0.425 \end{array} \right\}$$

This is the same b' obtained in Example 17-3 and will lead to the same final results for the bar forces.

Solution (2). It may be interesting to point out that when primary structure is used in analyzing an indeterminate structure, the same final results will be obtained if different primary structures are chosen in developing b_R and b_X.

To illustrate, let us first take member e as the redundant. The bar forces associated with the given primary structure due to external load $R_1 = 1$ are elements of b_R, as indicated in Fig. 17-13(a). Next, let member a be chosen as the redundant. Setting the redundant force equal to unity, we obtain a set of internal forces in equilibrium (Fig. 17-13(b)) which forms b_X.

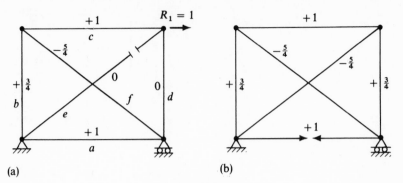

Fig. 17-13

$$b_R = \left\{ \begin{array}{c} 1 \\ \frac{3}{4} \\ 1 \\ 0 \\ 0 \\ -\frac{5}{4} \end{array} \right\} \qquad b_X = \left\{ \begin{array}{c} 1 \\ \frac{3}{4} \\ 1 \\ \frac{3}{4} \\ -\frac{5}{4} \\ -\frac{5}{4} \end{array} \right\}$$

$F_{XR} = b_X^T f b_R$

$$= \begin{bmatrix} 1 & \frac{3}{4} & 1 & \frac{3}{4} & -\frac{5}{4} & -\frac{5}{4} \end{bmatrix} \left(\frac{1}{E} \right) \left\{ \begin{array}{c} 1 \\ \frac{3}{4} \\ 1 \\ 0 \\ 0 \\ -\frac{5}{4} \end{array} \right\} = \frac{66}{16E}$$

$F_{XX} = b_X^T f b_X$

$$= \begin{bmatrix} 1 & \frac{3}{4} & 1 & \frac{3}{4} & -\frac{5}{4} & -\frac{5}{4} \end{bmatrix} \left(\frac{1}{E} \right) \left\{ \begin{array}{c} 1 \\ \frac{3}{4} \\ 1 \\ \frac{3}{4} \\ -\frac{5}{4} \\ -\frac{5}{4} \end{array} \right\} = \frac{100}{16E}$$

$$F_{XX}^{-1} = \frac{16E}{100}$$

The force transformation matrix b' is found to be

$$b' = b_R - b_X F_{XX}^{-1} F_{XR} = \begin{Bmatrix} 1 \\ \frac{3}{4} \\ 1 \\ 0 \\ 0 \\ -\frac{5}{4} \end{Bmatrix} - \begin{Bmatrix} 1 \\ \frac{3}{4} \\ 1 \\ \frac{3}{4} \\ -\frac{5}{4} \\ -\frac{5}{4} \end{Bmatrix} \left(\frac{16E}{100}\right)\left(\frac{66}{16E}\right) = \begin{Bmatrix} 0.340 \\ 0.255 \\ 0.340 \\ -0.495 \\ 0.825 \\ -0.425 \end{Bmatrix}$$

the same as previously obtained.

PROBLEMS

17-1. Use the force method to find the vertical deflection at each of the loaded points of the beam shown in Fig. 17-14. Assume constant *EI*.

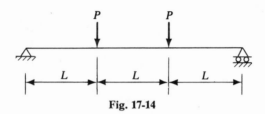

Fig. 17-14

17-2. Find, by the force method, all the bar forces and the vertical deflection at each of the loaded joints of the truss shown in Fig. 17-15. Assume that $A = 10$ in²., $E = 30,000$ kips per in². for all members.

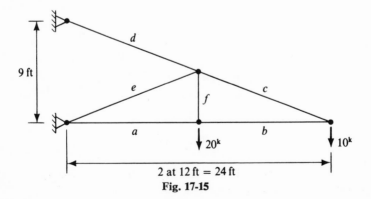

Fig. 17-15

17-3. Find, by the force method, all the member forces (end moments) and the nodal displacements corresponding to the applied loads for the frame in Fig. 17-16. Assume constant *EI*.

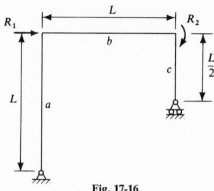

Fig. 17-16

17-4. Find, by the force method, the slope and deflection at the loaded end of the beam shown in Fig. 17-17. Assume constant EI.

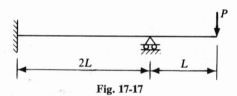

Fig. 17-17

17-5. Find, by the force method, the bar forces and the deflection components at the loaded point of the truss in Fig. 17-18. Assume that $A = 10\ \text{in}^2$. and $E = 30,000$ kips per in^2. for all members.

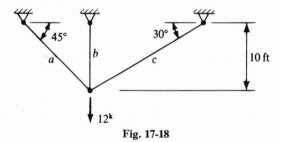

Fig. 17-18

17-6. Use the force method to obtain the member end moments for the frame shown in Fig. 17-19. Assume constant EI.

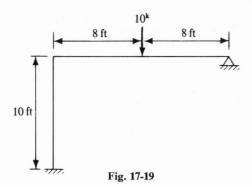

Fig. 17-19

17-7. Use the force method to find the member end moments for the frame shown in Fig. 17-20. Assume constant *EI*.

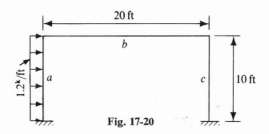

Fig. 17-20

17-8. Use the force method to find the member end moments for the gable bent in Fig. 17-21. Assume constant *EI*.

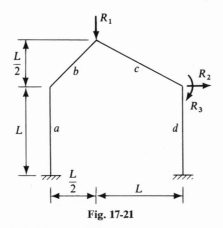

Fig. 17-21

17-9. Solve Prob. 17-5 disregarding the notion of primary structure.

17-10. Solve Prob. 17-6 by using different primary structures for developing the force transformation matrices b_R and b_X.

18

MATRIX ANALYSIS OF STRUCTURES
BY THE FINITE ELEMENT METHOD
PART II: THE DISPLACEMENT METHOD

18-1. GENERAL

As previously pointed out, the force method and the displacement method represent two different approaches to analyzing structures. The basic concepts of the structure remain the same (see Sec. 17-2). The fundamental difference between these two methods is that the force method chooses the member forces as the basic unknowns whereas the displacement method chooses the nodal displacements as the basic unknowns. Like the force method, the basic equations of the displacement method are derived from

1. The equilibrium of forces;
2. The compatibility of displacements;
3. The force-displacement relationship.

The compatibility condition is first satisfied by correlating the external nodal displacements to the end deformations of the members. The force-displacement relationship is then established between the member end forces and deformations and between the possible nodal forces and nodal displacements. Finally, using nodal equilibrium equations, we solve for the unknown nodal displacements and, therefore, for the member forces and deformations of the structure.

18-2. COMPATIBILITY, DISPLACEMENT
TRANSFORMATION MATRIX

The compatibility used in the displacement method is that the geometry of deformation must be such that the elements of structure fit together at the nodal points; i.e., the member deformations q should be consistently related to the nodal displacements r. Let a_{ij} represent the value of member deformation

q_i caused by a unit nodal displacement r_j. The total value of each member deformation caused by all the nodal displacements may be written as

$$q_1 = a_{11}r_1 + a_{12}r_2 + \cdots + a_{1n}r_n$$
$$q_2 = a_{21}r_1 + a_{22}r_2 + \cdots + a_{2n}r_n \tag{18-1}$$
$$\vdots$$
$$q_m = a_{m1}r_1 + a_{m2}r_2 + \cdots + a_{mn}r_n$$

in which $q_1 = q_i^a$, $q_2 = q_j^a$, $\ldots$, etc. represent the total set of member deformations and $r_1, r_2, \cdots, r_n$ the total set of nodal displacements. Note that no connection between the subscripts on q and r is implied. In matrix form

$$\begin{Bmatrix} q_1 \\ q_2 \\ \vdots \\ q_m \end{Bmatrix} = \begin{bmatrix} a_{11} & a_{12} & \cdots & a_{1n} \\ a_{21} & a_{22} & \cdots & a_{2n} \\ \vdots & & & \\ a_{m1} & a_{m2} & \cdots & a_{mn} \end{bmatrix} \begin{Bmatrix} r_1 \\ r_2 \\ \vdots \\ r_n \end{Bmatrix}$$

That is,

$$q = ar \tag{18-2}$$

where

$$a = \begin{bmatrix} a_{11} & a_{12} & \cdots & a_{1n} \\ a_{21} & a_{22} & \cdots & a_{2n} \\ \vdots & & & \\ a_{m1} & a_{m2} & \cdots & a_{mn} \end{bmatrix} \tag{18-2a}$$

called the *displacement transformation matrix* which relates the internal member deformations to the external nodal displacements. Matrix a is usually a rectangular matrix. It is simply a geometric transformation of coordinates representing the compatibility of the displacements of a system.

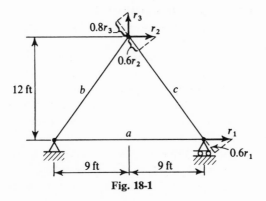

Fig. 18-1

For example, let us consider the truss in Fig. 18-1. The possible nodal displacements are one linear displacement at the roller support, denoted by r_1, and two linear displacement components at the top joint, denoted by r_2 and r_3. Also indicated in Fig. 18-1 are the member deformations of bar c as the result of the separate influences of r_1, r_2, and r_3, which gives the total q^c as

$$q^c = 0.6r_1 - 0.6r_2 + 0.8r_3$$

q^a and q^b can be similarly obtained. The result, put in matrix form, is

$$\begin{Bmatrix} q^a \\ q^b \\ q^c \end{Bmatrix} = \begin{bmatrix} 1 & 0 & 0 \\ 0 & 0.6 & 0.8 \\ 0.6 & -0.6 & 0.8 \end{bmatrix} \begin{Bmatrix} r_1 \\ r_2 \\ r_3 \end{Bmatrix}$$

$$a$$

Note that the displacement transformation matrix in this case happens to be a square matrix.

The relationship between the displacement transformation matrix and the force transformation matrix can be obtained by virtual work as follows:

$$R^T r = Q^T q$$

Because of 17-8, $$Q = bR$$

or $$Q^T = R^T b^T$$

and Eq. 18-2, $$q = ar$$

we have $$R^T r = R^T b^T ar$$

from which

$$b^T a = I \tag{18-3}$$

Similarly, we can prove

$$a^T b = I \tag{18-4}$$

18-3. FORCE-DISPLACEMENT RELATIONSHIP, STIFFNESS COEFFICIENT, STIFFNESS MATRIX

A *stiffness coefficient* k_{ij} is defined as the force developed at point i due to a unit displacement at point j, all other points (nodes) being fixed. Like the flexibility coefficient, the stiffness coefficient constitutes a relationship between force and displacement. Applying the principle of superposition, we may express the force component at any point of a system in terms of a set of prescribed displacements.

The first step in this analysis is to express the end forces in terms of the end deformations of an individual member. Using the stiffness coefficients and the notation defined in Fig. 17-2, we have

$$Q_i^a = k_{ii}^a q_i^a + k_{ij}^a q_j^a + k_{ik}^a q_k^a$$
$$Q_j^a = k_{ji}^a q_i^a + k_{jj}^a q_j^a + k_{jk}^a q_k^a \qquad (18\text{-}5)$$
$$Q_k^a = k_{ki}^a q_i^a + k_{kj}^a q_j^a + k_{kk}^a q_k^a$$

in which q_i^a, q_j^a, and q_k^a are the deformations of a particular member a and Q_i^a, Q_j^a, and Q_k^a the corresponding member forces. It is clear that the stiffness coefficient, say k_{ii}^a is defined as

$$k_{ii}^a = Q_i^a \quad \text{as} \quad q_i^a = 1 \qquad q_j^a = q_k^a = 0$$

The rest can similarly be defined.

Equation 18-5 in matrix form is

$$Q^a = k^a q^a \qquad (18\text{-}6)$$

in which

$$k^a = \begin{bmatrix} k_{ii}^a & k_{ij}^a & k_{ik}^a \\ k_{ji}^a & k_{jj}^a & k_{jk}^a \\ k_{ki}^a & k_{kj}^a & k_{kk}^a \end{bmatrix} \qquad (18\text{-}6a)$$

defined as the *element stiffness matrix*.

If Eq. 18-6 is premultiplied by $(k^a)^{-1}$,

$$(k^a)^{-1} Q^a = (k^a)^{-1} k^a q^a$$

or

$$q^a = (k^a)^{-1} Q^a \qquad (18\text{-}7)$$

Comparing Eq. 18-7 with Eq. 17-13,

$$q^a = f^a Q^a$$

we see that

$$f^a = (k^a)^{-1} \qquad (18\text{-}8)$$

Thus *the element flexibility matrix is the inverse of the element stiffness matrix and vice versa.*

The above descriptions refer to an individual element. For the entire assemblage composed of a, b, ... elements, we have

$$Q^a = k^a q^a$$
$$Q^b = k^b q^b$$
$$\vdots$$

Since

$$Q = \begin{Bmatrix} Q^a \\ Q^b \\ \vdots \end{Bmatrix} \qquad q = \begin{Bmatrix} q^a \\ q^b \\ \vdots \end{Bmatrix}$$

the above equations can be assembled as

$$Q = kq \tag{18-9}$$

where

$$k = \begin{bmatrix} k^a & & & \\ & k^b & & \\ & & \cdot & \\ & & & \cdot \end{bmatrix} \tag{18-9a}$$

which is a diagonal matrix with individual element stiffness matrices as its constituents.

Refer to Eqs. 18-5 and 18-6a, and consider a prismatic member with length L, cross-sectional area A, moment of inertia I, and modulus of elasticity E. The elements in the first column of k^a are, by definition, the member forces resulting from $q_i^a = 1$, i.e., a unit rotation at the left end of the member (see Fig. 17-2). Thus,

$$k_{ii}^a = \text{moment at the left end} = \frac{4EI}{L}$$

$$k_{ji}^a = \text{moment at the right end} = \frac{2EI}{L}$$

$$k_{ki}^a = \text{axial force of the member} = 0$$

Note that k_{ii}^a and k_{ji}^a can easily be obtained by the slope-deflection method; $k_{ki}^a = 0$ is apparent. All the other elements of k^a are similarly determined. Thus,

$$k^a = \begin{bmatrix} \dfrac{4EI}{L} & \dfrac{2EI}{L} & 0 \\[2mm] \dfrac{2EI}{L} & \dfrac{4EI}{L} & 0 \\[2mm] 0 & 0 & \dfrac{AE}{L} \end{bmatrix} \tag{18-10}$$

which is a symmetric matrix. As a check, we note that

$$f^a k^a = \begin{bmatrix} \dfrac{L}{3EI} & -\dfrac{L}{6EI} & 0 \\[2mm] -\dfrac{L}{6EI} & \dfrac{L}{3EI} & 0 \\[2mm] 0 & 0 & \dfrac{L}{AE} \end{bmatrix} \begin{bmatrix} \dfrac{4EI}{L} & \dfrac{2EI}{L} & 0 \\[2mm] \dfrac{2EI}{L} & \dfrac{4EI}{L} & 0 \\[2mm] 0 & 0 & \dfrac{AE}{L} \end{bmatrix} = \begin{bmatrix} 1 & 0 & 0 \\ 0 & 1 & 0 \\ 0 & 0 & 1 \end{bmatrix}$$

When the effect of axial forces is disregarded, as is usually done in rigid frame analysis,

$$k^a = \frac{EI}{L} \begin{bmatrix} 4 & 2 \\ 2 & 4 \end{bmatrix} \tag{18-11}$$

For a pin-connected truss

$$k^a = \left[\frac{AE}{L}\right] \tag{18-12}$$

The second step in the analysis is to express the nodal forces in terms of the corresponding nodal displacements. This can be accomplished easily by the method of vitual work. Let us start by Eq. 17-5

$$\bar{r}^T R = \bar{q}^T Q \tag{18-13}$$

From Eqs. 18-2 and 18-9, $\quad q = ar \quad$ and $\quad Q = kq$

we have $\qquad\qquad\qquad \bar{q} = a\bar{r}$

or

$$\bar{q}^T = \bar{r}^T a^T \tag{18-14}$$

and

$$Q = kar \tag{18-15}$$

Substituting Eqs. 18-14 and 18-15 in Eq. 18-13 gives

$$\bar{r}^T R = \bar{r}^T a^T kar$$

from which

$$R = a^T kar \tag{18-16}$$

or

$$R = Kr \tag{18-17}$$

if we make

$$K = a^T ka \tag{18-18}$$

K being called the *total stiffness matrix*, or *the stiffness matrix of structure*, which directly relates the nodal forces to the nodal displacements of a structure.

If we premultiply Eq. 18-17 with K^{-1} on both sides, we have

$$K^{-1}R = K^{-1}Kr = r$$

or

$$r = K^{-1}R \tag{18-19}$$

Comparing Eq. 18-19 with Eq. 17-23, i.e.,

$$r = FR$$

we have

$$F = K^{-1} \tag{18-20}$$

It is thus seen that *the total flexibility matrix is the inverse of total stiffness matrix and vice versa.*

18-4. EQUILIBRIUM

Refer to Eq. 18-17

$$R = Kr$$

If r denotes the elements of all possible unknown nodal displacements (not including the known support or boundary conditions), then R must denote all the corresponding nodal forces. The equilibrium of each node requires that the possible nodal forces, expressed in terms of unknown nodal displacements, must be equal to the applied loads. Thus if these nodal loads are given, we can solve for the unknown nodal displacements by Eq. 18-19

$$r = K^{-1}R$$

and for the member forces by Eqs. 18-2 and 18-9, i.e.,

$$Q = kq = kar$$

18-5. ANALYSIS OF STRUCTURES BY THE DISPLACEMENT METHOD

It is interesting that in the discussion of the displacement method the question of redundancy did not arise. The displacement method can apply with equal ease to statically determinate structures and statically indeterminate structures. The procedures of analysis by the displacement method are contained in the following steps:

1. Define all the possible unknown nodal displacements r.

Generally, a pin-connected node has two linear displacement components, with the rotation of the pin considered free of the connected members. A rigidly connected node has three displacement components, two linear and one rotational, but the linear displacements may be excepted if the axial deformations of the connected members are neglected and sidesway prevented. No displacements are assigned to the nodes that cannot move. Thus in a pin-connected truss the hinged support is considered completely restrained; the roller support has one linear movement. In a rigid frame the built-in support undergoes no displacement; a hinged support can have only an angular displacement whereas a roller support has one angular displacement and one linear displacement.

2. Determine the displacement transformation matrix a from geometric configuration.

3. From the individual stiffness matrices k^a, k^b, . . . obtain

$$k = \begin{bmatrix} k^a & & & \\ & k^b & & \\ & & \cdot & \\ & & & \cdot \\ & & & & \cdot \end{bmatrix}$$

4. Compute the stiffness matrix of structure K;

$$K = a^T k a$$

5. Obtain the inverse of K.

6. Compute nodal displacements r by

$$r = K^{-1} R$$

Note that R is in one-to-one correspondence with r. Some of the R are the actual loads; others are zero if no load is applied there. All of the R are known.

7. Compute the member forces Q by

$$Q = kar$$

8. Distributed loads are handled indirectly by the procedure outlined in Sec. 17-8.

Example 18-1. Compute the nodal displacements and bar forces for the truss shown in Fig. 18-2(a). Assume that $E = 30,000$ kips per in.^{2} and $L(\text{ft})/A(\text{in.}^2) = 1$ for all members.

The roller support has a possible displacement r_1, and the top joint has possible displacement components r_2 and r_3, as indicated in Fig. 18-2(b). The nodal forces R_1, R_2, and R_3 correspond to the nodal displacements. Note that $R_1 = R_2 = 0$ and $R_3 = -8$ kips in this problem.

Using the results of the example in Sec. 18-2, the displacement transformation matrix is

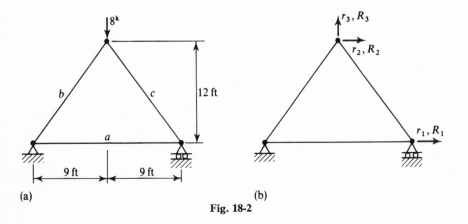

(a)   (b)

Fig. 18-2

$$a = \begin{bmatrix} 1 & 0 & 0 \\ 0 & 0.6 & 0.8 \\ 0.6 & -0.6 & 0.8 \end{bmatrix}$$

The individual member stiffness matrix is determined by Eq. 18-12,

$$\left[\frac{AE}{L}\right]$$

Since $A/L = 1$ for all members,

$$k^a = k^b = k^c = E$$

from which we form the diagonal matrix

$$k = E \begin{bmatrix} 1 & & \\ & 1 & \\ & & 1 \end{bmatrix}$$

Thus the total stiffness matrix is

$$K = a^T k a = \begin{bmatrix} 1 & 0 & 0.6 \\ 0 & 0.6 & -0.6 \\ 0 & 0.8 & 0.8 \end{bmatrix} (E) \begin{bmatrix} 1 & 0 & 0 \\ 0 & 0.6 & 0.8 \\ 0.6 & -0.6 & 0.8 \end{bmatrix}$$

$$= E \begin{bmatrix} 1.36 & -0.36 & 0.48 \\ -0.36 & 0.72 & 0 \\ 0.48 & 0 & 1.28 \end{bmatrix}$$

Using the procedures given in Sec. 16-7, we find the inverse of the total stiffness matrix to be

$$K^{-1} = \frac{1}{E} \begin{bmatrix} 1 & 0.5 & -0.375 \\ 0.5 & 1.639 & -0.188 \\ -0.375 & -0.188 & 0.922 \end{bmatrix}$$

The nodal displacements are then determined by $r = K^{-1} R$; i.e.,

$$\begin{Bmatrix} r_1 \\ r_2 \\ r_3 \end{Bmatrix} = \frac{1}{E} \begin{bmatrix} 1 & 0.5 & -0.375 \\ 0.5 & 1.639 & -0.188 \\ -0.375 & -0.188 & 0.922 \end{bmatrix} \begin{Bmatrix} 0 \\ 0 \\ -8 \end{Bmatrix} = \frac{1}{E} \begin{Bmatrix} 3 \\ 1.5 \\ -7.38 \end{Bmatrix}$$

Using $E = 30,000$ kips per in.^{2}, we obtain

$$\begin{Bmatrix} r_1 \\ r_2 \\ r_3 \end{Bmatrix} = \begin{Bmatrix} 0.0001 \\ 0.00005 \\ -0.00027 \end{Bmatrix} \text{ft}$$

Finally the bar forces are solved by $Q = kar$; i.e.,

$$
\begin{Bmatrix} Q^a \\ Q^b \\ Q^c \end{Bmatrix} = (E) \begin{bmatrix} 1 & 0 & 0 \\ 0 & 0.6 & 0.8 \\ 0.6 & -0.6 & 0.8 \end{bmatrix} \left(\frac{1}{E}\right) \begin{Bmatrix} 3 \\ 1.5 \\ -7.38 \end{Bmatrix} = \begin{Bmatrix} 3 \\ -5 \\ -5 \end{Bmatrix} \text{kips}
$$

Example 18-2. Solve the bar forces for the truss in Fig. 18-3(a). Assume constant E and $L(\text{ft})/A(\text{in.}^2) = 1$ for all members.

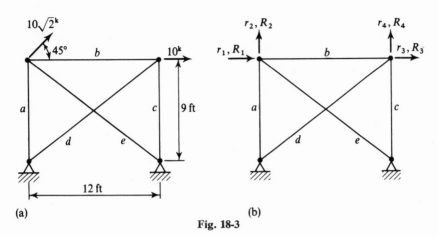

Fig. 18-3

The assigned nodal displacements and the corresponding nodal forces are shown in Fig. 18-3(b). In this problem $R_1 = R_2 = R_3 = 10$ kips, $R_4 = 0$.

From compatibility,

$$
\begin{array}{cccc} r_1 = 1 & r_2 = 1 & r_3 = 1 & r_4 = 1 \end{array}
$$

$$
\begin{Bmatrix} q^a \\ q^b \\ q^c \\ q^d \\ q^e \end{Bmatrix} = \begin{bmatrix} 0 & 1 & 0 & 0 \\ -1 & 0 & 1 & 0 \\ 0 & 0 & 0 & 1 \\ 0 & 0 & 0.8 & 0.6 \\ -0.8 & 0.6 & 0 & 0 \end{bmatrix} \begin{Bmatrix} r_1 \\ r_2 \\ r_3 \\ r_4 \end{Bmatrix}
$$

$$a$$

Since $L/A = 1$ for all members, the individual member stiffness matrices are found to be

$$
k^a = k^b = k^c = k^d = k^e = E
$$

from which
$$
k = E \begin{bmatrix} 1 & & & & \\ & 1 & & & \\ & & 1 & & \\ & & & 1 & \\ & & & & 1 \end{bmatrix}
$$

Thus we have the total stiffness matrix;

$$K = a^T ka$$

$$= \begin{bmatrix} 0 & -1 & 0 & 0 & -0.8 \\ 1 & 0 & 0 & 0 & 0.6 \\ 0 & 1 & 0 & 0.8 & 0 \\ 0 & 0 & 1 & 0.6 & 0 \end{bmatrix} (E) \begin{bmatrix} 0 & 1 & 0 & 0 \\ -1 & 0 & 1 & 0 \\ 0 & 0 & 0 & 1 \\ 0 & 0 & 0.8 & 0.6 \\ -0.8 & 0.6 & 0 & 0 \end{bmatrix}$$

$$= E \begin{bmatrix} 1.64 & -0.48 & -1 & 0 \\ -0.48 & 1.36 & 0 & 0 \\ -1 & 0 & 1.64 & 0.48 \\ 0 & 0 & 0.48 & 1.36 \end{bmatrix}$$

Using the procedures in Sec. 16-7, we find the inverse of K to be

$$K^{-1} = \frac{1}{2.150E} \begin{bmatrix} 2.721 & 0.960 & 1.850 & -0.653 \\ 0.960 & 1.920 & 0.653 & -0.230 \\ 1.850 & 0.653 & 2.721 & -0.960 \\ -0.653 & -0.230 & -0.960 & 1.296 \end{bmatrix}$$

The nodal displacements are given by $r = K^{-1}R$; i.e.,

$$\begin{Bmatrix} r_1 \\ r_2 \\ r_3 \\ r_4 \end{Bmatrix} = \frac{1}{2.150E} \begin{bmatrix} 2.721 & 0.960 & 1.850 & -0.653 \\ 0.960 & 1.920 & 0.653 & -0.230 \\ 1.850 & 0.653 & 2.721 & -0.960 \\ -0.653 & -0.230 & -0.960 & 1.296 \end{bmatrix} \begin{Bmatrix} 10 \\ 10 \\ 10 \\ 0 \end{Bmatrix} = \frac{1}{E} \begin{Bmatrix} 25.7 \\ 16.4 \\ 24.3 \\ -8.57 \end{Bmatrix}$$

The bar forces are then obtained from $Q = kar$.

$$\begin{Bmatrix} Q^a \\ Q^b \\ Q^c \\ Q^d \\ Q^e \end{Bmatrix} = (E) \begin{bmatrix} 0 & 1 & 0 & 0 \\ -1 & 0 & 1 & 0 \\ 0 & 0 & 0 & 1 \\ 0 & 0 & 0.8 & 0.6 \\ -0.8 & 0.6 & 0 & 0 \end{bmatrix} \left(\frac{1}{E}\right) \begin{Bmatrix} 25.7 \\ 16.4 \\ 24.3 \\ -8.57 \end{Bmatrix}$$

$$= \begin{Bmatrix} 16.40 \\ -1.40 \\ -8.57 \\ 14.30 \\ -10.72 \end{Bmatrix} \text{ kips}$$

Example 18-3. Find all the end moments for the frame shown in Fig. 18-4(a). Assume constant EI.

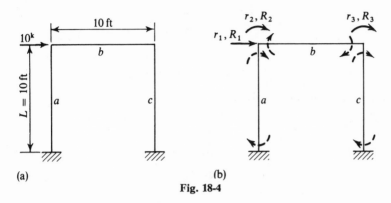

Fig. 18-4

We assume unknown nodal displacements r_1, r_2, and r_3 and their corresponding forces R_1, R_2, and R_3, as shown in Fig. 18-4(b). Note that r_1 represents the sidesway of the frame and that r_2 and r_3 represent the rotations at the joints. In the present case $R_1 = 10$ kips, $R_2 = R_3 = 0$. The dotted lines in Fig. 18-4 (b) indicate the assumed directions for the member moments, $Q_i^a, Q_j^a, \ldots$, etc., and their corresponding member rotations, $q_i^a, q_j^a, \ldots$, etc. From compatibility

$$r_1 = 1 \quad r_2 = 1 \quad r_3 = 1$$

$$\begin{Bmatrix} q_i^a \\ q_j^a \\ q_i^b \\ q_j^b \\ q_i^c \\ q_j^c \end{Bmatrix} = \begin{bmatrix} -\dfrac{1}{10} & 0 & 0 \\[2mm] -\dfrac{1}{10} & 1 & 0 \\[2mm] 0 & 1 & 0 \\[2mm] 0 & 0 & 1 \\[2mm] -\dfrac{1}{10} & 0 & 1 \\[2mm] -\dfrac{1}{10} & 0 & 0 \end{bmatrix} \begin{Bmatrix} r_1 \\ r_2 \\ r_3 \end{Bmatrix}$$

$$a$$

Since the members are identical, the individual member stiffness matrices are the same; i.e.,

$$k^a = k^b = k^c = \frac{EI}{L} \begin{bmatrix} 4 & 2 \\ 2 & 4 \end{bmatrix}$$

See Eq. 18-11.

Using Eq. 18-9a, we have

$$k = \frac{EI}{L}\begin{bmatrix} 4 & 2 & & & & \\ 2 & 4 & & & & \\ & & 4 & 2 & & \\ & & 2 & 4 & & \\ & & & & 4 & 2 \\ & & & & 2 & 4 \end{bmatrix}$$

Thus the total stiffness matrix is:

$$K = a^T k a$$

$$= \begin{bmatrix} -\dfrac{1}{10} & -\dfrac{1}{10} & 0 & 0 & -\dfrac{1}{10} & -\dfrac{1}{10} \\ 0 & 1 & 1 & 0 & 0 & 0 \\ 0 & 0 & 0 & 1 & 1 & 0 \end{bmatrix}$$

$$\cdot \left(\frac{EI}{L}\right)\begin{bmatrix} 4 & 2 & & & & \\ 2 & 4 & & & & \\ & & 4 & 2 & & \\ & & 2 & 4 & & \\ & & & & 4 & 2 \\ & & & & 2 & 4 \end{bmatrix}\begin{bmatrix} -\dfrac{1}{10} & 0 & 0 \\ -\dfrac{1}{10} & 1 & 0 \\ 0 & 1 & 0 \\ 0 & 0 & 1 \\ -\dfrac{1}{10} & 0 & 1 \\ -\dfrac{1}{10} & 0 & 0 \end{bmatrix}$$

$$= \frac{EI}{L}\begin{bmatrix} 0.24 & -0.6 & -0.6 \\ -0.6 & 8 & 2 \\ -0.6 & 2 & 8 \end{bmatrix}$$

Using the procedure stated in Sec. 16-7, we obtain

$$K^{-1} = \frac{L}{EI}\left(\frac{1}{10.08}\right)\begin{bmatrix} 60 & 3.6 & 3.6 \\ 3.6 & 1.56 & -0.12 \\ 3.6 & -0.12 & 1.56 \end{bmatrix}$$

The nodal displacements are then obtained from $r = K^{-1}R$;

$$\begin{Bmatrix} r_1 \\ r_2 \\ r_3 \end{Bmatrix} = \frac{L}{EI}\left(\frac{1}{10.08}\right)\begin{bmatrix} 60 & 3.6 & 3.6 \\ 3.6 & 1.56 & -0.12 \\ 3.6 & -0.12 & 1.56 \end{bmatrix}\begin{Bmatrix} 10 \\ 0 \\ 0 \end{Bmatrix} = \frac{L}{EI}\left(\frac{1}{10.08}\right)\begin{Bmatrix} 600 \\ 36 \\ 36 \end{Bmatrix}$$

Finally, the end moments are determined by $Q = kar$. Using the values of k, a, and r previously obtained gives

$$\begin{Bmatrix} Q_i^a \\ Q_j^a \\ Q_i^b \\ Q_j^b \\ Q_i^c \\ Q_j^c \end{Bmatrix} = \begin{Bmatrix} -28.6 \\ -21.4 \\ 21.4 \\ 21.4 \\ -21.4 \\ -28.6 \end{Bmatrix} \text{ft-kips}$$

Example 18-4. The end moments of the rigid frame shown in Fig. 18-5

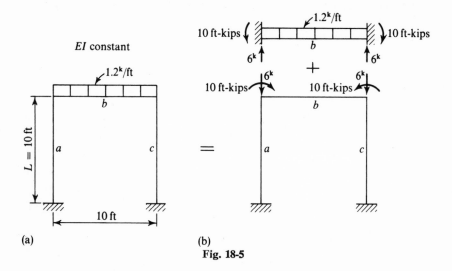

Fig. 18-5

were solved by the force method in Example 17-5 and will now be re-solved by the displacement method.

Neglecting the effect of axial deformations, the frame prepared for nodal force analysis is shown in Fig. 18-6 where r_1 and r_2 denote the joint rotations and R_1 and R_2 the corresponding moments. In this problem $R_1 = -R_2 = 10$ ft-kips. The dotted lines indicate assumed directions for member end moments, Q_i^a, $Q_j^a, \ldots$, etc., and their corresponding member end rotations, $q_i^a, q_j^a, \ldots$, etc. These can be expressed in terms of r_1 and r_2 as

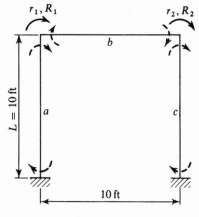

Fig. 18-6

$$r_1 = 1 \quad r_2 = 1$$

$$
\begin{Bmatrix} q_i^a \\ q_j^a \\ q_i^b \\ q_j^b \\ q_i^c \\ q_j^c \end{Bmatrix}
=
\begin{bmatrix}
0 & 0 \\
1 & 0 \\
1 & 0 \\
0 & 1 \\
0 & 1 \\
0 & 0
\end{bmatrix}
\begin{Bmatrix} r_1 \\ r_2 \end{Bmatrix}
$$

$$a$$

The diagonal matrix k is the same as that obtained in the preceding example.

The stiffness matrix of structure is obtained by

$$K = a^T k a$$

$$
= \begin{bmatrix} 0 & 1 & 1 & 0 & 0 & 0 \\ 0 & 0 & 0 & 1 & 1 & 0 \end{bmatrix}
\left(\frac{EI}{L}\right)
\begin{bmatrix}
4 & 2 & & & & \\
2 & 4 & & & & \\
& & 4 & 2 & & \\
& & 2 & 4 & & \\
& & & & 4 & 2 \\
& & & & 2 & 4
\end{bmatrix}
\begin{bmatrix}
0 & 0 \\
1 & 0 \\
1 & 0 \\
0 & 1 \\
0 & 1 \\
0 & 0
\end{bmatrix}
$$

$$
= \frac{EI}{L} \begin{bmatrix} 8 & 2 \\ 2 & 8 \end{bmatrix}
$$

Thus
$$K^{-1} = \frac{L}{EI} \frac{\begin{bmatrix} 8 & -2 \\ -2 & 8 \end{bmatrix}^T}{60} = \frac{L}{30EI} \begin{bmatrix} 4 & -1 \\ -1 & 4 \end{bmatrix}$$

and
$$r = K^{-1} R,$$

$$
\begin{Bmatrix} r_1 \\ r_2 \end{Bmatrix} = \frac{L}{30EI} \begin{bmatrix} 4 & -1 \\ -1 & 4 \end{bmatrix} \begin{Bmatrix} 10 \\ -10 \end{Bmatrix} = \frac{L}{30EI} \begin{Bmatrix} 50 \\ -50 \end{Bmatrix}
$$

Then the member end moments based on the nodal force analysis (Fig. 18-6) are determined by $Q = kar$. Using the values of $k, a,$ and r, previously found, gives

$$
\begin{Bmatrix} Q_i^a \\ Q_j^a \\ Q_i^b \\ Q_j^b \\ Q_i^c \\ Q_j^c \end{Bmatrix}
=
\begin{Bmatrix}
3.33 \\
6.67 \\
3.33 \\
-3.33 \\
-6.67 \\
-3.33
\end{Bmatrix} \text{ft-kips}
$$

The end moments of member b, Q_i^b and Q_j^b, must be corrected by adding the fixed-end moments shown in the upper part of Fig. 18-5(b). The final result is

$$
\begin{Bmatrix} Q_i^a \\ Q_j^a \\ Q_i^b \\ Q_j^b \\ Q_i^c \\ Q_j^c \end{Bmatrix} = \begin{Bmatrix} 3.33 \\ 6.67 \\ -6.67 \\ 6.67 \\ -6.67 \\ -3.33 \end{Bmatrix} \text{ft-kips}
$$

Example 18-5. Obtain the end moments for the frame shown in Fig. 18-7 (a). Use the equivalent form in Fig. 18-7(b) for the analysis so that the size of the matrices will be reduced. Assume constant EI.

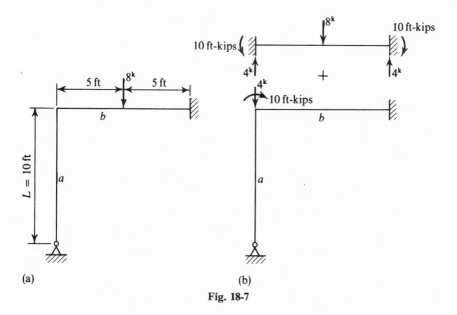

Fig. 18-7

If the effect of axial deformations is neglected, the frame prepared for nodal force analysis is shown in Fig. 18-8. The frame is subjected to the joint rotations r_1 and r_2 and the corresponding nodal moments R_1 and R_2. In this case, $R_1 = 0$ and $R_2 = 10$ ft-kips.

As before, the dotted lines indicate the assumed directions for the member end moments Q_i^a, Q_j^a, Q_i^b, and Q_j^b and their corresponding end rotations q_i^a, q_j^a, q_i^b, and q_j^b. The latter can be expressed in terms of r_1 and r_2 from compatibility as

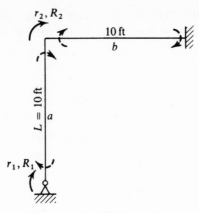

Fig. 18-8

$$r_1 = 1 \quad r_2 = 1$$

$$\begin{Bmatrix} q_i^a \\ q_j^a \\ q_i^b \\ q_j^b \end{Bmatrix} = \begin{bmatrix} 1 & 0 \\ 0 & 1 \\ 0 & 1 \\ 0 & 0 \end{bmatrix} \begin{Bmatrix} r_1 \\ r_2 \end{Bmatrix}$$

$$a$$

Since the members are identical,

$$k = \frac{EI}{L} \begin{bmatrix} 4 & 2 & & \\ 2 & 4 & & \\ & & 4 & 2 \\ & & 2 & 4 \end{bmatrix}$$

The total stiffness matrix is then determined.

$$K = a^T k a = \begin{bmatrix} 1 & 0 & 0 & 0 \\ 0 & 1 & 1 & 0 \end{bmatrix} \left(\frac{EI}{L}\right) \begin{bmatrix} 4 & 2 & & \\ 2 & 4 & & \\ & & 4 & 2 \\ & & 2 & 4 \end{bmatrix} \begin{bmatrix} 1 & 0 \\ 0 & 1 \\ 0 & 1 \\ 0 & 0 \end{bmatrix} = \frac{EI}{L} \begin{bmatrix} 4 & 2 \\ 2 & 8 \end{bmatrix}$$

from which $\quad K^{-1} = \dfrac{L}{EI} \dfrac{\begin{bmatrix} 8 & -2 \\ -2 & 4 \end{bmatrix}^T}{28} = \dfrac{L}{14EI} \begin{bmatrix} 4 & -1 \\ -1 & 2 \end{bmatrix}$

The nodal displacements are expressed by

$$\begin{Bmatrix} r_1 \\ r_2 \end{Bmatrix} = K^{-1} R = \frac{L}{14EI} \begin{bmatrix} 4 & -1 \\ -1 & 2 \end{bmatrix} \begin{Bmatrix} 0 \\ 10 \end{Bmatrix} = \frac{L}{14EI} \begin{Bmatrix} -10 \\ 20 \end{Bmatrix}$$

Thus the end moments from nodal force analysis, based on Fig. 18-8, are found
to be

$$
\begin{Bmatrix} Q_i^a \\ Q_j^a \\ Q_i^b \\ Q_j^b \end{Bmatrix} = kar = \frac{EI}{L}
\begin{bmatrix} 4 & 2 & & \\ 2 & 4 & & \\ & & 4 & 2 \\ & & 2 & 4 \end{bmatrix}
\begin{bmatrix} 1 & 0 \\ 0 & 1 \\ 0 & 1 \\ 0 & 0 \end{bmatrix}
\left(\frac{L}{14EI} \right)
\begin{Bmatrix} -10 \\ 20 \end{Bmatrix} =
\begin{Bmatrix} 0 \\ 4.28 \\ 5.72 \\ 2.86 \end{Bmatrix} \text{ft-kips}
$$

After adding the fixed-end moments (see the upper part of Fig. 18-7(b)) to Q_i^b
and Q_j^b, we obtain the final solution as

$$
\begin{Bmatrix} Q_i^a \\ Q_j^a \\ Q_i^b \\ Q_j^b \end{Bmatrix} =
\begin{Bmatrix} 0 \\ 4.28 \\ -4.28 \\ 12.86 \end{Bmatrix} \text{ft-kips}
$$

18-6. USE OF THE MODIFIED MEMBER STIFFNESS
MATRIX

Referring to Eq. 18-11, we find the stiffness matrix for a uniform member
in frame analysis is given by

$$
k^a = \frac{EI}{L} \begin{bmatrix} 4 & 2 \\ 2 & 4 \end{bmatrix}
$$

Note that the elements in the first column of the matrix are the member end
moments obtained by producing in this end (i end) a unit rotation, the other
end (j end) being fixed. The procedure described for obtaining these values is
exactly the same as that for finding the stiffness of the i end and its carry-over

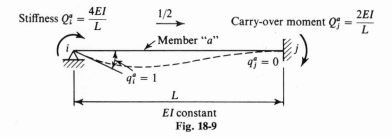

Stiffness $Q_i^a = \dfrac{4EI}{L}$ 1/2→   Carry-over moment $Q_j^a = \dfrac{2EI}{L}$

Member "a"

$q_i^a = 1$ $q_j^a = 0$

L

EI constant

Fig. 18-9

value to the j end in the method of moment distribution, as illustrated in Fig.
18-9. Recall, in the moment-distribution procedures, that if the actual condition of the other end is known, then the computation can be simplified by using
the modified stiffness of this end and omitting the presentation of the other

portion of structure in the analysis. This technique can also be applied to the displacement method by introducing the *modified member stiffness matrix*.

With reference to Sec. 13-7 (modified stiffness in moment distribution), we note the following three special cases:

1. When the other end is simply supported, then the moment needed to produce a unit rotation in this end is $3EI/L$, i.e., $Q_i^a = 3EI/L$ for $q_i^a = 1$ as indicated in Fig. 18-10.

Modified stiffness

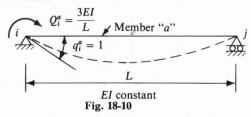

EI constant
Fig. 18-10

The modified member stiffness matrix, if we disregard the factors of Q_j^a and q_j^a, is then given by

$$(k^a)' = \left[\frac{3EI}{L}\right] \qquad (18\text{-}21)$$

where $(k^a)'$ denotes the modified member stiffness matrix for member a.

2. When the other end rotates an equal but opposite angle to that of this end (the case of symmetry),

$$(k^a)' = \left[\frac{2EI}{L}\right] \qquad (18\text{-}22)$$

3. When the other end rotates the same angle as that of this end (the case of antisymmetry),

$$(k^a)' = \left[\frac{6EI}{L}\right] \qquad (18\text{-}23)$$

Application of the procedures described above is illustrated by re-solving Examples 18-3, 18-4, and 18-5 of the preceding section as follows:

Example 18-6. Re-solve Example 18-3 by using the modified stiffness matrix.

The frame in Fig. 18-4(a) may be put in the form of Fig. 18-11(a). The structure and its loading represent a case of antisymmetry for which we may assume the nodal displacements and their corresponding nodal forces as shown in Fig. 18-11(b). Observe that $R_1 = 5$ kips and $R_2 = 0$.

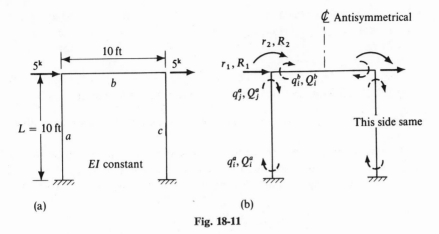

Fig. 18-11

The analysis can be simplified by working with only half the structure if the modified stiffness matrix for member b is used; i.e.,

$$(k^b)' = \left[\frac{6EI}{L}\right]$$

Hence

$$k = \begin{bmatrix} k^a & \\ & (k^b)' \end{bmatrix} = \frac{EI}{L}\begin{bmatrix} 4 & 2 & 0 \\ 2 & 4 & 0 \\ 0 & 0 & 6 \end{bmatrix}$$

From compatibility

$$\begin{matrix} & r_1 = 1 & r_2 = 1 \\ \begin{Bmatrix} q_i^a \\ q_j^a \\ q_i^b \end{Bmatrix} = & \begin{bmatrix} -\frac{1}{10} & 0 \\ -\frac{1}{10} & 1 \\ 0 & 1 \end{bmatrix} & \begin{Bmatrix} r_1 \\ r_2 \end{Bmatrix} \end{matrix}$$

$$a$$

Using the values already found for a and k, we have

$$K = a^T k a = \frac{EI}{L}\begin{bmatrix} 0.12 & -0.6 \\ -0.6 & 10 \end{bmatrix}$$

$$K^{-1} = \frac{L}{0.84EI}\begin{bmatrix} 10 & 0.6 \\ 0.6 & 0.12 \end{bmatrix}$$

$$r = K^{-1}R = \frac{L}{0.84EI}\begin{bmatrix} 10 & 0.6 \\ 0.6 & 0.12 \end{bmatrix}\begin{Bmatrix} 5 \\ 0 \end{Bmatrix} = \frac{L}{0.84EI}\begin{Bmatrix} 50 \\ 3 \end{Bmatrix}$$

Thus, $$\begin{Bmatrix} Q_i^a \\ Q_j^a \\ Q_i^b \end{Bmatrix} = kar = \frac{EI}{L}\begin{bmatrix} 4 & 2 & 0 \\ 2 & 4 & 0 \\ 0 & 0 & 6 \end{bmatrix}\begin{bmatrix} -\frac{1}{10} & 0 \\ -\frac{1}{10} & 1 \\ 0 & 1 \end{bmatrix}\left(\frac{L}{0.84EI}\right)\begin{Bmatrix} 50 \\ 3 \end{Bmatrix}$$

$$= \left\{ \begin{matrix} -28.6 \\ -21.4 \\ 21.4 \end{matrix} \right\} \text{ft-kips}$$

The results, $Q_i^a = Q_j^c = -28.6$ ft-kips, $Q_j^a = Q_i^c = -21.4$ ft-kips, and $Q_i^b = Q_j^b = 21.4$ ft-kips, are the same as previously found in Example 18-3.

Example 18-7. Re-solve Example 18-4 by using the modified stiffness matrix.

The portal frame shown in Fig. 18-5 is symmetrical about the center line of the beam. The frame assumed for nodal force analysis may be given as in Fig. 18-12. Referring to Fig. 18-5(b), we note that $R_1 = 10$ ft-kips. The

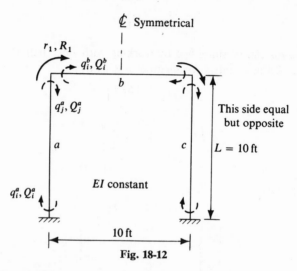

Fig. 18-12

analysis can be simplified by working with only half the frame and using the modified stiffness matrix for member b; i.e.,

$$(k^b)' = \left[\frac{2EI}{L} \right]$$

Thus

$$k = \begin{bmatrix} k^a & \\ & (k^b)' \end{bmatrix} = \frac{EI}{L} \begin{bmatrix} 4 & 2 & 0 \\ 2 & 4 & 0 \\ 0 & 0 & 2 \end{bmatrix}$$

From compatibility

$$\left\{ \begin{matrix} q_i^a \\ q_j^a \\ q_i^b \end{matrix} \right\} = \left\{ \begin{matrix} 0 \\ 1 \\ 1 \end{matrix} \right\} [r_1]$$

$$a$$

With the values of a and k found, we obtain

$$K = a^T k a = \left[\frac{6EI}{L}\right]$$

It follows that
$$K^{-1} = \left[\frac{L}{6EI}\right]$$

$$r_1 = K^{-1} R_1 = \left[\frac{L}{6EI}\right][10] = \left[\frac{10L}{6EI}\right]$$

With the values of a, k, and r_1 found, we obtain the end moments from nodal force analysis as

$$\begin{Bmatrix} Q_i^a \\ Q_j^a \\ Q_i^b \end{Bmatrix} = kar_1 = \begin{Bmatrix} 3.33 \\ 6.67 \\ 3.33 \end{Bmatrix} \text{ft-kips}$$

Adding the fixed-end moment of -10 ft-kips (see the upper part of Fig. 18-5(b)) to O_i^b in the above result and using symmetry, we obtain the final solution as

$$Q_i^a = -Q_j^c = 3.33 \text{ ft-kips}$$
$$Q_j^a = -Q_i^c = 6.67 \text{ ft-kips}$$
$$Q_i^b = -Q_j^b = -6.67 \text{ ft-kips}$$

These are the same as previously obtained in Example 18-4.

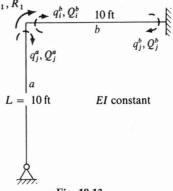

Fig. 18-13

Example 18-8. Re-solve Example 18-5 by using modified member stiffness.

Refer to Fig. 18-7. Since the moment at the hinged end is zero, we may simplify the computation by using the modified stiffness matrix $[3EI/L]$ for member a and by assuming the frame for nodal force analysis as in Fig. 18-13. Note that in present case $R_1 = 10$ ft-kips.

$$k = \begin{bmatrix} (k^a)' & \\ & k^b \end{bmatrix} = \frac{EI}{L}\begin{bmatrix} 3 & 0 & 0 \\ 0 & 4 & 2 \\ 0 & 2 & 4 \end{bmatrix}$$

From compatibility
$$\begin{Bmatrix} q_j^a \\ q_i^b \\ q_j^b \end{Bmatrix} = \begin{Bmatrix} 1 \\ 1 \\ 0 \end{Bmatrix}[r_1]$$

$$a$$

With k and a determined, we have

$$K = a^T ka = \frac{EI}{L} [7]$$

$$r_1 = K^{-1} R_1 = \frac{L}{EI} \begin{bmatrix} 10 \\ 7 \end{bmatrix}$$

Thus the end moments from nodal force analysis are given by

$$\begin{Bmatrix} Q_j^a \\ Q_i^b \\ Q_j^b \end{Bmatrix} = kar_1 = \frac{EI}{L} \begin{bmatrix} 3 & 0 & 0 \\ 0 & 4 & 2 \\ 0 & 2 & 4 \end{bmatrix} \begin{Bmatrix} 1 \\ 1 \\ 0 \end{Bmatrix} \left(\frac{L}{EI}\right) \begin{bmatrix} 10 \\ 7 \end{bmatrix} = \begin{Bmatrix} 4.28 \\ 5.72 \\ 2.86 \end{Bmatrix} \text{ft-kips}$$

After the values of Q_i^b and Q_j^b are corrected by adding the fixed-end moment (see the upper part of Fig. 18-7(b)), we obtain the final solution as previously found in Example 18-5; i.e.,

$$\begin{Bmatrix} Q_j^a \\ Q_i^b \\ Q_j^b \end{Bmatrix} = \begin{Bmatrix} 4.28 \\ -4.28 \\ 12.86 \end{Bmatrix} \text{ft-kips}$$

18-7. THE DIRECT STIFFNESS METHOD

In the previous discussion the establishment of total stiffness matrix was performed by

$$K = a^T ka$$

in which a and k were assembled on the basis of the entire structure. Alternatively, we may treat each member or element as a structure (member-structure) and obtain the *stiffness matrix for the member-structure* by using the smaller matrices a, a^T, and k of the member. The total stiffness matrix of the entire structure is then established by superimposing the stiffness matrix of the individual structural members. This method, referred to as the *direct stiffness method* is particularly convenient for larger structures, for it eliminates the necessity of establishing all the large matrices except one, the final K matrix of the structure. To illustrate, let us consider the frame in Fig. 18-4.

The total stiffness matrix was found to be

$$\frac{EI}{L} \begin{bmatrix} 0.24 & -0.6 & -0.6 \\ -0.6 & 8 & 2 \\ -0.6 & 2 & 8 \end{bmatrix}$$

in Example 18-3 through the multiplication of $a^T ka$ in which k is of order (6 × 6), a is of order (6 × 3), and a^T is of order (3 × 6). The same result may be achieved by the direct stiffness procedures, which enable the use of smaller size matrices, as contained in the following steps:

1. Treat member a as a structure. To find the total stiffness matrix of the member, we note that if only the flexural deformation is considered,

$$\begin{Bmatrix} Q_i^a \\ Q_j^a \end{Bmatrix} = \frac{EI}{L}\begin{bmatrix} 4 & 2 \\ 2 & 4 \end{bmatrix}\begin{Bmatrix} q_i^a \\ q_j^a \end{Bmatrix}$$

and from compatibility $\quad \begin{Bmatrix} q_i^a \\ q_j^a \end{Bmatrix} = \begin{bmatrix} -\frac{1}{10} & 0 \\ -\frac{1}{10} & 1 \end{bmatrix}\begin{Bmatrix} r_1 \\ r_2 \end{Bmatrix}$

since the end rotations of member a is not affected by r_3 under the condition that all other nodes are fixed.

Thus we have

$$a = \begin{bmatrix} -\frac{1}{10} & 0 \\ -\frac{1}{10} & 1 \end{bmatrix} \qquad a^T = \begin{bmatrix} -\frac{1}{10} & -\frac{1}{10} \\ 0 & 1 \end{bmatrix} \qquad k = \frac{EI}{L}\begin{bmatrix} 4 & 2 \\ 2 & 4 \end{bmatrix}$$

for the member-structure, all of order (2×2). If we use K^a to denote the total stiffness matrix of member a, we have

$$
K^a = a^T ka = \begin{bmatrix} -\frac{1}{10} & -\frac{1}{10} \\ 0 & 1 \end{bmatrix}\left(\frac{EI}{L}\right)\begin{bmatrix} 4 & 2 \\ 2 & 4 \end{bmatrix}\begin{bmatrix} -\frac{1}{10} & 0 \\ -\frac{1}{10} & 1 \end{bmatrix} = \frac{EI}{L}\begin{bmatrix} \overset{r_1}{0.12} & \overset{r_2}{-0.6} \\ -0.6 & 4 \end{bmatrix}
$$

2. In the same way, we treat member b as a structure. We find that

$$\begin{Bmatrix} Q_i^b \\ Q_j^b \end{Bmatrix} = \frac{EI}{L}\begin{bmatrix} 4 & 2 \\ 2 & 4 \end{bmatrix}\begin{Bmatrix} q_i^b \\ q_j^b \end{Bmatrix}$$

and

$$\begin{Bmatrix} q_i^b \\ q_j^b \end{Bmatrix} = \begin{bmatrix} 1 & 0 \\ 0 & 1 \end{bmatrix}\begin{Bmatrix} r_2 \\ r_3 \end{Bmatrix}$$

Thus, $\qquad a = a^T = I \qquad k = \frac{EI}{L}\begin{bmatrix} 4 & 2 \\ 2 & 4 \end{bmatrix}$

from which $\qquad K^b = a^T ka = \frac{EI}{L}\begin{bmatrix} \overset{r_2}{4} & \overset{r_3}{2} \\ 2 & 4 \end{bmatrix}$

3. Member c is in a condition similar to that of member a except that it is subjected to the influences of r_1 and r_3. Thus,

$$K^c = \frac{EI}{L}\begin{bmatrix} \overset{r_1}{0.12} & \overset{r_3}{-0.6} \\ -0.6 & 4 \end{bmatrix}$$

4. The above matrices K^a, K^b, and K^c are not immediately subjected to superposition since the columns in the three matrices are under different influences of r. To make them compatible for addition, we expand each of the

three to the order of (3×3) by adding a column and a row of zeros for the nodal displacement, which is irrelevant for the member in question. In this manner, we obtain

$$K^a = \frac{EI}{L} \begin{bmatrix} & r_1 & r_2 & r_3 \\ & 0.12 & -0.6 & 0 \\ & -0.6 & 4 & 0 \\ & 0 & 0 & 0 \end{bmatrix} \qquad K^b = \frac{EI}{L} \begin{bmatrix} r_1 & r_2 & r_3 \\ 0 & 0 & 0 \\ 0 & 4 & 2 \\ 0 & 2 & 4 \end{bmatrix}$$

$$K^c = \frac{EI}{L} \begin{bmatrix} r_1 & r_2 & r_3 \\ 0.12 & 0 & -0.6 \\ 0 & 0 & 0 \\ -0.6 & 0 & 4 \end{bmatrix}$$

The order of the total stiffness matrix for each member is now the same as the order of K for the entire structure. By directly adding K^a, K^b, and K^c, we obtain the total structure stiffness matrix K as

$$K = K^a + K^b + K^c = \frac{EI}{L} \begin{bmatrix} 0.24 & -0.6 & -0.6 \\ -0.6 & 8 & 2 \\ -0.6 & 2 & 8 \end{bmatrix}$$

After K is found, the rest of analysis is the same as previously discussed in the displacement method. Thus the direct stiffness method differs from the displacement method only in the manner by which the total stiffness matrix of structure is constructed. For complex structures this difference is important, since the direct stiffness method is a direct and easier way to establish the overall stiffness matrix by the superposition of total stiffness matrices of individual elements.

18-8. COMPARISON OF THE FORCE METHOD AND
THE DISPLACEMENT METHOD

The force method and the displacement method represent two parallel ways of analyzing structures. The basic procedures for the two methods may be briefly recapitulated as in Table 18-1.

The duality between the two methods is apparent; one is the inverse of the other. The choice of the methods mainly lies in the accuracy of the solution and the ease of computation which in turn would depend upon the idealization of the structure, the rounding off of error, and the type of and formulation of the problem. Generally, except for structures that involve many joint displacements but few redundants, the displacement method is often preferred. Some of the reasons are as follows:

TABLE 18-1

Force Method	Displacement Method
(1) Select member forces as basic unknowns.	(1) Select nodal displacements as basic unknowns.
(2) Establish the force transformation matrix	(2) Establish the displacement transformation matrix.
(3) Evaluate member flexibility matrices	(3) Evaluate member stiffness matrices.
(4) Obtain the total flexibility matrix.	(4) Obtain the total stiffness matrix.
(5) Express the nodal displacement in terms of the nodal forces	(5) Express the nodal forces in terms of the nodal displacements

1. In the displacement method the irrelevancy of redundancy enables the use of the same procedures for analyzing statically determinate structures and statically indeterminate structures.

2. It is much easier to form the displacement transformation matrix than the force transformation matrix, since the effects of displacements are often localized.

3. It is convenient to establish the stiffness matrix of structure by the direct stiffness method for complicated structures.

4. It is found that the displacement method usually produces a well conditioned stiffness matrix of structure; whereas in the force method, a well conditioned flexibility matrix of the structure depends upon a good choice of redundants.

Note that a *well conditioned matrix* is the one for which the largest terms lie on the main diagonal and is, thus, most suitable for computer operation.

PROBLEMS

18-1. Solve Prob. 17-4 by the displacement method.

18-2. Solve Prob. 17-5 by the displacement method.

18-3. Solve Prob. 17-6 by the displacement method.

18-4. Find, by the displacement method, the moment at support B of the beam shown in Fig. 18-14.

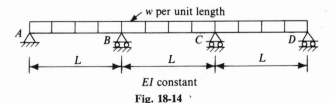

EI constant

Fig. 18-14

18-5. Obtain, by the displacement method, all the end moments for the rigid frame shown in Fig. 18-15.

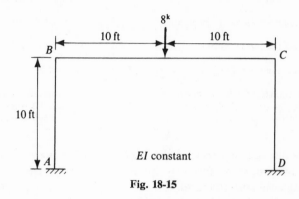

Fig. 18-15

18-6. Obtain, by the displacement method, all the end moments for the frame of Fig. 18-16, using the modified member stiffness matrix for the center beam due to antisymmetry.

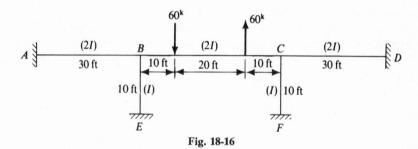

Fig. 18-16

18-7. Use the displacement method to obtain the end moments for the frame shown in Fig. 18-17. Assume constant EI.

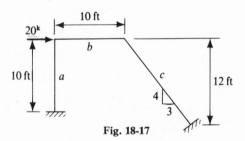

Fig. 18-17

18-8. Use the direct stiffness method to obtain the total stiffness matrix for the frame of Fig. 18-17.

SELECTED REFERENCES

Andersen, P., *Statically Indeterminate Structures*. New York: The Ronald Press Co., 1953.

Andersen, P., and G. M. Nordby, *Introduction to Structural Mechanics*. New York: The Ronald Press Co., 1960.

Argyris, J. H., *Recent Advance in Matrix Methods of Structural Analysis*. New York: Pergamon Press, 1964.

Argyris, J. H., and S. Kelsey, *Energy Theorems and Structural Analysis*. London: Butterworth & Co. (Publishers), Ltd., 1960.

Asplund, S. O., *Structural Mechanics: Classical and Matrix Methods*. Englewood Cliffs, N.J.: Prentice-Hall, Inc., 1966.

Borg, S. F., and J. J. Gennaro, *Advanced Structural Analysis*. Princeton, N.J.: D. Van Nostrand Company, Inc., 1959.

Cross, H., and N. D. Morgan, *Continuous Frames of Reinforced Concrete*. New York: John Wiley & Sons, Inc., 1954.

Fung, Y. C., *Foundations of Solid Mechanics*. Englewood Cliffs, N.J.: Prentice-Hall, Inc., 1965.

Gaylord, E. H., and C. N. Gaylord, *Structural Engineering Handbook*. New York: McGraw-Hill Book Company, 1968.

Gere, J. M., *Moment Distribution*. Princeton, N.J.: D. Van Nostrand Company, Inc., 1963.

Grinter, L. E., *Theory of Modern Steel Structures* (rev. ed.), Vol 2. New York: The Macmillan Company, Publishers, 1949.

Hall, A. S., and R. W. Woodhead, *Frame Analysis* (2nd ed.). New York: John Wiley & Sons, Inc., 1967.

Kinney, J. S., *Indeterminate Structural Analysis*. Reading, Mass.: Addison-Wesley Publishing Co., Inc., 1957.

Laursen, H. I., *Structural Analysis*. New York: McGraw-Hill Book Company, 1969.

Lightfoot, E., *Moment Distribution*. London: E. & F. M. Spon, 1961.

Martin, H. C., *Introduction to Matrix Methods of Structural Analysis*. New York: McGraw-Hill Book Company, 1966.

Norris, C. H., and J. B. Wilbur, *Elementary Structural Analysis* (2nd ed.). New York: McGraw-Hill Book Company, 1960.

Sutherland, H., and H. L. Bowman, *Structural Theory* (4th ed.). New York: John Wiley & Sons, Inc., 1958.

419

Timoshenko, S. P., and D. H. Young, *Theory of Structures* (2nd ed.). New York: McGraw-Hill Book Company, 1965.

Wang, C. K., and C. L. Eckel, *Elementary Theory of Structures.* New York: McGraw-Hill Book Company, 1957.

Wang, P. C., *Numerical and Matrix Methods in Structural Mechanics.* New York: John Wiley & Sons, Inc., 1966.

Willems, N., and W. M. Lucas, Jr., *Matrix Analysis for Structural Engineers.* Englewood Cliffs, N.J.: Prentice-Hall, Inc., 1968.

ANSWERS TO SELECTED PROBLEMS

CHAPTER 2

2–1. (a) Stable and indeterminate to the fifth degree
 (b) Unstable
 (c) Stable and determinate
 (d) Stable and indeterminate to the second degree
2–2. (b) Stable and indeterminate to the third degree
 (c) Unstable externally
 (d) Unstable internally
2–3. (b) Stable and indeterminate to the fifth degree
 (e) Stable and indeterminate to the fourth degree
 (f) Unstable externally
 (g) Stable and indeterminate to the 102nd degree

CHAPTER 3

3–2. $R_A = 8.33$ kips (down) $R_B = 23.33$ kips (up)
 $R_D = 5$ kips (up) $M_B = -80$ ft-kips
 $V_C = 5$ kips
3–3. $R_A = 18.33$ kips (up) $M_A = -123.33$ ft-kips
 $R_F = 11.67$ kips (up) $V_B = 18.33$ kips
 $M_B = -31.67$ ft-kips

CHAPTER 4

4–1. (e) Simple   (g) Simple   (h) Complex
 The rest are compound trusses.
4–2. (b) $S_{ab} = S_{bc} = +14.5$ kips $S_{aB} = +9.25$ kips
 $S_{Bb} = +17.4$ kips $S_{Bc} = -18.8$ kips
 (d) $S_{Ab} = S_{Bb} = S_{Bc} = S_{Cc} = S_{Cd} = S_{Dd} = 0$
 $S_{Ae} = +36$ kips $S_{ae} = -37$ kips
4–3. (a) $S_a = -32$ kips $V_b = +12$ kips $S_c = +24$ kips
 (b) $H_a = +18.75$ kips $V_b = -15$ kips $S_c = -6.25$ kips
 (c) $V_a = +11.6$ kips $H_b = +116.7$ kips $S_c = -20$ kips
 (d) $S_a = +20$ kips $V_b = +100$ kips $V_c = -45$ kips
 (e) $S_a = -50$ kips

421

4–4. $V_a = -20$ kips $V_b = +10$ kips $V_c = -20$ kips $S_d = 0$

4–5. $S_{AB} = S_{BC} = S_{CD} = S_{BF} = S_{CF} = 0$ $S_{AD} = +15.4$ kips

   $S_{AE} = +11.2$ kips $S_{DE} = -13.6$ kips $S_{EF} = -10$ kips

4–6. $S_a = -34.8$ kips $S_b = -1.95$ kips

4–8. $S_{ad} = +7.85$ kips

CHAPTER 5

5–1. (d) $M_c = +108$ ft-kips

   (e) $M_b = +31.25$ ft-kips

   (f) $M_b = -144$ ft-kips

5–4. $S_{ch} = S_{hi} = S_{ei} = 0$ $S_{cd} = -15$ kips

   $S_{de} = +5$ kips $V_{bd} = -V_{df} = 8.33$ kips

CHAPTER 6

6–1. $V = 49.33$ kips $M = 458.67$ ft-kips

6–4. $R_A = 12$ kips $R_C = 46.67$ kips $R_E = 8$ kips

   $V_B = -13.67$ kips $M_B = -64$ ft-kips $M_C = -128$ ft-kips

CHAPTER 7

7–1. (a) $R = 22.78$ kips

   (b) $V = 14.44$ kips $M = 216.67$ ft-kips

7–2. $V_{1-2} = 27.5$ kips $M_2 = 690$ ft-kips

7–3. (a) $V = 83.75$ kips

   (b) $V = 57.5$ kips

   (c) $M = 975$ ft-kips

   (d) $M = 978.5$ ft-kips

7–4. $V_a = +29.68$ kips (maximum tension)

   $V_a = -4.4$ kips (maximum compression)

   $S_b = +52.75$ kips

CHAPTER 8

8–1. $\Delta_c = 5wl^4/384EI$ (down) $\theta_a = wl^3/24EI$ (clockwise)

8–2. $\Delta_P = 0.0147Pl^3/EI$ (down) Δ(at midspan) $= 0.0236Pl^3/EI$ (down)

8–3. $\Delta = 0.54$ in. (down) $\theta = 0.006$ rad (counter-clockwise)

8–4. $\Delta_h = 0.43$ ft (right) $\Delta_v = 0.696$ ft (down)

   $\Delta_r = 0.006$ rad (counter-clockwise)

8–5. (a) $\Delta_B = 0.00746$ ft (down)

   (b) $\Delta_C = 0.00278$ ft (right)

   (c) $\Delta_{bC} = 0.0002$ ft (toward each other)

   (d) $\theta = 0.000433$ rad (clockwise)

8–14. $\theta_b = 0.006$ rad (clockwise, at the left side)
$\theta_b = 0.001$ rad (counter-clockwise, at the right side)
$\Delta_b = 0.48$ in. (down)

CHAPTER 9

9–1. (a) $R_b = 50$ kips (up)
(b) $M_b = -100$ ft-kips
9–2. $R_b = 10$ kips (up)
9–3. $R_b = 7.25$ kips (up)
9–4. $M_a = wl^2/20$ (counter-clockwise)
$M_b = wl^2/30$ (clockwise)
9–5. $H_a = 1.25$ kips (right) $V_a = 5$ kips (up)
$M_a = 4.16$ ft-kips (clockwise)
9–6. (a) $S_{bC} = -0.76$ kip
(b) $H_d = 30$ kips (left)
,   (c) $S_{bC} = -5.4$ kips $H_d = 31.1$ kips (left)
9–7. $S_{Bc} = S_{bC} = +8.75$ kips

CHAPTER 10

10–7. $S_{ac} = +4.25$ kips $M_a = -3.25$ ft-kips
The effect of axial force in beam is neglected.
10–8. 21.75 kips, 17.4 kips, and 13.05 kips

CHAPTER 11

11–1. The influence ordinates are 1.3, 1, 0.704, 0.432, 0.208, 0.056, 0
11–2. The influence ordinates are 0, 2.09, 2.67, 2.25, 1.33, 0.417, 0
11–3. (a) The influence ordinates are 1, 0.578, 0.222, 0, -0.05, 0
(b) The influence ordinates are 0, -0.532, -0.668, 0, -0.30, 0
(c) The influence ordinates are 0, -0.422, 0.222, 0, -0.05, 0
11–5. (a) The influence ordinates for S_{cd} are 0, 0.320, 0.666, 0.293, 0, -0.039, -0.038, 0
(b) $S_{BC} = -S_{cd}$

CHAPTER 12

12–1. $M_{ab} = -22$ ft-kips $M_{ba} = -M_{bc} = 28$ ft-kips
$M_{cb} = 31$ ft-kips
12–2. $M_{ab} = -23.55$ ft-kips $M_{ba} = 16.89$ ft-kips
$M_{bd} = -8.89$ ft-kips $M_{db} = -4.44$ ft-kips
12–3. $M_{ab} = -22.1$ ft-kips $M_{ba} = -M_{bc} = 68.3$ ft-kips
$M_{cb} = 20$ ft-kips
12–5. $M_{ab} = -44.8$ ft-kips $M_{ba} = -M_{bc} = -34.4$ ft-kips

12–6. $M_{ab} = 4.9$ ft-kips $M_{ba} = -M_{bc} = -0.9$ ft-kip
       $M_{cb} = -M_{cd} = 2.24$ ft-kips $M_{dc} = -1.76$ ft-kips

12–7. (a) $M_{ab} = -14.9$ ft-kips $M_{ba} = -M_{bc} = 14.9$ ft-kips
             $M_{cb} = 39.4$ ft-kips $M_{ce} = -29.8$ ft-kips

     (b) $M_{ab} = -120.8$ ft-kips $M_{ba} = -M_{bc} = -112.6$ ft-kips
         $M_{cb} = -M_{cd} = 104$ ft-kips $M_{dc} = -103.6$ ft-kips

     (c) $M_{ab} = 8.52$ ft-kips $M_{ba} = -M_{bc} = 9.58$ ft-kips
         $M_{cb} = -M_{cd} = -1$ ft-kip $M_{dc} = 26.74$ ft-kips

     (d) $M_{ba} = M_{ef} = -150$ ft-kips $M_{be} = M_{eb} = 156.25$ ft-kips
         $M_{bc} = M_{ed} = -6.25$ ft-kips $M_{cb} = M_{de} = -43.75$ ft-kips
         $M_{cd} = M_{dc} = 43.75$ ft-kips

     (e) $M_{ab} = -118.1$ ft-kips $M_{ba} = -M_{bc} = -82$ ft-kips
         $M_{cb} = -M_{cd} = 11.5$ ft-kips $M_{dc} = -M_{de} = 25.8$ ft-kips
         $M_{ed} = -74.1$ ft-kips

CHAPTER 13

13–5. End moment $= 14.6$ ft-kips
13–6. Moments at interior supports are 36.5 ft-kips and 27.3 ft-kips
13–7. $M_{ba} = -M_{bc} = 11.4$ ft-kips $M_{cb} = -M_{cd} = 0.6$ ft-kip

CHAPTER 14

14–1. $M_{ba} = -M_{bc} = -160$ ft-kips
14–2. $M_{ba} = -M_{bc} = 13.55$ ft-kips $M_{cb} = -M_{cd} = 13.55$ ft-kips
14–3. $M_{ab} = -2.6$ ft-kips $M_{ba} = -6.4$ ft-kips
     $M_{bc} = -9.6$ ft-kips $M_{cb} = -M_{cd} = -4.2$ ft-kips
     $M_{dc} = 3.8$ ft-kips
14–4. End moment $= 1.52$ ft-kips
14–6. $M_{ad} = -50.9$ ft-kips $M_{be} = -12.2$ ft-kips
           $M_{cg} = -35.6$ ft-kips $M_{da} = -M_{de} = 85.8$ ft-kips
           $M_{ed} = 119.8$ ft-kips $M_{eb} = -48.5$ ft-kips
           $M_{ef} = -71.3$ ft-kips $M_{fe} = -M_{fg} = 108.8$ ft-kips
     $M_{gf} = -M_{gc} = 78.1$ ft-kips
14–7. The influence ordinates for the left half of structure are 0, 0.268, 0.428, 0.375,
     0, -0.697, -1.286, -1.232, 0

CHAPTER 15

15–2. $C_{ab} = 0.743$ $C_{ba} = 0.462$ $S_{ab} = 4.62(EI_a/l)$ $S_{ba} = 7.43(EI_a/l)$
15–3. $M_{ab}^F = 50$ ft-kips (counter-clockwise)
     $M_{ba}^F = 96$ ft-kips (clockwise)
15–4. $M_{ba} = -M_{bc} = 135$ ft-kips $M_{cb} = 148.5$ ft-kips

15–5. $M_{ba} = -M_{bc} = -1,077$ ft-kips $M_{cb} = 1,988$ ft-kips
15–6. $M_{ba} = -M_{bc} = 126.8$ ft-kips
15–8. (a) $M_{ad} = 109$ ft-kips $M_{da} = -M_{de} = 295.2$ ft-kips
   $M_{ed} = 420.4$ ft-kips $M_{be} = 0$
   (b) $M_{ad} = -137$ ft-kips $M_{da} = -M_{de} = -20.5$ ft-kips
   $M_{ed} = 26.9$ ft-kips $M_{eb} = -67.2$ ft-kips
   $M_{be} = -131.6$ ft-kips $M_{ef} = 40.3$ ft-kips
   $M_{fe} = -M_{fc} = 39.8$ ft-kips
   $M_{cf} = -52.9$ ft-kips

CHAPTER 16

16–3.
$$A = \frac{1}{2}\begin{bmatrix} 4 & 11 & -3 \\ 11 & 8 & 2 \\ -3 & 2 & 6 \end{bmatrix} + \frac{1}{2}\begin{bmatrix} 0 & 5 & -5 \\ -5 & 0 & 2 \\ 5 & -2 & 0 \end{bmatrix}$$

16–5.
$$A^{-1} = \begin{bmatrix} -\frac{3}{4} & \frac{3}{2} & -2 \\ \frac{7}{16} & -\frac{5}{8} & 1 \\ \frac{1}{4} & -\frac{1}{2} & 1 \end{bmatrix}$$

CHAPTER 17

17–1. $r_P = 5PL^3/6EI$ (down)
17–2. r(under 10 kips load) $= 0.01608$ ft (down)
   r(under 20 kips load) $= 0.00548$ ft (down)
17–3. $r_1 = (4L^3 R_1 - L^2 R_2)/6EI$ $r_2 = (-L^2 R_1 + 2LR_2)/6EI$
17–4. Deflection $= 5PL^3/6EI$ Slope $= PL^2/EI$
17–5. $$\begin{Bmatrix} Q_a \\ Q_b \\ Q_c \end{Bmatrix} = \begin{Bmatrix} 3.36 \\ 8.28 \\ 2.76 \end{Bmatrix} \text{kips} \qquad \begin{Bmatrix} r_1 \\ r_2 \end{Bmatrix} = \frac{1}{AE}\begin{Bmatrix} 83.2 \\ -15.72 \end{Bmatrix} \text{ft}$$
17–6. Moment at fixed support $= 10.2$ ft-kips
17–7. $Q_i^a = -29.5$ ft-kips $Q_j^a = -Q_i^b = -6.5$ ft-kips
   $Q_j^b = -Q_i^c = 8.4$ ft-kips $Q_j^c = -15.5$ ft-kips

CHAPTER 18

18–4. $M_B = wL^2/10$
18–5. $M_{AB} = 8$ ft-kips $M_{BA} = 16$ ft-kips

18–6. $M_{AB} = 31$ ft-kips $M_{BA} = 62$ ft-kips
$M_{BE} = 93.2$ ft-kips $M_{EB} = 46.6$ ft-kips
$M_{BC} = -155.2$ ft-kips

18–7. $Q_i^a = -46.48$ ft-kips $Q_j^a = -Q_i^b = -42.25$ ft-kips
$Q_j^b = -Q_i^c = 33.80$ ft-kips $Q_j^c = -30.98$ ft-kips

18–8.

$$K = \frac{EI}{L} \begin{bmatrix} 0.243 & -0.150 & 0.117 \\ -0.150 & 8 & 2 \\ 0.117 & 2 & 6.667 \end{bmatrix}$$

INDEX